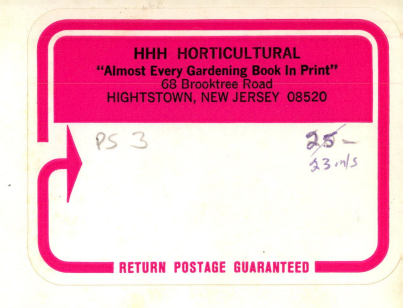

Viability of Seeds

Mutant phenotypes produced from aged seeds (see Chapter 9, pp. 261–3). Peas (*Pisum sativum*): 1A, Yellowish-edges; 1B, Maculata; 1C, Xantha. Broad beans (*Vicia faba*): 2A, Greenish-yellow; 2B, Maculata; 2C, Xantha; 3A Chlorina. Barley (*Hordeum vulgare*): 3B, Albino; 3C, Striata. All photographs (except 1B and 1C) include examples of the normal phenotypes for comparison.

Viability of Seeds

Edited by

E. H. Roberts

Professor of Crop Production,
Department of Agriculture, University of Reading

SYRACUSE UNIVERSITY PRESS

1972

First published 1972
© 1972 Chapman & Hall Ltd
Published in the U.S.A. by
Syracuse University Press
ISBN 0–8156–5033–7
Library of Congress Catalog Card Number 73–39736

Printed in Great Britain

Contents

Contributors

R. B. AUSTIN *National Vegetable Research Station, Welles-bourne, Warwick, UK.*

C. M. CHRISTENSEN *Professor of Plant Pathology, Institute of Agriculture, University of Minnesota, St Paul, Minnesota 55101, USA.*

W. HEYDECKER *Lecturer in Horticulture, Department of Agriculture and Horticulture, School of Agriculture, University of Nottingham, Sutton Bonington, Loughborough LE12 5RD, UK.*

HIROSHI ITO *Plant Breeder in charge of genetic stocks, Division of Genetics, Department of Physiology and Genetics, National Institute of Agricultural Sciences, Hiratsuka, Kanagawa, 254 Japan.*

EDWIN JAMES *Head, National Seed Storage Laboratory, Crops Research Division, Agricultural Research Service, US Department of Agriculture, Fort Collins, Colorado 80521, USA.*

D. B. MACKAY *Chief Officer, The Official Seed Testing Station for England and Wales, Huntingdon Road, Cambridge, CB3 0LE, UK.*

R. P. MOORE *Professor of Research-Crop Stands, North Carolina Agricultural Experiment Station, Raleigh, North Carolina, 27607, USA.*

BRUCE M. POLLOCK *PO Box 187, Allens Park, Colorado 80510, USA.*

E. H. ROBERTS *Professor of Crop Production, Department of Agriculture, University of Reading, Earley Gate, Reading RG6 2AT, UK.*

H. THOMAS *Department of Botany, University College of Wales, Aberystwyth, SY23 3DA, UK.*

Preface

From prehistoric times man has had a special relationship with seed plants – as a source of food, materials for tools, buildings, clothing and pharmaceuticals, and for ornamenting his surroundings for his own delight (probably in that chronological order which, incidentally, also gives some indication of the priorities of life). Today man's most important staple foods are derived directly from seeds as they have been since neolithic times. (It is a sobering thought, as Harlan* has pointed out, that nothing significant has been added to his diet since then.) From those times he must have learned to collect, conserve and cultivate seeds; and the accumulated experience has been handed down. This book then is part of an ancient tradition, for here we are still primarily concerned with these skills.

Seeds are plant propagules comprised of embryos in which growth has been suspended, usually supplied with their own food reserves and protected by special covering layers. Typically they are relatively dry structures compared with other plant tissues and, in this condition, they are resistant to the ravages of time and their environment. But resistant is a relative term and seeds do deteriorate: the type, the extent and the rapidity of the deterioration, and the factors which control it are important to agronomists, horticulturalists, plant breeders, seedsmen, seed analysts, and those concerned with the conservation of genetic resources. In addition the changes in seeds which lead to loss of viability may be relevant to the wider biological problems of senescence and ageing.

It is not surprising that a subject of such practical importance and theoretical interest has gathered a vast technical and scientific literature – not to mention a great deal of folk-lore. The literature is not only large but also diffuse, with isolated pieces of information scattered among a variety of journals. As a result, the accumulated knowledge has been difficult for the research worker to grasp or the practitioner to employ. Nevertheless, although the subject still contains much undigested material and many mysteries, general prin-

*Harlan, J. R., 1956. Distribution and utilization of natural variability in cultivated plants. *Brookhaven Symp. Biol.*, 9, 191–208.

ciples which can guide both types of worker have recently begun to emerge; and it was this realisation which suggested the time was ripe for a book which takes a fresh look at the problems of seed viability.

Department of Agriculture,   E. H. Roberts
University of Reading

April 1971

CHAPTER I

Introduction

E. H. Roberts

'*Probably few sections of human knowledge contain a larger percentage of contradictory, incorrect and misleading observations than prevail in the works dealing with this subject, and, although such fables as the supposed germination of mummy wheat have long since been exploded, equally erroneous records are still current in botanical physiology. In addition there are considerable differences of opinion as to the causes which determine the longevity of seeds in the soil or air. The works of Candolle, Duvel and Becquerel are the most accurate and comprehensive dealing with the question, and, in addition, Vilmorin has published very useful data in regard to the seeds of culinary vegetables. The subject is still, however, in an incomplete and fragmentary condition,*' – Alfred J. Ewart (1908). On the longevity of Seeds. *Proc. Roy. Soc. Victoria, 21* (1), 1–210.

A great deal of work has been done since Ewart dispaired of the fragmentary nature of our knowledge on matters of seed viability and our lack of understanding of the causes of loss of viability; yet, although it was written at the turn of the century, one could be forgiven if it were thought that this is a contemporary quotation: there are still contradictions, errors, and differences of opinion, and the subject is still incomplete and fragmentary. But progress has been made: more fragments exist and some of them begin to tell a story. The main purpose of this book is to pick out the more significant of these and attempt to interpret them; and it is the main task of this chapter to survey the relevance of viability studies and to act as a guide to the book as a whole.

The relevance of seed viability studies

The problems of seed viability are important in a number of applied fields. The factors affecting viability before harvest (Chapters 4 and 5) are a special concern of seed producers, and the problems encountered after sowing (Chapter 6) are important to farmers, agrono-

mists and horticulturalists. The problems of maintaining viability in
storage (Chapters 2 and 3) have always been an important concern of
seedsmen but more recently they have become particularly relevant
to a much wider group of interests because of the developing importance of long-term storage systems in a number of fields. For example,
plant breeders have become increasingly interested in such systems
not only because they provide a cheap and labour-saving device for
maintaining genotypes which are not currently being used, but also
because it is often difficult to maintain the genetic integrity of a particular cultivar by continued multiplication. This development can be
illustrated in Japan where in 1959 Ito and Hayashi introduced a comprehensive rice-breeding system based on long-term seed storage at
the General Assembly of Rice Breeders, and this has been adopted in
its essentials as part of the Japanese government breeding system
since 1965 (Ito, 1970).

Many breeders have been concerned that a number of cultivars
which were common at the beginning of this century have now disappeared completely, even though they may have contained genes or
gene combinations which might have been of value in some current
or future breeding programme. Concern with this problem has led
to the establishment of a few national seed storage laboratories. Prominent among these are the laboratories of the United States and Japan
(described in Appendices 1 and 2). But these are relatively recent
developments and there is a great need for additional laboratories.
Important specialist seed collections are maintained in a number of
centres – e.g. in the United Kingdom at the Welsh Plant Breeding
Station at Aberystwyth and at the Forestry Commission's Research
Station at Farnborough. A number of others, established or developing, have been listed (Anon, 1970), but many do not have facilities
for long-term storage. Frankel and Bennett (1970) have emphasized
that 'it should be recalled that first-rate seed storage laboratories are
few in number, and they are devoted to national or specialist tasks.
In consequence, some of the world's most valuable collections are
without the use of adequate storage facilities, hence they have to be
grown every few years, and are thus exposed to the process of "genetic
erosion" . . .; losses due to inadequate storage can also be alarming.'

Furthermore, it has now been recognised that in addition to losing
recently produced cultivars there is the more urgent danger of loss of
genetic material from the natural centres of genetic diversity. This
is the result of the spread of modern breeding technology which is
rapidly leading to a narrowing of the genetic base of cultivars – inten-
sified by other developments in agricultural technology (the use of

fertilizers, plant protection measures, irrigation, etc.) which make it possible, and desirable in many senses, to use the same cultivar over wide areas. Such techniques are now spreading rapidly to the less developed parts of the world in the lower latitudes where many of the gene centres are to be found.

Because of a growing concern with these developments (which has eloquently been discussed by Bennett, 1965), the International Biological Programme (IBP), at its inception in 1964, set up a subcommittee to study ways and means for collecting and conserving plant genetic resources which were threatened by agricultural development in many centres of diversity. In 1961 the Food and Agriculture Organisation of the United Nations (FAO), Division of Plant Production and Protection held a Technical Meeting at which they considered the same subject; out of this arose a further conference in 1967 which was the result of collaboration between the FAO, IBP, and ICSU (International Council of Scientific Unions). The papers from this conference, extended and revised, were published recently (Frankel and Bennett, 1970); they indicate that the principles of exploration and collection are well understood – even if financial support for such endeavours is not always available. But it is also evident that the principles of seed storage – which is an essential part of genetic conservation – are not so well understood and need to be clarified and developed. Furthermore, there are insufficient seed storage facilities for the growing needs of genetic conservation. To paraphrase the situation, the software needs to be taught and developed and the hardware needs building. Frankel and Bennett (1970) point out 'There is, however, one need of general and overriding importance, and that is the need for international seed storage facilities. An international gene bank, at least in the form of a "clearing house" available to all nations and complemented by agreements on the reconstitution, or "rejuvenation", would give strong encouragement to long-term conservation.' In recognition of the urgency of the problem of genetic conservation a new Crop Ecology and Genetic Resources Branch was established in 1968 as part of the Plant Production and Protection Division within the framework of FAO. The detailed aims, which have been set out by the Chief of the Branch (Pichel, 1969) include the 'promotion of international action in the establishment of national and regional gene banks for the conservation and use of valuable genetic resources.' Obviously much remains to be done in the field of genetic conservation but it is hoped that this book may contribute in two ways to the practice of long-term seed storage: first it provides a survey of the principles and problems as they appear at present, and secondly it attempts to signpost those areas in which research is needed and likely to yield useful results.

In addition to the use of long-term seed storage in plant breeding and genètic conservation, Thompson (1970) has pointed out recently that such methods would have a number of advantages in providing the material for seed lists published by botanical gardens. This material is used primarily by taxonomists and bio-systematists. Thompson says 'Complaints concerning seed lists refer most frequently to the poor quality (i.e. low viability) of seed, inaccurate naming, and lack of information about the sample offered. Suggestions for improving quality and information vary in detail but all would require, as they are bound to, a very great deal both of additional labour and expenditure of professional time on each sample of seed offered . . . Suggestions, however necessary and constructive, which require still more labour are unlikely to be widely adopted unless ways can also be found of reducing the time spent on the samples of individual species. One way of doing this is to store seed under conditions which will prolong its viability for the longest possible time.' Thompson goes on to point out that although considerable work has been done on species of agricultural or horticultural interest, much less work has been done on the wide variety of species of interest to botanic gardens. Apparently experiments at the Royal Botanic Gardens, Kew, aim to provide methods for long-term storage of seventy-five per cent or more of the seeds listed in the *Index Seminum*.

The discussion above has dealt with some of the major applications of seed viability studies but there are a number of other technical applications which should be mentioned briefly. Heat treatments can be used to some extent as a method of control of plant pathogens. The principle of the treatment is to use conditions which will destroy the pathogen without significantly affecting the viability or vigour of the seed. In the past the treatments have generally been arrived at empirically, but a particularly interesting example has been described recently in which careful consideration was given to the effect of heat treatments on the viability of tomato seeds in developing a system for the control of Tobacco Mosaic Virus (Rees, 1970).

In the technology of grain storage for food it has long been known that percentage viability is a good criterion of grain quality (Oxley, 1948; Zeleny, 1954); and in grain drying, even when the product is destined for food or feed and the germination capacity is of no direct significance as it would be for seed or malting, it has been found that seed viability is still one of the more useful criteria of the quality of the product from the driers. For example in high-temperature wheat drying studies it has been found that germination percentage is a satisfactory indicator of damage since loaf volume, dough extensibility, and turbidity are closely correlated with this value (Bailey, 1970).

The problems raised by the survival of seed populations of wild species in the soil is of interest to plant ecologists and particularly to those concerned with the practical problems of weed control. Seeds of many wild species can survive for very long periods in the soil. Harrington (1970) has recently brought together an impressive list of species which have shown an ability to survive 30 years or more in the soil. Apparently some seeds can survive for extremely long periods: a claim of 1,700 years' survival, based on archaeological dating has now been made for two common weeds (*Chenopodium album* and *Spergula arvensis*) (Ødum, 1965) but, if we are looking for records, the main present contender is *Lupinus arcticus*, found buried in permanently frozen silt of Pleistocene age in central Yukon; it is claimed that these seeds which germinated rapidly and produced normal looking plants were more than 10,000 years old (Posild, Harrington, and Mulligan, 1967). Although these extreme records are fascinating, from the practical point of view we are more concerned with the very large number of species whose seeds can survive at least several years in the soil. One of the main points of interest about such seeds is that they are capable of germination immediately they are removed from interment – even though the soil in which they have been buried may, for the most part, have been moist and contain sufficient oxygen for germination. Thus in the soil environment, in addition to the other qualities necessary for longevity in 'dry' storage, an additional physiological factor, dormancy, becomes vital; otherwise the seed could not survive for any length of time when buried. If methods could be found for breaking the dormancy of buried seeds *in situ*, the problem of annual weeds could be largely overcome.

In addition to technical applications there is one further aspect of seed-viability studies which may have wider implications, and that is their relevance to the general problem of ageing. This book does not concern itself with the overall problem of senescence in plants, for at this time it would seem that 'ageing' or loss of viability of seeds in storage is a different phenomenon from most other aspects of senescence in plants. Many workers believe that senescence in the whole plant is generally the result of programmed physiological changes (Carr and Pate, 1967; Wareing and Seth, 1967; Woolhouse, 1967) as also appears to be the case for the senescence of the individual organs of the growing plant, e.g. in cotyledons (Cherry, 1967), leaves

(Leopold, 1967; Osborne, 1967; Simon, 1967; Wollgiehn, 1967), and fruits (Sacher, 1967); the ageing of meristems of excised roots may be a special phenomenon, but again this seems to be a physiologically controlled process depending on hormone balance (Street, 1967). In many cases these types of ageing are reversible to some extent, but generally speaking seed ageing is not and the evidence presented in Chapter 9 is more compatible with the notion that it is the result of the accumulation of deleterious unprogrammed changes.

But although the 'ageing' of seeds may not be comparable with many of the wider problems of senescence in plants, paradoxically it may have greater relevance to ageing processes in other organisms where the possibility remains that the changes are the result of the accumulation of unprogrammed events at the cellular level. In most cases in ageing in higher animals it is still not clear whether ageing is the result of programmed or unprogrammed changes, but some of the more popular theories favour the idea of the accumulation of random cellular damage – some of which at least is nuclear – particularly in tissue which is free of cell division. Tissues with undividing cells are specified because in a mitotically active tissue those cells containing aberrations are likely to be removed by diplontic selection (Comfort, 1954, 1955; Curtis, 1963; Medvedev, 1967; Strehler, 1962). Such theories are not without their difficulties but it has been pointed out (Roberts, Abdalla, and Owen, 1967) that seeds provide convenient experimental material for investigating these ideas because they have several notable advantages: (1) during storage no cell division takes place so there is no possibility of loss of damage by selection following cell division; (2) it is easy to initiate cell division for the cytological investigation of chromosomes in the surviving population; (3) because a number of crop species are self-pollinating, it is easy to obtain homozygous material and this simplifies investigations into the induction of genetic mutations during ageing; (4) a number of these self-pollinating species have few but relatively large chromosomes, and this facilitates cytological investigations; (5) unlike many other organisms, the rate of ageing can be altered over extremely wide limits by altering temperature, moisture content, or oxygen pressure; seeds are also convenient material for investigating 'unnatural' ageing factors such as ionising radiation; (6) very large numbers can be conveniently handled in experiments; (7) in animals the investigation of mortality curves is confused by the fact that two types of death may be involved – that due to senescence and that due to accidents (e.g. in man, disease and road accidents); at first it might seem that these two categories can be clearly distinguished but, as Benjamin (1959) has pointed out, this is not so since there is a correlation of road accidents with age which suggests that one's susceptibility to this type of accident is increased by the progress of senescence, and no doubt analogous parallels could be drawn for other organisms; with seeds, however, it is possible to use storage conditions in which it is reasonable to assume that similar accidental deaths have been eliminated.

The structure of the book

A guide to the organisation of this book may be helpful because there are many diverse aspects to the subject and investigations have involved many disciplines. To put the subject matter together in a coherent whole is not easy since a book, of necessity, has a linear structure, whereas the various aspects indicated by the chapter titles interrelate in a multi-dimensional fashion; and it is not always obvious which topic should be considered first. However, a decision had to be made and it was decided to begin with the environmental effects – physical and biotic – on seed deterioration: those occurring during storage are dealt with in Chapters 2 and 3; then the factors important during harvesting are considered in Chapter 4, those before harvesting in Chapter 5, and those after sowing in Chapter 6. Thus the first half of the book deals primarily with the effects of external factors on seed deterioration and loss of viability.

The second half of the book begins with a consideration of the methods of measuring seed viability (Chapter 7) and this is followed by a discussion of the various properties of seeds, collectively known as 'vigour', which decline before a seed completely loses the ability to germinate (Chapter 8). This chapter on vigour is closely related to the previous one on the methods of measuring viability because of the increasing tendency for seed analysts to introduce some aspects of seed vigour into their tests. The book then moves on to deal with some of the consequences of seed deterioration associated with loss of viability: Chapter 9 describes the cytological, genetical and metabolic changes that occur as seeds deteriorate and also discusses the current theories of loss of viability. Chapter 10 gives a short account of the effects of seed deterioration on plant growth and crop yield. Chapter 11 is rather different from the preceding topics in that it deals with the physiological properties which are required for the survival of wild populations of seeds in the soil, whereas the previous chapters are more concerned with cultivated species in which long periods of survival in the soil are of little significance. And finally, so far as the

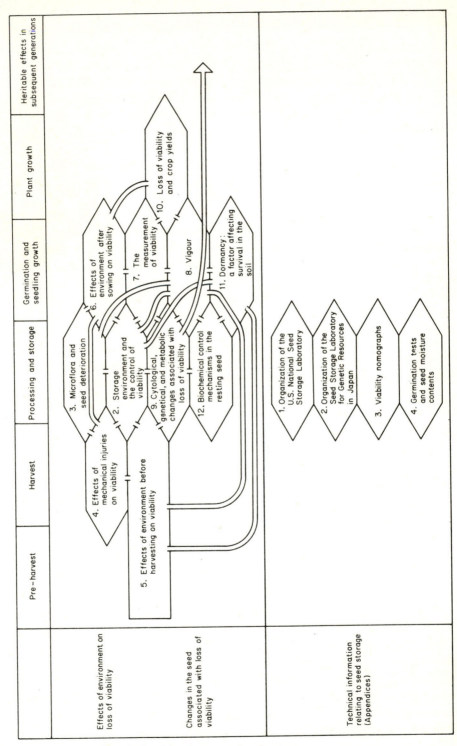

FIGURE I.I   The relationships between the chapters.

main body of the book is concerned, Chapter 12 deals with the bio-chemical control mechanisms in the resting seed. This has been left to the end, not because it is unimportant, but because it is one of the newer and more speculative areas – of great relevance to research workers but, so far, less important to those concerned with practical applications.

The appendices contain descriptions of the aims, organisation and methods of two of the most important national seed storage laboratories – in the United States (Appendix 1) and Japan (Appendix 2), nomographs for relating the period of viability of some common agricultural species to the storage environment (Appendix 3), the international rules for carrying out germination tests and seed moisture determinations, a scale for converting between the two systems of expressing moisture content, data on the relationships between relative humidity and seed moisture content, and recommended safe drying temperatures (Appendix 4).

Figure 1.1 attempts a synoptic view of the overall problems of viability. It explains the context of each chapter in relation to the subject matter as a whole. The individual chapters are represented as hexagons placed in a field described by two coordinates. The axis running from left to right indicates the time sequence in the life of a seed from before harvest until it has developed into a plant and produced its own progeny. The position of a chapter on this scale indi-

TABLE 1.1 *A guide to the essential chapters relating to various topics.*

Topic	Chapters												Appendices			
	1	2	3	4	5	6	7	8	9	10	11	12	1	2	3	4
Ageing and senescence	★	★						★	★		★			★		
Agronomic factors			★	★	★			★		★						
Genetic conservation	★	★	★	★	★				★				★	★	★	★
Grain drying and storage	★	★	★													★
Phytosanitary heat treatment of seeds		★	★				★	★	★							★
Plant breeding	★	★	★					★	★				★	★	★	★
Seed drying		★	★	★			★	★	★	★					★	★
Seed production		★	★	★	★	★	★	★							★	★
Seed storage		★	★	★			★	★	★	★			★	★	★	★
Seed testing		★	★	★	★	★	★	★			★					★
Weed ecology and control		★				★				★						

cates the main stage in this sequence with which it deals – either because of the environmental effects at that stage on seed deterioration, or because of the consequences of previous deterioration which may be manifest at that stage. The vertical scale indicates the emphasis of the subject matter: chapters towards the top of the diagram deal mainly with the effects of environment on seed deterioration and loss of viability, whereas chapters towards the bottom are concerned primarily with the changes brought about in the seed which are associated with loss of viability. The main relationships between the subject matter of the chapters are indicated by passageways between the hexagons. To complete the picture the appendices are grouped separately in a position to show that their contents are primarily concerned with processing and storage.

Although the chapters interconnect the book has been designed so that each chapter is essentially self-supporting and intelligible without reference to the rest. Consequently it is not imperative to read the book through in numerical order of the chapters; neither is it essential to read all. Accordingly it has been possible to recognise that we live in busy times and include Table 1.1 as a guide to the essential reading relating to the topics surveyed in the first section of this introduction – though it is hoped that this will not dissuade anyone from dipping into all we have to offer.

Definitions

SEED

Strictly speaking an angiosperm seed is the structure which develops from an ovule, usually after fertilisation. However, for the purposes of this book the strict botanical definition is not helpful, and we use the term in the agricultural or horticultural sense, i.e. it is a 'dry' dispersal unit which develops from the ovule or ovule and associated tissues: it includes not only true seeds but also any fruits (e.g. achenes and caryopses) which consist largely of seed tissue. In this sense a seed is an embryonic plant in a resting stage usually – though not always – supplied with a food reserve in cotyledons (seed leaves), endosperm (tissue normally derived from the triple fusion of two polar nuclei of the ovule with the second sperm nucleus of the pollen tube), or perisperm (tissue derived from the nucellus); all this usually contained within protective structures consisting of the testa (derived from the integuments of the mother plant), and possibly other structures formed in a variety of ways. The embryo typically results from a zygote derived from the fertilisation of the egg cell in the embryo sac by one of the male nuclei from a pollen tube; but in some plants it may also arise by apomixis or non-sexual processes.

SEED LOT

'Seed lot' is a term used in seed-testing circles and in the seed trade. It denotes a nominally uniform consignment of seeds on which representative tests can be made on random samples. Different seed lots may differ in their place of production, time of harvest, method of harvesting, drying, threshing, cleaning, handling, and storing.

DORMANCY

The term 'dormancy' refers to a condition in a viable seed which prevents it from germinating when supplied with the factors normally considered adequate for germination – i.e. at a suitable temperature in a medium providing adequate water and access to a gaseous environment approximating to air. Dormancy may be the result of an impermeable barrier in the seed coat to water or possibly oxygen (e.g. in hard-seeded legumes) or, more commonly, it may be caused by a specific physiological block to germination. The term can be qualified further into different types of dormancy, and these are defined in Chapters 6 (p. 158), 7 (p. 181), 11 (p. 321) and 12 (p. 370).

VIABILITY

Sometimes when a seed is given suitable conditions for germination it fails to germinate. This inability may be due to dormancy – which is a temporary condition in living seeds which can often be removed artificially, or to loss of viability – a degenerative change which is irreversible and generally considered to represent the death of the seed. Unless otherwise defined, a non-viable seed will be considered to be one which could not germinate when given near optimal conditions, even when it is non-dormant. Thus a viable seed is one which can germinate under favourable conditions, providing any dormancy that may be present is removed.

However, in this book we decided not to abide rigidly by a single definition of viability: sometimes a different definition is more useful in a particular context; it will be made clear where this occurs in the text. For example it will be apparent that the term is used in a rather different sense in Chapter 4; there, in order for a seed to be classed as viable, it is considered necessary for it to be capable of establishing a seedling in the field under conditions which may not be ideal, and this is a rather more demanding requirement. The difficulties of defining viability rigidly are further discussed in Chapters 7 and 8.

MOISTURE CONTENT

Percentage moisture content may be expressed in two ways, either as a percentage of the weight of the material before water is removed (wet weight basis), or as a percentage of the dry weight of the seed (dry weight basis). In seed technology moisture content is usually expressed on a wet weight basis and this is the standard practice adopted in this book, unless specifically stated otherwise. A scale for converting between the two methods of expression and extracts from the International Seed Testing Association rules for determining moisture contents are included in Appendix 4.

References to Chapter 1

ANON, 1970. *Report of the Fourth Session of the FAO Panel of Experts on Plant Exploration and Introduction.* FAO, Rome.

BAILEY, P. H., 1970. A feasibility study of high temperature grain drying. *Home-grown Cereals Authority Prog. Rep. on Research and Development 1969–70*, 34–36. Home-grown Cereals Authority, London.

BENJAMIN, B., 1959. The actuarial aspects of human life-spans. *CIBA Foundation Colloquia on Ageing*, **5**, *The Lifespan of Animals*, 2–20. J. & H. Churchill, London.

BENNETT, E., 1965. Plant introduction and genetic conservation: genecological aspects of an urgent world problem. *Scottish Pl. Breed. Stat. Record, 1965*, 27–113.

CARR, D. J., and PATE, J. S., 1967. Ageing in the whole plant. *Symp. Soc. exp. Biol.*, **21**, 559–99.

CHERRY, J. H., 1967. Nucleic acid metabolism in ageing cotyledons. *Symp. Soc. exp. Biol.*, **21**, 247–68.

COMFORT, A., 1964. *Ageing: the Biology of Senescence.* Routledge and Kegan Paul, London.

COMFORT, A., 1965. *The Process of Ageing.* Weidenfeld and Nicolson, London.

CURTIS, H. J., 1963. Biological mechanisms underlying the ageing process. *Science*, **141**, 686–94.

FRANKEL, O. H., and BENNETT, E. (edits.), 1970. *Genetic Resources in Plants – their Exploration and Conservation.* IBP Handbook No. 11. F. A. Davis Co., Philadelphia.

HARRINGTON, J. F., 1970. Seed and pollen storage for conservation of plant gene resources. Pp. 501–21 in Frankel and Bennett, 1970.

ITO, H., 1970. A new system of cereal breeding based on long term seed storage. *SABRAO (Soc. Adv. Breed. Res. in Asia and Oceana) Newsletter*, **2** (1), 65–70.

LEOPOLD, A. C., 1967. The mechanism of foliar abscission. *Symp. Soc. exp. Biol.*, **21**, 507–16.

MEDVEDEV, ZH. A., 1967. Molecular aspects of ageing. *Symp. Soc. exp. Biol.*, **21**, 1–28.

OSBORNE, D. J., 1967. Hormonal regulation of leaf senescence. *Symp. Soc. exp. Biol.*, **21**, 305–21.

OXLEY, T. A., 1948. *The Scientific Principles of Grain Storage.* Northern Publishing Co. Ltd, Liverpool.

φDUM, S., 1965. Germination of ancient seeds. Floristical observations and experiments with archaeologically dated soil samples. *Dansk. bot. Ark.*, **24**, 1–70.

PICHEL, R. J., 1969. Activities of the Crop Ecology and Genetic Resources Branch. *Plant Introd. Newsletter, No. 22, July, 1969*, 5–6. FAO, Rome.

PORSILD, A. E., HARINGTON, C. R., and MULLIGAN, G. A., 1967. *Lupinus arcticus* Wats. grown from seeds of the Pleistocene age. *Science*, **158**, 113–14.

REES, A. R., 1970. Effect of heat-treatment for virus attenuation on tomato seed viability. *J. hort. Sci.*, **45**, 33–40.

ROBERTS, E. H., ABDALLA, F. H., and OWEN, R. J., 1967. Nuclear damage and the ageing of seeds. *Symp. Soc. exp. Biol.*, **21**, 65–100.

SACHER, J. A., 1967. Studies of permeability, RNA and protein turnover during ageing of fruit and leaf tissues. *Symp. Soc. exp. Biol.*, **21**, 269–303.

SIMON, E. W., 1967. Types of leaf senescence. *Symp. Soc. exp. Biol.*, **21**, 215–30.

STREET, H. E., 1967. The ageing of root meristems. *Symp. Soc. exp. Biol.*, **21**, 517–42.

STREHLER, B. L., 1962. *Time, Cells, and Ageing*. Academic Press, New York.

THOMPSON, P. A., 1970. Seed banks as a means of improving the quality of seed lists. *Taxon*, **19**, 59–62.

WAREING, P. F., and SETH, A. K., 1967. Ageing and senescence in the whole plant. *Symp. Soc. exp. Biol.*, **21**, 543–58.

WOLLGIEHN, R., 1967. Nucleic acid and protein metabolism of excised leaves. *Symp. Soc. exp. Biol.*, **21**, 231–46.

WOOLHOUSE, H. W., 1967. The nature of senescence in plants. *Symp. Soc. exp. Biol.*, **21**, 179–213.

ZELENY, L., 1954. Chemical, physical and nutritive changes during storage. In *Storage of Cereal Grains and their Products*, ed. J. A. Anderson and A. W. Alcock, 46–76. Amer. Assoc. Cereal Chemists, St Paul.

CHAPTER 2

Storage Environment and the Control of Viability

E. H. Roberts

It has long been known that the major factors which influence the longevity of seeds in storage are temperature, moisture content and oxygen pressure (Owen, 1956; Barton, 1961; James, 1967). There have been a multitude of empirical investigations on the effects of temperature and moisture content on the viability period of seeds. In the vast majority of cases it has been shown that the lower the temperature and the lower the moisture content the longer the period of viability. It would be unnecessary, and in any case impossible, to attempt to catalogue all the publications substantiating this statement but some of the more comprehensive recent investigations on a wide variety of species include Anon. (1954), Boswell, Toole, Toole and Fisher (1940), Brett (1953), Gane (1948), Toole, Toole and Gorman (1948), Ching, Parker and Hill (1959). Rather less work has been done on the effects of oxygen and some of the earlier work was conflicting (Owen, 1956; Roberts, 1961b; Touzard, 1961). Nevertheless, it can now be said that for most species, the higher the oxygen pressure, the shorter the period of viability (Roberts, Abdalla and Owen, 1967; Roberts and Abdalla, 1968).

There do appear to be a number of exceptions to these generalisations since in some species there is evidence that the optimum moisture content for maximum retention of viability is relatively high. For example, seeds of many of the large-seeded hardwoods such as *Quercus, Fagus, Aesculus, Castanea* and *Acer saccharinum* (Holmes and Buszewicz, 1958), many species of *Citrus* (Barton, 1943), oil palm (*Elaeis guineensis*) (Rees, 1963), coffee (*Caffea robusta*) (Huxley, 1964) and *Caffea arabica* (Bacchi, 1955, 1956) have all been reported to have relatively high optimum moisture contents for maximum longevity. It has also been reported that cocoa (*Theobroma cacao*) not only has a high optimum moisture content (46 per cent dry weight basis) but a storage temperature of 10°C was more deleterious than 30°C (Barton, 1965). However, the viability response to the environment of these exceptional species is still not entirely clear and in some cases at least there is the possibility that low moisture content has been confounded with deleterious effects of rapid drying. Further work on these species is obviously necessary and it is not possible to generalise about their responses at present. Consequently, in what follows, I shall concentrate entirely on the much more common response in which viability period is increased by decreasing all three factors – temperature, moisture content and oxygen.

Survival curves

Even within a single population of homozygous seeds stored under constant conditions there is a marked variation from seed to seed in period of viability. The frequency distribution of the viability periods of the individual seeds in a population will determine the shape of the survival curve of the population (percentage germination of samples periodically taken from the population plotted against time). All this may appear obvious but it needs mentioning because in descriptions of survival curves in the literature many statements which could lead to misunderstandings have appeared. For example it has often been assumed that the slope of a survival curve is a measure of the rate of loss of viability, whereas the slope is actually a measure of the spread of the distribution of the individual viability periods amongst the seeds of the population – the shallower the gradient, the greater the variation between seeds. At the other extreme, if all seeds had an identical viability period, the survival curve would begin as a horizontal line at 100 per cent viability, then drop vertically at the point in time when all seeds lost viability.

There is no doubt that under constant environmental conditions the form of the seed survival curve is sigmoid. Gane (1948), working with seeds of *Festuca rubra* var. *commutata*, suggested that his observed survival curves agreed well with a modified Gompertz equation:

$$\log R = A - Be^t \qquad (2.1)$$

where R = retention of viability as percentage, t = time of storage in years, and A and B are constants. Because of its sigmoid shape the equation does provide an approximate fit to seed survival curves. But because it assumes that the logarithm of the relative chance of death increases linearly with time, it is asymmetrical, whereas it has been shown that over a very wide range of conditions seed survival curves are symmetrical since the observations conform well to negative cumulative normal distributions or ogives. In other words, when seeds are stored under constant conditions, the periods of viability of the individual seeds in a population are randomly distributed around some mean value. This normal distribution of seed deaths in time has been shown in all species which have been examined in detail – in wheat (*Triticum aestivum*) (Roberts, 1961a), rice (*Oryza sativa*) (Roberts, 1961b), broad beans (*Vicia faba*), peas (*Pisum sativum*), barley (*Hordeum distichon*) (Roberts and Abdalla, 1968), and tomato (*Lycopersicon esculentum*) (Rees, 1970).

A convenient way of dealing with data which fit a cumulative normal distribution is to plot the probit of percentage germination against time or, what amounts to the same thing, to plot percentage germination on a probability scale against time. When this is done, a cumulative normal distribution results in a straight line. Figure 2.1 shows some examples of such survival curves for samples of the same population of broad bean seeds stored under various conditions. In this figure two horizontal lines have been drawn, one at the 50 per cent viability level and one at the 15.9 per cent level. It has been pointed out (Roberts, 1961a) that the distance along the abscissa to the point

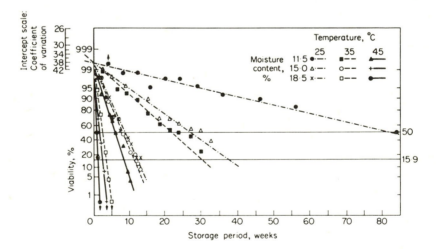

FIGURE 2.1   Survival curves of broad bean seeds stored under various combinations of temperature and moisture content. Percentage viability is plotted on a probability scale. Since values of 100 and 0 per cent cannot be plotted on a probability scale, where such values occurred experimentally they are indicated by points at 99.5 and 0.5 per cent with arrows pointing up and down respectively. As explained in the text, the intercept of the survival curves at zero time is a function of the Coefficient of Variation of the distribution of deaths in time ($\sigma/p \times 100$). The relationship between the intercept and the Coefficient of Variation is shown by the intercept scale.

where the survival curve intersects the 50 per cent level estimates the mean viability period ($\bar{p}$) for the treatment; and it can be shown from the area under a normal curve that the distance along the abscissa from this point to the point where the survival curve intersects the 15.9 per cent level is an estimate of the standard deviation (σ) of the distribution. Since all the survival curves in Fig. 2.1 have a common origin at zero time, it can then be demonstrated by the simple geometry of equiangular triangles that the ratio ($\sigma/\bar{p} \times 100$) – i.e. the Coefficient of Variation – remains the same, irrespective of how the mean viability period has been altered by the storage treatment. This has been shown to be true in all the species mentioned previously which have been examined in detail.

It may be concluded that when samples of seed are stored under a variety of constant conditions and the data are plotted in this way, a given population generates a family of curves of common intercept on the ordinate at time zero indicating, as we have seen, a constant Coefficient of Variation. The ratio $\sigma:\bar{p}$ determines the value of the intercept and so it is possible to indicate the Coefficient of Variation of the population directly by an intercept scale on the ordinate as shown in Fig. 2.1*

Although the distribution of seed deaths in time remains normal under most conditions, when the storage conditions are so severe that

*At first sight it may seem strange that the intercept should be a measure of the degree of variability amongst the viability periods of the individual seeds of the population. One might think that intercepts would always tend to occur at 100 per cent or values close to it. And so they do, but a probability scale reflects a normal distribution and stretches to infinity. Thus sample data near 100 per cent are on a part of the scale which is rapidly expanding to infinity and, accordingly, when data are transformed in this way, substantial differences in intercept appear. However, on this part of the scale it would be unwise to take the actual sample values at the beginning of the survival curves as an indication of the intercept value for the whole population since sample errors here are at their greatest: the behaviour of only a single seed in a sample of 100 in this region of the curve has an enormous effect. It is, therefore, much better to estimate the value of the intercept by extrapolation from the sample values found near the centre of the distribution. In fact, when the curves are fitted by the most appropriate statistical treatment, probit analysis, greatest weight is placed on the more central values of the distribution and very little weight on the more extreme values (Finney, 1952).

It might also be thought strange to use a scale on which 100 per cent viability cannot be indicated since this is a sample value that turns up frequently in practice. However, it is probably the case that a population would seldom, if ever, have a viability of 100 per cent, although it may have a value which approaches it very closely; in such cases samples drawn from the population would frequently show a value of 100 per cent.

the mean viability period is of the order of a week or less, the distribution may become skewed so that the longer surviving seeds in a distribution lose their viability somewhat sooner than would be expected (Roberts and Abdalla, 1968). Examples of such survival curves are shown in Fig. 2.2. Although these deviations from the normal distribution are of theoretical interest, for most practical purposes they can be ignored since the deviations are not great and, in any case, one does not normally have to deal with such severe storage conditions.

It is necessary to mention here that recently Watson (1970) has suggested a different treatment of survival curves. He has described survival curves for wheat and rice seed subjected to relatively high temperatures (he also comments that survival curves for soya beans, *Glycine Max*, at lower temperatures are very similar) in which percentage germination was plotted on a log scale against time on a linear scale. It was suggested that these graphs 'clearly indicate two logarithmic rates of loss of germination capacity'. This statement is based on the fact that two straight lines can be drawn through the survival curves – the first of shallow gradient during the period when the shorter-lived members of the population die, and the second of

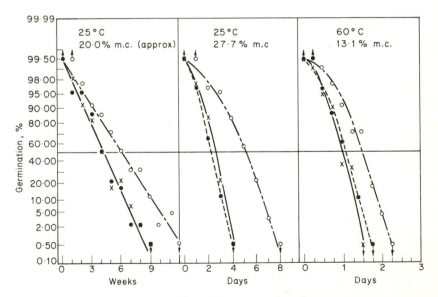

FIGURE 2.2   Survival curves of broad beans stored under constant gaseous environment at the temperatures and moisture contents (m.c.) indicated.
o nitrogen; × air; • oxygen. Percentage germination is plotted on a probability scale. (From Roberts and Abdalla, 1968.)

much steeper gradient during the period when the majority of the seeds die. Watson's results are in fact very similar to the data illustrated by the survival curves shown in this chapter: the first part of the survival curve has a shallow gradient – particularly when percentage germination is plotted on a log scale – but later becomes steeper as the mean viability period (and the mode of the distribution) is reached. These results therefore do not indicate two rates of loss of germination capacity, but represent the typical normal distribution of viability periods of individual seeds. When seeds are subjected to conditions in which there is a rapid loss of viability, the resulting skewness, in which the tail of the distribution tends to be lost, will make Watson's pattern agree quite well with those observed in practice. However, it must be emphasised that the presentation of data on the distribution of deaths in this way does not necessarily indicate two rates of loss of viability.

The relationship between temperature, moisture content and period of viability

In the varied literature on seed storage and grain-drying technology a number of investigators have attempted to see whether it is possible to define quantitatively the relationship between environmental factors and loss of viability. Often these essays have been made without acquaintance with previous similar attempts and so different approaches and symbols have been used.

One of the earliest attempts to define quantitative relationships was undertaken by Groves (1917) who investigated the viability of wheat seeds stored at temperatures between 50° and 100°C. He described the relationship between period of viability and temperature as:

$$T = a - b \log Z \tag{2.2}$$

where T = temperature (°C), Z = the time taken for 75 per cent of the seeds to be killed, and a and b are constants.

More recently Touzard (1961), who worked on ten species of horticultural seeds over a range of temperature from 4–30°C and at various relative humidities from 10–75 per cent, described the relationship between viability and temperature by the equation:

$$L_m = L_{mo} a^{(t_o - t)/10} \tag{2.3}$$

where L_m = mean viability period at temperature t, L_{mo} = the mean viability period at temperature t_o and a (the Q_{10} for rate of loss of viability) is of the order of 2.5–3.0. Although the end-point used as a

measure of period of viability is different from that used by Groves –
50 per cent death as against 75 per cent death – it will be seen that the
relationship is the same type, although expressed in a different way.

These equations do not take into account the effect of the moisture
content of the seeds and therefore, without altering the value of a
constant, they would only be correct for one particular moisture
content. A number of other investigators have developed equations
in which the effect of moisture as well as temperature is accounted for.
Hutchinson (1944) investigated the effects of temperature (46–114°C)
over a range of moisture contents from 14–35 per cent on wheat seed.
For defining the period of viability he used two alternative end-points,
either the point at which all seeds had died (100 per cent death) or the
point at which no death had occurred but at which a definite delay in
germination was observed when the seeds were set to germinate
(which probably represents the point at which germination is about
to begin to fall). For the first end-point (100 per cent death) the
following equation was given:

$$\Theta = 130.3 - 5.4 \log_{10}t - 43.87 \log_{10}m \qquad (2.4)$$

where Θ = temperature (°C), t = time of heating in minutes, and
m = percentage moisture content. For the second end-point the same
equation was found to apply if the first constant were altered from
130.3 to 122.0. However, at moisture contents below 14 per cent,
Hutchinson found that these relationships no longer applied and his
results showed that viability was held longer than would be predicted
by the equations.

Using results on wheat and other temperate cereals stored at temperatures from 15–25°C and moisture contents from 11–23 per cent
published by other workers, it was suggested (Roberts, 1960) that the
following equation provided a good fit to the data:

$$\log p_{50} = K_v - C_1 m - C_2 t \qquad (2.5)$$

where p_{50} = the half-viability period (time taken for 50 per cent of
the seed to lose viability),* m = moisture content, t = temperature
(°C) and K_v, C_1 and C_2 are constants. This equation was shown later

---

*In previous publications the symbol p has been used for the half-viability period.
It is now suggested that p_{50} is used for the half-viability period and $\bar{p}$ for the mean
viability period. As discussed earlier, under most circumstances the distribution of
the viability periods amongst the individual seeds of a population is normal and
therefore symmetrical, in which case p_{50} and $\bar{p}$ are identical. There are some advantages, however, in retaining the distinction and being clear as to which measure of
viability is adopted for particular circumstances.

to apply to rice which was investigated over the range 12–14.5 per cent moisture content and 27–47°C, although the values of the constants were different from wheat. Further experiments on barley (*Hordeum distichon*), broad beans (*Vicia faba*), and peas (*Pisum sativum*) over the ranges 25–45°C and 12–18 per cent moisture content showed that equation (2.5) provided an excellent fit in all cases, although the constants are slightly different for the three species (Roberts and Abdalla, 1968). An equation based on the relationship expressed by equation (2.4) was also tested. It will be seen that this relationship could be written in terms of equation (2.5) as:

$$\log p_{100} = K_v - C_1 \log m - C_2 t \qquad (2.6)$$

although since the end-point (p_{100}) is different and since the moisture-content constant refers to a log value, the values of the constants K_v and C_1 would be different. The essential difference between the two equations, however, resides in the moisture-content term. Curves were fitted to the data on barley, broad beans, and peas according to each equation by the method of least squares (Roberts and Abdalla, 1968). The residual variance in all cases was extremely small and failed to show that either equation was superior to the other. Hutchinson's equation (2.4) was originally derived from results over a wider range of moisture contents, and a moisture-content term of his type might prove ultimately to be preferable. On the other hand, although equations were not fitted, data have been provided by McFarlane, Hogan and McLemone (1955) on rice treated at temperatures from 50–80°C over the wide moisture-content range from 5–25 per cent. These results suggest that the moisture-content term of equation (2.5) would fit very well over this range (which covers the normal range of storage conditions). For the time being, then, we prefer to use equation (2.5) which also has the advantage of being slightly simpler.

Hukill (1963) approached the problem of defining the relationship between viability and environmental factors in rather a different way: using the data of Toole and Toole (1946) on the viability of soya beans he developed the concept of an 'age index'. This could be considered as an index of physiological age and he defined it as follows:

$$\text{age index} = \text{months of storage} \times 10^{0.143mc} \times 10^{0.0645 T} \qquad (2.7)$$

where mc = moisture content (per cent) and T = temperature (°C). Hukill's graphs indicated that germination percentage is a fairly precise though complex function of the age index.

It was pointed out previously (Roberts, Abdalla and Owen, 1967) that equation (2.7) shows the same relationship as equation (2.5).

This can be demonstrated as follows. Taking the log of equation (2.7) and rearranging (using AI to represent age index), we get:

$$\text{log months of storage} = \log AI - 0.143\,mc - 0.0645\,T$$

In equation (2.5), $K_v = \log AI$ when germination has dropped to 50 per cent it follows therefore that:

$$\log p_{50} = K_v - 0.143\,mc - 0.0645\,T$$

which, with adjustment of constants according to species, is the same as equation (2.5).

Harrington (1963) has also implied the same type of relationship as equation (2.5), but using a different end-point as a measure of viability period, by his 'rule of thumb' of seed storage which states: (1) For each 1 per cent reduction in moisture content the storage life of seed is doubled; and (2) for each 5°C lowering of the storage temperature the storage life of the seed is doubled.

In this context the work of Burgess, Edwards, Burrell and Commell (1963) should also be mentioned. They published data on barley in which the time taken for viability to drop to 95 per cent was determined at combinations including 32° and 40°C and 12.8 and 13.3 per cent moisture content, and these results were compared with similar data derived from other workers over the range 10–30°C and 12–24 per cent moisture content. They did not describe these results in terms of equations but it is evident from their graphs that these data tend to show the same relationships as equation (2.5).

Using his own data on wheat at high temperatures (60–68.3°C) and other workers' data on wheat, rice and soya beans at lower temperatures, Watson (1970) has shown that, at any constant moisture content, the log of the specific reaction rate (measured as the log of the reciprocal of the time taken for germination to drop to 90 per cent) is a negative linear function of the reciprocal of the absolute temperature. In other words, since 90 per cent viability implies 10 per cent death, we could express his relationship as:

$$\log \frac{1}{p_{10}} = a - b/T \qquad (2.8)$$

where p_{10} is the time taken for viability to drop to 90 per cent, T is temperature (°K) and a and b are constants.

Over a relatively narrow range of temperature at ambient or higher, there is an approximately negative linear relationship between temperature (°C) and the reciprocal of absolute temperature ($1/T$) and, as a result, either equation (2.8) or (2.5) would be equally satisfactory for describing the temperature-viability relationship. Since equation (2.8) implies an Arrhenius relationship, it might be thought that it would be preferred since it would emphasise a conformity of the phenomenon of loss of viability with the laws of physical chemistry. However, equation (2.8) implies a decrease in Q_{10} for rate of loss of viability with increase in temperature whereas, if there is a variation in Q_{10} over wide ranges of temperature, the available evidence, which will be discussed shortly, indicates an *increase* in Q_{10} with increase in temperature.

It is quite evident from this survey that a number of investigators, working independently on various species, have concluded that the period of viability of a population of seeds under constant conditions can be defined quite accurately providing the temperature and moisture content are known. Although there is still some doubt as to the *precise* relationship between moisture content and viability, there is apparently good agreement on the type of relationship between temperature and viability. Furthermore, where both factors have been investigated simultaneously, there is no evidence of a mathematical interaction between temperature and moisture content.

If this work is taken in conjunction with the work on survival curves, it is possible to define precisely the whole pattern of loss of viability in relation to temperature and moisture content. The pattern is most clear and usefully expressed in terms of three equations; for convenience I shall refer to these as the three basic viability equations. They are as follows:

(a) Seed survival curves under constant environmental conditions are negative cumulative normal distributions. Such curves cannot be explicitly expressed in terms of an equation, but they result from the fact that the distribution of deaths in time is described by the normal distribution. The distribution of deaths for the whole population can be expressed as:

$$ y = \frac{1}{\sigma\sqrt{2\Pi}} e^{-\frac{(p-\bar{p})^2}{2\sigma^2}} \tag{2.9}$$

where y is the relative frequency of deaths occurring at time p, $\bar{p}$ is the mean viability period, and σ is the standard deviation of the distribution of deaths in time.

(b) The spread of the distribution in time is proportional to the mean viability period. This has been expressed (Roberts 1960, 1961a; Roberts and Abdalla, 1968) as:

$$\sigma = K_{\dot\sigma}\,\bar p \qquad\qquad (2.10)$$

where $K_{\dot\sigma}$ is a constant for the species.

(c) The mean viability period may be related to temperature and moisture content by equation (2.5) in which the term $\bar p$ is substituted for p_{50}:

$$\log \bar p = K_v - C_1 m - C_2 t \qquad\qquad (2.11)$$

These basic viability equations have been shown to be reasonably accurate for predicting percentage viability from a few days to several

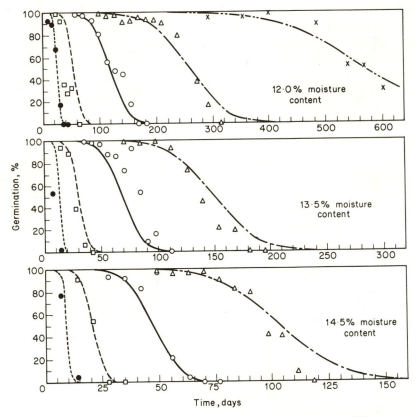

FIGURE 2.3   Survival curves for rice hermetically sealed in air at different temperatures and moisture contents. Note that the time scale is different for each moisture content: reading from top to bottom, each scale is a 2 × magnification of the previous one. Storage temperatures were as follows: 27°C (×); 32°C (△); 37°C (○); 42°C (□); 47°C (●). The curves were fitted to the points according to the three basic equations mentioned in the text using the following values for the constants in equations (2.10) and (2.11): $K_{\dot v} = 5.686$, $C_1 = 0.159$, $C_2 = 0.069$ and $K_\sigma = 0.210$. (From Roberts, 1961b.)

years (Roberts and Abdalla, 1968). Furthermore Ito and his associates have found that the equations are satisfactory for predicting the viability of rice for a period of at least 15 years (see Appendix 2).

It is possible to express the relationships shown by the three equations simultaneously in graphic form by a number of different conventions. For example Fig. 2.3 shows the theoretical curves fitted to some experimental data on rice. Figure 2.4 shows the relationships for broad beans using a different convention.

When the relationships are expressed in the form of Fig. 2.4 it is easy to see why different workers have found essentially the same relationship between temperature, moisture content and period of viability irrespective of which level of germination they adopted as a measure of the end-point of viability period. In summary the following end-points have been used: the time taken for viability to begin to fall (p_0), the 95 per cent level (p_5), the 75 per cent level (p_{25}), the 50 per cent level (p_{50}), and finally the time taken for viability to be lost completely (p_{100}). As pointed out previously the end-points at the extremes of the distribution (p_0 and p_{100}) will be subject to grave sampling errors and that is why p_{50} was adopted in equation (2.5). Nevertheless, because the relationship between p_{50} and any other end-point remains constant, as indicated by equation (2.10), any end-point would be expected to yield the same slope constants (C_1 and C_2 in equations 2.5 or 2.9), although the value of the intercept constant, K_v, would depend on the end-point adopted. Thus when log viability period is plotted on a vertical scale against temperature and moisture content on horizontal scales, all levels of viability are represented by planes of common slope (compare the 10, 50 and 90 per cent levels of viability shown in Fig. 2.4).

Although Figs. 2.3 and 2.4 help one to visualise the relationships described by the basic viability equations, a nomograph (sometimes called nomogram) is probably the most useful form for presenting the data for a given species in a way they can be most conveniently used in practice (for interesting discussions on the use and construction of nomographs see, for example, Allcock, Jones and Michel, 1962, or Davis, 1962). Nomographs for all species for which sufficient data have been obtained at the present time are shown in Appendix 3.

The nomographs in Appendix 3 can be considered as no more than a beginning to the cataloguing of viability data and its presentation in a way which would be useful to those concerned with the practical problems of seed storage and the preservation of genetic stocks. They provide a convenient method of predicting percentage viability after any time at any given combination of storage conditions or, alterna-

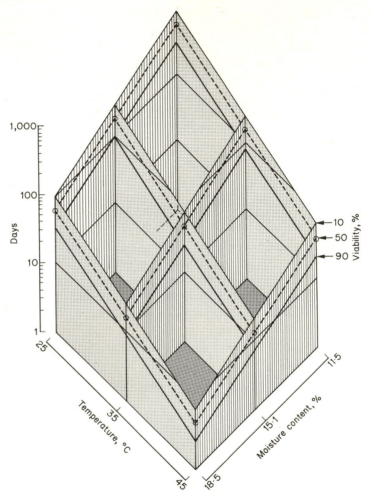

FIGURE 2.4   Isometric three-dimensional graph showing the relationship
between temperature, moisture content and the period taken for the viability of
broad beans to drop to various levels. Time is plotted on a log scale. The small
circles represent the experimental points for the observed times for germination to
drop to 50 per cent. A plane (broken lines) was fitted to these points by the
method of least squares according to equation (2.11). The values for the constants
thus determined were: $K_v = 5.766$, $C_1 = 0.139$, and $C_2 = 0.056$. The value K_σ in
equation (2.10) was derived from the survival curves and found to have a value of
0.379. Using this constant in conjunction with tables showing the area under a
normal curve, positions of the 90 per cent viability and the 10 per cent viability
planes were fixed. For any given combination of conditions shown on the
horizontal axes of the graph, 80 per cent of the population may be expected to die
in the period of time represented by the space between these two planes.
(From Roberts, Abdalla and Owen, 1967.)

tively, they enable one to determine the alternative combinations of temperature and moisture content to maintain viability above a given level for a given period. Before going on to deal with the limitations of the basic viability equations which will, of course, restrict the extent to which the nomographs can be used in practice, it is important to emphasise that only a small amount of work is necessary to construct a nomograph for a species which conforms to the basic viability equations. One can be reasonably optimistic that the basic viability equations will be capable of widespread application since all of the first five species which have been investigated from this point of view have conformed. If it is known that a species complies with the basic viability equations it is only necessary to determine the values of the four constants K_v, C_1, C_2 and K_σ in order to define its complete pattern of loss of viability. And although it would be neither prudent nor statistically desirable to restrict the experimental work to the minimum, it is interesting to note that the determination of the value of these constants theoretically demands only four germination tests. Two germination tests carried out at different times for seed stored under one set of conditions would determine the slope of the survival curve, from which $\bar{p}$ and σ could be determined for those conditions; from these values K_σ could be determined. Two further germination tests, each for a different set of conditions from the first two tests and from each other, carried out any time after the start of the experiment would fix the survival curve for these conditions since the survival curves would have a common origin at zero time with the survival curve derived from the first two tests. Thus two further values of $\bar{p}$ could be determined from the latter tests. Finally using the three values of $\bar{p}$, three simultaneous equations (equation 2.11) would determine the values of K_v, C_1 and C_2.

Limitations of the basic viability equations

POSSIBLE INACCURACIES IN THE BASIC EQUATIONS

The fact that the frequency distribution of deaths may become skewed when the storage conditions are severe and the mean viability period is consequently a week or less means that equation (2.9) does not apply under such conditions. But this is a complication which is of no consequence within the normal range of storage conditions and may be ignored. There is no reason to suppose that conditions arise when equation (2.10) does not apply. Thus any inaccuracies of practical consequence are more likely to involve equation (2.11).

It has already been mentioned that there is some question as to whether the moisture-content term of equation (2.11) is strictly accurate. If the relationship should ultimately prove to be that indicated by equation (2.6), then the prediction of viability periods using equation (2.11) outside the range of moisture contents from which the value C_1 was calculated would lead to an underestimate of tne period of viability.

With regard to the temperature term, although it is evident that all workers are in agreement that the relationships is of the type indicated by equation (2.11) it has been pointed out (Roberts and Abdalla, 1968) that this may not be strictly correct when an attempt is made to encompass a very wide range of temperature. A measure of the rate of loss of viability is the reciprocal of the time taken for viability to be lost. On this basis equation (2.11) indicates that the Q_{10} for the rate of loss of viability for all temperature ranges is the same, and this is what the experimental results of any individual investigator suggest. Generally speaking, however, when a comparison is made between the various temperature coefficients for rate of loss of viability which can be derived from the literature, the Q_{10} value appears to be greater the higher the temperature range which was investigated. For example, Touzard (1961), working on ten species of vegetable and flower seeds over the range 4–30°C, reported a Q_{10} of 2.5–3.0 for all the seeds he studied; the results for wheat over the range 15–25°C showed a Q_{10} of 3.3 (Roberts, 1960); for rice between 27° and 47°C it was 4.9 (Roberts, 1961b) and for barley, beans and peas between 25° and 45°C the values found were 5.6, 3.6 and 4.5 respectively (Roberts and Abdalla, 1968); for wheat between 50° and 100°C the results of Groves (1917) suggest a Q_{10} of about 10 (Roberts, 1961b) and for wheat between 63° and 70°C the results of Watson (1970) suggests a Q_{10} of 14, whereas the constants for the equations produced by Hutchinson (1944) for wheat over the range 46–114°C indicate the extremely high Q_{10} of 71. Thus there is indirect evidence that the Q_{10} values are not constant as indicated by the published equations, but increase with increase in temperature. Burges *et al.* (1963), in trying to relate their own work on barley between 32° and 40°C with that of other workers' data between 10° and 30°C, suggested that there may be an increase in the rate of loss of viability above 35°C. The inference from all these observations would be that the temperature term in equation (2.11) would be inappropriate for dealing with a very wide range of temperatures. If the equation were used to predict viability above or below the temperature range over which C_2 were calculated it would be expected to result in an over-estimate of viability period.

In summary, there is a possibility of inaccuracies in both the temperature and moisture-content terms: the first could lead to an under-estimate of viability period at extreme values whereas the second could lead to an over-estimate. If these errors exist, it would be difficult to know whether or not they would tend to cancel each other when used to predict conditions suitable for long-term storage. However, it has been possible to obtain at least a very rough check on equation (2.11) for an extremely long period of storage, and the evidence suggests that in practice, the equation can give at least a rough guide for extremely long periods of storage. This conclusion is based on a consideration (Roberts, 1961a) of data reported by Aufhammer and Simon (1957) on various species of seeds which were found sealed in glass tubes within the foundation stone of the Nuremburg City Theatre where they had lain for 123 years. The effective temperature during storage was estimated (using equation 2.13) to have been 10.6°C. Unfortunately, relatively few seeds were available for test – between 20 and 32 for the species considered here; nevertheless the results are at least indicative of the potential applicability of the equations to such conditions. Although all the wheat (8.3 per cent moisture content) was dead, the barley (7.3 per cent moisture content) gave 12 per cent germination, and the oats (8.0 per cent moisture content) gave 22 per cent germination. When these figures were being examined the constants for barley had not been worked out and on the basis of the wheat constants it was suggested at that time (Roberts, 1961a) that the barley appeared to be lasting better than would be expected. However, it is now possible to make the proper comparisons with the extrapolations from the short-term experiments on the appropriate species. Calculations from extrapolating the nomograph in Appendix 3 suggest that barley would be expected to give between 50–60 per cent germination after 123 years under the Nuremburg conditions, or alternatively to give 12 per cent germination after about 165 years. In other words, the seeds did not last as well as would have been predicted by the basic viability equations. However, considering the very small size of sample of barley from the foundation stone (25 seeds) this discrepancy is small enough to give reasonable confidence in the application of the basic viability equations for long-term seed storage. All that can be said for wheat from the foundation stone (0 per cent germination) is that this would be predicted by the appropriate nomograph in Appendix 3, but of course it is not known how long the seeds in the stone had been non-viable.

GENOTYPIC VARIATIONS IN VIABILITY PERIOD

So far attempts to apply the basic viability equations have only been made on five species. It is difficult not to be impressed by the fact that in all these species the experimental data provide a very good fit; and one is sometimes tempted to wonder whether the basic viability equations could be of almost universal application. However, there are a number of reports which provide a salutary corrective to unwarranted optimism. For example, an examination of the data published by Ching, Parker and Hill (1959) on *Lolium perenne* and *Trifolium incarnatum* suggests that although equations (2.9) and (2.10) would apply, and although decreasing temperature and decreasing moisture content increase viability period, equation (2.11) would provide only a very rough fit to the data. The main reasons for this are that the seeds lasted longer than would be predicted at the lowest temperature (3°C), and also at the lower moisture contents (6 and 8 per cent) in the case of the ryegrass.

The question then arises, in those cases where it has been shown that the basic viability equations do apply, to what extent are the same constants applicable to different species? Unfortunately the evidence is somewhat conflicting. A re-examination (Roberts, 1960) of some work by Robertson, Lute and Gardner (1939) on three cereal species – wheat, barley and oats – suggested that these species behave very similarly. But more recent work has indicated differences between barley and wheat, particularly with regard to the constant K_v (see Appendix 3). On the other hand, an analysis by MacKay and Tonkin (1967) of an extensive series of investigations on material stored at the Official Seed Testing Station at Cambridge, England, has indicated that barley and wheat behave almost identically whereas oats has a mean viability period which is almost twice as great as these two cereals. On balance it would be best to assume that species differ, even when they have, like the cereals, similar types of seed.

The next question to ask is to what extent would the same constants apply to different varieties or cultivars within a species? Here again, unfortunately, the evidence is not entirely consistent. For example, although the earlier work was not very critical, there were many suggestions in the literature on rice, that different cultivars show marked differences in their ability to retain viability (e.g. Jones, 1926; Sahadevan, 1953; Mudalier and Sundararaj, 1954; Dore, 1955; Ghose, Chatge and Subrahmanyan, 1956; Nagai, 1959; Anthanakrishna Rao, 1959). However, when special care was taken to store six widely different cultivars (including representatives of two species,

Oryza sativa and *O. glaberrima*, and two sub-species of *O. sativa*, *indica* and *japonica*) under *identical* conditions, the survival curves were also found to be identical (Roberts, 1963) (see Fig. 11.1, p. 325).

It may well be that some of the difficulties of comparison arise because of different techniques employed by different investigators. One of the major difficulties is the measurement of moisture content. An examination of the nomographs in Appendix 3 shows that a small discrepancy in the determination of moisture content could lead to quite large differences in predicted viability period. Different methods of estimation of seed moisture content give different values and most methods cannot claim to measure moisture content to a greater accuracy than about 0.5 per cent;* and different types of seeds of the same species can give different results even though they contain the same amount of water, e.g. hard and soft wheats (Oxley and Pixton, 1960).

Although there have been a number of reports of differences in longevity between different cultivars or varieties, in much of the work described one cannot be certain that the moisture contents of the seeds which were being compared were identical: in many cases the moisture contents were neither controlled nor recorded. Nevertheless, even if such cases are ignored, there are still many reports which show convincingly that significant differences in viability periods between genotypes within a species do exist and, furthermore, that these differences are heritable. Lindstrom (1942) investigated the inheritance of seed longevity in maize; he concluded that the degree of longevity was inherited but not simply, and reciprocal crosses showed pronounced maternal effects. Sayre (1947) and Haber (1950) also demonstrated heritable differences in longevity amongst cultivars of this species. Toole and Toole (1954) reported a varietal difference in beans. One of the most extreme cases of intraspecific differences investigated under controlled conditions is that reported by Harrison (1966) who investigated 20 cultivars of lettuce (*Lactuca sativa*) between 5 and 6 per cent moisture content at 18°C: when seeds were stored in carbon dioxide a threefold difference was found between the longest and shortest lived cultivars. James, Bass and Clark (1967) carried out an investigation on 8 tomato, 8 bean, 5 pea, 15 watermelon, 11 cucumber and 5 sweet corn cultivars under controlled conditions; significant differences in longevity between cultivars were found within all species, but the differences between the best and worst cultivars of a species were apparently not very great. Although

*Because of these difficulties it is strongly recommended that research workers should adhere to the rules of the International Seed Testing Association (see Appendix 4).

the moisture contents were not strictly controlled, interesting comparisons were made by MacKay and Tonkin (1967) between cultivars of a number of different species from an extensive series of trials. In three cultivated forms of beet (*Beta vulgaris*) – mangel, garden beet-root and sugar beet – the mean viability period (as indicated by the time taken for viability to fall to 50 per cent) was identical for all three, although the Coefficient of Variation of the distribution of deaths (as indicated by the time taken for viability to fall to 80 per cent) was much smaller in mangel than in the other two forms. In the case of oats, wheat and barley, although differences were shown between different cultivars, these were not very great.

THE INFLUENCE OF PROVENANCE ON VIABILITY

In Chapters 4 and 5, a number of the factors, operating before and during harvest, which can affect viability are discussed. It is not surprising, then, that samples of seed obtained from different sources may show differences in viability behaviour. It is not always easy to assess what the causes of these differences are or even sometimes to know how important they are. One of the major difficulties in assessing the variation in viability between samples from different sources is that the experimental investigations on the subsequent behaviour of the seeds have not been carried out under controlled storage conditions and, to make matters worse, the different samples have not been dried to identical moisture contents. Nevertheless, the seed begins its existence before it is harvested and it is only to be expected that seeds harvested in different climates or at different times will have been subjected to different pre-harvest conditions which will have caused different amounts of deterioration by the time the seeds are harvested.

MacKay and Tonkin (1967) have published data which indicate the range of intra-cultivar differences in viability period which may be expected in practice in a number of common agricultural species. For example it was found that the half-viability period of red clover derived from English sources was about 18 per cent less than that derived from Canadian and New Zealand sources. A number of other examples of differences between seed of various species obtained from different geographical sources are also mentioned.

Haferkamp, Smith and Nolan (1953) in the United States, reported reduced longevity in seed of several wheat varieties harvested in a year of high summer rainfall. MacKay and Tonkin (1967) followed up this observation by examining the longevity of wheat, barley and oats harvested in England in seasons when sunshine was above average and rainfall below average as compared with seasons with below average sunshine and above average rainfall. In each case the viability of seeds from the years when ripening and harvesting conditions were poor declined more rapidly. The decrease in half-viability period in seeds derived from poor harvest years as compared with those from good harvests was about 8 per cent in oats, 14 per cent in wheat and 24 per cent in barley. More disconcerting than this is the data they present showing that the half-viability periods between extreme years may vary by a factor of almost two in both wheat and barley.

Within a given season there is evidence that the time of harvest can also affect viability. Bass (1965) found that mature seed of Kentucky bluegrass (*Poa pratensis*) remained viable longer than immature seed. Shands, Janisch and Dickson (1967) showed that harvesting barley prematurely was slightly deleterious to viability. In addition these workers showed that delaying harvest by about three weeks was also deleterious to subsequent viability in experiments where the grain was stored at high moisture content (18 per cent).

In Chapter 4 it is shown that mechanical damage can have a considerable effect on viability in certain species. In that chapter attention is concentrated on those seeds, particularly some of the legumes, which are most susceptible to such damage. However it is comforting to note that Webster and Dexter (1961) have shown that there are some species which are practically immune from normal mechanical abuse. Of the four species they investigated, barley was found to be in this category, whereas wheat was slightly susceptible to mechanical damage, maize more so, and beans were the most susceptible. Although I have no data, from my own experience with both rice and barley, I would suggest that they behave very similarly in this respect, as I imagine would any seed similarly well-protected by a husk. So far as the application of the nomographs in Appendix 3 is concerned it would seem that problems of mechanical abuse are not likely to affect their application in rice and barley, but severe mechanical damage may affect their application in wheat and particularly in peas and beans. Mechanical damage in wheat is likely to be greatest during harvest if the moisture content is within the range 19–25 per cent (Mitchell and Caldwell, 1962).

In summary there is evidence that harvest and pre-harvest conditions can affect subsequent viability. Consequently although any particular sample may obey the basic viability equations, the constants may be affected to some extent by the previous history of the seeds. For this reason, the nomographs in Appendix 3 should not be considered more than a useful guide at this stage. It would be interesting to know how many of the four constants are affected by pre-storage conditions. A critical examination of pre-storage conditions in relation to their effects on these constants might well enable more precise viability predictions to be made.

THE EFFECTS OF FLUCTUATING ENVIRONMENTAL CONDITIONS ON VIABILITY

The data which led to the development of the basic viability equations were obtained under constant conditions of temperature and moisture content. It would be useful to know whether the equations could be extended to include fluctuating conditions. There have been a few reports in the literature that fluctuating conditions are harmful, but in no case has the effect of fluctuating conditions been compared with what would be expected from integration of the environmental parameters according to the relationships indicated by equation (2.11). At present there is no *a priori* reason to suppose that *change* in temperature or moisture content would itself be deleterious save, possibly, for very rapid changes in moisture content.

In the absence of evidence to the contrary it would seem reasonable to suppose that equation (2.11) could be applied to fluctuating conditions. However, it would not be sufficient to take arithmetic mean values for the fluctuating conditions since viability period is related logarithmically to both parameters. As an aid to applying equation (2.11) to fluctuating conditions the following equation was suggested (Roberts, 1963) for estimating the 'effective temperature' of a fluctuating temperature regime:

$$T_E = \frac{\log\left\{\dfrac{\Sigma[\text{antilog }(tC_2) \times w]}{100}\right\}}{C_2} \tag{2.12}$$

where T_E = effective temperature (substituting for t in equation (2.11)), t = recorded temperature in °C, w = the percentage of time spent at each temperature t, and C_2 is the value of the constant applied in equation (2.11).

If each temperature (t) is applied for the same length of time, equation (2.12) may be simplified as follows:

$$T_E = \frac{\log\left\{\dfrac{\Sigma[\text{antilog }(tC_2)]}{n}\right\}}{C_2} \tag{2.13}$$

where n = the number of different temperatures given.

(A similar approach could be made to deal with fluctuating moisture contents.)

MacKay and Flood (1968) used equation (2.13) for applying equation (2.5) to wheat stored under fluctuating temperatures. At 21 per cent moisture content the actual half-viability period was less than expected; at 20, 17 and 16 per cent there was very close agreement; but at lower moisture contents the actual half-viability periods were substantially greater than the calculated ones.

In our present imperfect state of knowledge it may be taken that equations (2.10) and (2.11) may be used with reasonable confidence for designing long-term storage systems for fluctuating conditions, since under fluctuating but otherwise favourable storage conditions it seems that period of viability tends to be underestimated. However, it is evident that more critical investigations are needed on the effects of fluctuating environmental conditions.

SPECIAL EFFECTS OF EXTREME STORAGE CONDITIONS ON PERIOD OF VIABILITY

There is evidence that there are three sets of conditions of temperature and moisture content which limit the applicability of the basic viability equations. First, at very high moisture contents, say above about 30 per cent in cereals, providing the temperature is suitable, germination will take place and thus the seeds will be lost. Secondly, if the temperature is sufficiently low a special type of damage, freezing injury, will result in loss of viability when seeds are very moist. Thirdly, there is evidence that if seeds are subjected to extreme dessication, their period of viability may be less than would be predicted by the basic viability equations.

Little needs to be said here about the first limitation – the conditions under which germination will take place at high moisture contents. Obviously the details will vary from species to species. For example, the range of temperature at which this can take place generally increases as seeds lose dormancy (Vegis, 1963). Two examples will illustrate interspecific differences: non-dormant rice is capable of germinating over a range of temperature from about 17–45°C (Roberts, 1962) whereas in barley although the upper limit might be slightly less, the lower limit for germination is very much lower, i.e. about 3°C.

With few exceptions, there is no evidence that extremely low temperatures are anything but beneficial to the maintenance of viability, providing the moisture content is not high. For example, seeds of many species have been reduced to a temperature of 1.35°K without any harmful effects (Lipman, 1936). Providing the moisture content is not high enough to allow freezing injury, there is a great deal of evidence for many species to show that very low temperatures, −20°C or less, are beneficial to the maintenance of maximum viability (Barton, 1966; Von Senbusch, 1955; Weibull, 1953, 1955).

Sometimes a species may develop an undeserved reputation for being particularly susceptible to injury at low temperature, e.g. onion seed. In this case the reputation derives from a report by Barton (1966) that a temperature of −18°C was more deleterious than one of −2°C. But an examination of her data shows that the deleterious effect was shown only at 23.3 per cent moisture content: at 15.9 per cent for example −22°C was better than −2°C. In fact the deleterious effect of −22°C only occurred at a moisture content high enough for the additional factor of freezing injury to operate. A number of specific investigations have been made into freezing injury in other species. For example Robbins and Porter (1946) found that sorghum was reduced in viability when frozen at −20°F (−28.5°C) at moisture contents as low as 22 per cent, whereas soya beans at this temperature were not affected by moisture contents as high as 32 per cent. Carlson and Atkins (1960) showed that loss of viability in sorghum occurs at −3°C and below when the moisture content is 25 per cent or greater, but no injury was caused at 20 per cent moisture content. Rosenow, Casady and Heyne (1962), also working on sorghum, reported that greatly reduced germination was caused at 34 per cent moisture content or higher by temperatures of 25°F (−4.5°C) and below; a temperature of 28°F (−2°C) had no effect. Seed with 23 per cent moisture content or less showed little injury at temperatures as low as 22°F (−5.5°C). They suggested that the critical moisture content for freezing injury is between 23 and 34 per cent, probably close to 30 per cent. Similar results were obtained by Kantor and Webster (1967) who showed that freezing at −1.6°C had no effect on sorghum at moisture contents up to 40 per cent; at −3.3°C seed was injured above 33 per cent moisture content but not below. McRostie (1939) reported that seeds of maize with moisture contents of above 15 per cent would suffer severe damage from freezing temperatures and such damage was greater under fluctuating temperatures. However, Rossman (1949) found that maize grains were not injured by sub-freezing point temperatures at moisture contents less than about 20 per cent, a finding which was confirmed by Rall' (1961). The data of Kiesselbach and Ratcliffe (1920) on this species indicate that 75 per cent death can be expected after one hour's exposure to slightly less than 20 per cent moisture content at −18°C; at about −5°C the same degree of injury is only produced if the moisture content is about 33 per cent (See also p. 140).

These results are not completely consistent and there appear to be differences between species in that maize is apparently more sensitive to freezing injury than sorghum. But as a working rule for cereals, at any rate, it would seem that it is unsafe to store seeds at temperatures below −2°C when the moisture content is 20 per cent or greater, and for deep-freezing temperatures of about −20°C or less it would be safer to recommend 15 per cent moisture content or less. In normal practice, of course, it is unlikely that seeds will be stored under conditions which combine very high moisture contents with very low temperatures, but such damage may be relevant in considering deterioration which can occur in the field before harvest; it may be worthwhile then considering these data in relation to the other environmental factors which operate before harvest discussed in Chapter 5. However, it should be pointed out that there is some evidence that grain maturing on plants is somewhat less susceptible to freezing injury than would be indicated by studies on harvested grain (Rossman, 1949; Carlson and Atkins, 1960).

Equation (2.11) implies that for any given temperature the maximum period of viability would be obtained at 0 per cent moisture content. This is almost certainly not true. There is no clear distinction between physically bound and chemically bound water and attempts to achieve 0 per cent moisture content would probably result in chemical changes and consequent damage to the seed. Nevertheless it is often possible to reduce seed moisture contents to less than 1 per cent without resort to drastic techniques and it is important to know whether such severe drying is likely to have any detrimental effect on viability.

In some of the work which has been carried out on very dry seeds it is difficult to assess whether reported deleterious effects are due to the low moisture content or to damage which occurred on drying, particularly if high temperatures were used. H. M. Roberts (1959) reported that seed of Timothy (*Phleum pratense*) did not maintain its viability quite so well over a three-year period when dried to 5 per cent moisture content as compared with seed dried to 7 per cent. In this case, although the effects of moisture content are confounded with period of initial drying, the drying treatments were relatively mild – 16 and 9 hours respectively at 100°F (38°C) – and would not be expected to have a great effect on viability. Then again the results of Ching, Parker and Hill (1959) showed that the mean viability period of *Lolium perenne* was not improved by storage at 6 per cent moisture content as compared with 8.3 per cent although at all moisture contents down to 8.3 per cent there was increase in viability period with decrease in moisture content. These results all suggest that at 5 or 6 per cent moisture content there may be a limit to the rule that decreasing moisture content increases viability. However, other work suggests that if there is a critical moisture content at which a further reduction ceases to be an advantage, it is at a much lower value than this. For example Evans (1957) reported that heat-drying seeds of *Lolium perenne* (in which the temperature was raised in steps to 100°C) to 0.66 per cent moisture content did not injure the germination capacity except that after a period of 7 years, viability began to fall in this treatment before the most favourable treatment in which the seeds had a moisture content of 1.6 per cent. Harrington and Crocker (1918) by using dessicants found that reducing the moisture content of barley, wheat, Kentucky bluegrass (*Poa pratensis*), Sudan grass (*Holcus halepensis sudanensis*) and Johnson grass (*Sorghum halepense*) to less than 1 per cent did not adversely affect viability. The percentage germination of Kentucky bluegrass or Johnson grass was not affected when the moisture content was further reduced to 0.1 per cent, although the vigour of the Kentucky bluegrass seedlings was greatly reduced.

One of the most comprehensive investigations on the effects of extremely low moisture contents has been carried out by Nutile (1964) on 9 kinds of vegetable seeds: cabbage (*Brassica oleracea*), carrot (*Daucus carota*), celery (*Apium graveolens*), cucumber (*Cucumis sativa*), eggplant (*Solanum melongena*), lettuce (*Lactuca sativa*), onion (*Allium cepa*), pepper (*Capsicum frutescens*), and tomato (*Lycopersicon esculentum*), and three kinds of grasses: Highland bent grass (*Agrostis tenuis*), Kentucky bluegrass (*Poa pratensis*), and red fescue (*Festuca rubra*). These were dried over concentrated sulphuric acid in desiccators to moisture contents of 4, 2, 1 and 0.3–0.4 per cent. By the end of 5 years' storage at laboratory temperature (23–30°C) the viability of seeds of celery, eggplant, pepper and Kentucky bluegrass at 1 and 0.4 per cent moisture content, and carrot, tomato and red fescue sealed at 0.4 per cent moisture content was seriously impaired. The germination of cabbage, cucumber, lettuce, onion and Highland bent grass was not seriously affected by the lowest moisture content.

It has been reported that drying seeds of *Vicia faba* below 13 per cent moisture content reduces viability (Klingmüller and Lane, 1960). However, this is not our experience, at least down to 11.5 per cent moisture content (Roberts and Abdalla, 1968). It may be that the results of Klingmüller and Lane are complicated by the fact that their germination test involved a 16-hour soaking period in water and there are some reports that such a soaking treatment itself is capable of leading to chromosome breakages (Levan and Lofty, 1951; Rieger and Michaelis, 1958), and, as is shown in Chapter 9, any treatment which leads to an increase in chromosome damage decreases viability.

From this review it would seem that in most species one can expect an improvement in viability period with reduction in moisture content down to at least 5 per cent; in many species there would be a further improvement on drying down to about 2 per cent; but drying below this value may be deleterious to some species (see also p. 222).

THE RELATIONSHIP BETWEEN OXYGEN PRESSURE AND PERIOD OF VIABILITY

For carrying our storage experiments under controlled conditions, a convenient way to control seed moisture content at a constant value is to use sealed ampoules. Most of the work described so far in the development of the basic viability equations adopted this approach. However, this system has the disadvantage that the gaseous composition of the atmosphere alters with time because of the respiration of the seeds and their associated micro-flora. For example it was shown (Roberts and Abdalla, 1968) that when pea seeds at 18.4 per cent moisture content were stored in sealed ampoules at 25°C there was a more or less linear increase in carbon dioxide content and decrease in oxygen content such that after 11.3 weeks (the period required for viability to drop to 50 per cent under these conditions) the oxygen content in the ampoule had dropped from 21 to 1.4 per cent while the CO_2 content had risen from 0.03 to about 12 per cent. Thus the nomographs shown in Appendix 3, strictly speaking refer to hermetic storage in which the environment is becoming more anaerobic as the seeds lose viability. Consequently it is important to know to what extent the viability period may differ in open storage where the composition of the inter-seed atmosphere remains approximately the same as air.

In much of the earlier work on the effect of gaseous environment on seed viability, which has been reviewed by Owen (1956) and Barton (1961), the evidence is often not critical since the other factors affecting viability were not closely controlled. Because of this lack of control it is not surprising that there were conflicting views: for example suggestions have sometimes been made (e.g. Curtis and Clark, 1950; Milner and Geddes, 1954) that anaerobic conditions lead to rapid loss of viability. In some experiments on Chewings fescue seed (*Festuca rubra* var. *commutata*), Gane (1948) reported that in storage for periods of as long as 4 years there was little to choose between storage in air or nitrogen under conditions where temperature and moisture were closely controlled. In some work on wheat, Glass, Ponte, Christensen and Geddes (1959) stored seed under conditions where temperature, moisture content and gaseous composition were all controlled. In one trial at 30°C no difference was shown between the effects of air and nitrogen in the range 13–18 per cent moisture content, but in a second trial nitrogen was shown to be slightly more effective for retaining viability. In a trial at 20°C storage in nitrogen was shown to cause a distinctly greater retention of viability, particularly at higher moisture contents. Peterson, Schlegel, Hummel, Cuendet, Geddes and Christensen (1956) had shown previously that, in short trials at 30°C and 18 per cent moisture content in controlled oxygen-nitrogen mixtures, decreasing the oxygen content led to an increased germination percentage when tested after 16 days' storage. In mixtures containing 21 per cent oxygen and variable amounts of nitrogen, increasing the carbon dioxide content also led to an increase in percentage viability. In some experiments at various moisture contents from 12.0–14.5 per cent and at temperatures from 32–45°C, rice was sealed in ampoules in oxygen, air and nitrogen (Roberts, 1961b). Under these conditions it was found that, although the results were not entirely consistent, there was a tendency for the period of viability to be increased by decrease in the partial pressure of oxygen, particularly at the lower temperatures and moisture contents. In some long-term experiments on lettuce and onion seeds lasting up to 18 years, Harris (1966) showed that sealed storage in carbon dioxide extended the period of viability over that of seed sealed in air; for example in 20 varieties of lettuce stored at 5–6 per cent moisture content sealed in air his data showed that the mean viability period was about 8 years, whereas when sealed in carbon dioxide it was more than 9 years. In onion seed at 8.5 per cent moisture content the corresponding mean viability periods were about 4 years for seed sealed in air and 5 years for seed sealed in carbon dioxide.

From the more recent of this work on gaseous environment, it has emerged that there is now general agreement that increases in the pressure of oxygen tend to decrease the period of viability. This conclusion has been derived from two types of experiment. In the first type all three major factors which affect viability – temperature, moisture content and partial pressure of oxygen – were all held constant throughout; but these experiments were short-term, i.e. they employed combinations of conditions which led to a relatively rapid loss of viability. The second type of experiment was long-term but carried out in sealed ampoules so that while temperature and moisture content were maintained at constant values, there would have been changes in the gaseous composition during the experiment.

Further short-term experiments on barley, broad beans and peas in which all three factors have been kept constant throughout have confirmed that in all three species the period of viability is decreased by the presence of oxygen (Roberts and Abdalla, 1968). Figure 2.2 only shows the results obtained for broad beans, but the results for the other two species showed very similar survival patterns. In all cases the effect of oxygen was relatively more pronounced under conditions where viability was lost most rapidly, irrespective of whether this rapid loss was due primarily to high moisture content or high temperature. The results also showed that most of the dele-terious effects of oxygen are produced at relatively low partial pressures. This is illustrated by the fact that in all cases a pronounced acceleration of loss of viability was produced by increasing the oxygen level from 0 to 21 per cent whereas a further increase to 100 per cent had little or sometimes no further effect. It should be mentioned here that it is difficult to reconcile our results with those of Kreyger (1958, 1959) who found that hermetic storage of barley in nitrogen was slightly more deleterious than similar storage in air; however, his results were obtained at the very high moisture content of 22 per cent.

From time to time there have been suggestions that gases other than oxygen have an effect on viability. For example the work of Peterson *et al.* (1956) in which carbon dioxide was found to be bene-ficial as compared with nitrogen when the oxygen level was held constant. However, in general, it would seem that carbon dioxide itself has no special beneficial effect on longevity. In this connection recent work by Harrison (1966) is illuminating. In some experiments on 10 varieties of lettuce stored for 3 years at approximately 18°C and 6 per cent moisture content he found that when the seeds were sealed in oxygen the mean viability at the end of one experiment was 8 per cent; in air it was 57 per cent; whereas in nitrogen, or argon, or carbon dioxide the viability gave the identical result of 78 per cent, and when stored in vacuum the value was 77 per cent. Similar results were obtained using onion stored at 8 per cent moisture content at the same temperature for 4 years. In oxygen the viability was 3 per cent; in air it was 36 per cent; in carbon dioxide, nitrogen and argon the results were, 80, 75 and 79 per cent (i.e. not significantly different). With onion, however, the vacuum treatment seemed to be less effective in that the viability was 51 per cent.

This discussion on the effects of oxygen raises the question of how far can the data obtained under hermetic storage be applied to condi-tions of open storage? In other words what would be the errors

Viability of Seeds

TABLE 2.1 *Mean viability periods (days) obtained under constant storage conditions compared with values predicted from conditions of sealed storage where the gaseous environment became more anaerobic with time (from Roberts and Abdalla, 1960).*

Treatment			Barley		Broad beans		Peas	
Temp. (°C)	Moist. cont. (%)	Gas	Sealed* ampoules	Constant condi- tions	Sealed* ampoules	Constant condi- tions	Sealed* ampoules	Constan condi- tions
25	27	N₂	2.6	3.0	3.3	5.2	3.9	6.3
		air		1.5		2.6		3.5
		O₂		1.3		2.2		3.5
25	18	N₂	50.7	56.7	38.5	42.0	82.3	91.0
		air		45.5		29.4		56.0
		O₂		44.1		29.4		56.0

*The gas indicated for the sealed ampoules is 'air' since the seeds were initially enclosed in air. However, as indicated in the text, the gas composition in the sealed ampoules would have changed during the experiment; for example in peas stored at 25°C and 18 per cent moisture content, the O_2 and CO_2 composition in the ampoules after half the seeds had lost their viability would have been about 1.4 per cent O_2 and 12 per cent CO_2.

involved in applying the data presented in the nomographs in Appendix 3 to conditions of open storage? Table 2.1 compares some results obtained under constant gaseous conditions with results obtained from hermetic storage. It is evident that, as would be expected, the results obtained in hermetic storage fall between those obtained under constant conditions in air and constant conditions in nitrogen. It is impossible, at this time, to say anything more precise since it is evident both from work of Glass *et al.* (1959) already mentioned and from Fig. 2.2 and Table 2.1 that interactions occur between oxygen and the other factors affecting viability. There is evidently a need for more comprehensive experiments under conditions with constant gaseous environments to enable viability to be accurately predicted in open storage. Until such investigations are made, however, the nomographs shown in Appendix 3 may be used as a guide, providing it is realised that they may over-estimate the period of viability in open storage; the available evidence suggests, however, that the error will not be very large at the lower moisture contents which are normally used in practice.

The discussion here has been concerned primarily with the effects of gaseous environment insofar as modifications of the environment in hermetic storage will influence the applicability of the basic viability equations in open storage. However, this is also a convenient point to discuss briefly whether it would be possible to prevent the deleterious effects of oxygen by treatments in which no attempt is made to control the gaseous environment of the seeds. Very little work appears to have been carried out on the use of anti-oxidants but Siegal (1953) showed that the heat injury (100–103°C) of kidney bean (*Phaseolus vulgaris*) embryos was decreased in reduced oxygen pressures and that application of cysteine overcame the injury to some extent. Kaloyereas, Mann and Miller (1961) have reported some interesting work in which seeds of onion (*Allium cepa*) and okra (*Hibiscus esculentus*) were treated with either starch phosphate or alpha-tocopherol and then stored at the same temperature and moisture content as untreated control seeds. The results presented suggested that starch phosphate was very effective in prolonging the viability of both species and alpha-tocopherol had some beneficial effect on onion seeds. It is evident that there is a need for this work to be confirmed and applied to other species and extended to other anti-oxidants.

The effects of environmental factors on the activity of organisms associated with seeds in storage

There are five main types of organisms associated with seeds in storage – bacteria, fungi, mites, insects and rodents. The activity of all these organisms can lead to damage resulting in loss of vigour or viability or, particularly in the case of rodents, to complete loss of seed. It has often been suggested that fungi are a primary cause of loss of viability; this problem is considered in greater detail in Chapters 3 and 9. Rodents come into a different category from the other organisms since their activity is not dependent on the temperature and moisture content of the seed; the control of rodents has been reviewed by Parkin (1963) and measures include construction of rodent-proof stores, trapping, fumigating and poison baiting. The activity of the remaining four types of organism depends on temperature, moisture content – or, more probably, the relative humidity of the inter-seed atmosphere – and gaseous environment. There is no doubt that the activity of all types of organism (except for some bacteria) is reduced as conditions become more anaerobic, particularly that of insects and mites. Hermetic storage of damp grain, which utilises the rapid production of anaerobic conditions by the respiration of the microflora, is sometimes used to reduce the activity of the microflora and other organisms in the storage of grain for feed but, under these conditions, the seed rapidly loses viability because of the high moisture content (Hyde and Oxley, 1960). The present discussion, then, will concentrate on the effects of temperature and relative humidity under aerobic conditions.

BACTERIA AND FUNGI

The microflora of seed in storage is dealt with in detail in Chapter 3; here it will be sufficient to indicate the main constraints that are put on the activity of the microflora by the seed environment. It has been shown that the important moisture consideration in the control of the seed microflora is the relative humidity of the inter-seed atmosphere and not the moisture content of the seed itself (Milner and Geddes, 1954). Of course this means that the activity of the microflora can be related to seed moisture content since, as will be discussed shortly, there is an equilibrium relationship between seed moisture content and the relative humidity of the inter-seed atmosphere. However, it is important to recognise that the microflora activity is related to relative humidity rather than seed moisture content because different types of seeds have different equilibrium relationships between these two factors (see Appendix 4). The evidence indicates that all storage fungi are completely inactive below 62 per cent relative humidity (Semenuik, 1954) and that there is very little activity below about 75 per cent relative humidity (Milner and Geddes, 1954). From 75 per cent relative humidity upwards the amount of fungi in a seed often shows an exponential relationship with relative humidity (Bottomley, Christensen and Geddes, 1950). It is hardly necessary to consider storage bacteria because they require at least 90 per cent relative humidity for growth (Semenuik, 1954) and they therefore only become significant under conditions in which fungi are already very active.

With regard to the effect of temperature on the growth of the microflora, certain organisms can grow at temperatures as low as $-8°C$ and others at temperatures as high as 80°C (Semenuik, 1954); consequently, since high temperatures rapidly decrease seed viability, the only practical method of controlling microfloral activity by temperature alone is by deep-freezing. At this time there are no satisfactory chemical methods of control of these organisms in storage (Christensen and López, 1963).

INSECTS AND MITES

There is no insect activity at seed moisture contents below 8 per cent but, if the grain is infected, increased activity may generally be

TABLE 2.2　*Optimum temperature for rapid insect growth and the temperature ('safe temperature') at which the developmental cycle needs 100 days on one of the best foods for each species (from Burges and Burrell, 1964).*

	Optimum temperature, °C	Safe temperature, °C
Saw-toothed grain beetle, *Oryzaephilus surinamensis* L.	34	19
Grain weevil, *Sitophilus (Calandra) granarius* L.	28–30	17
Rust-red grain beetle, *Cryptolestes (Laemophoeus) ferrugineus*	36	20
Rust-red flour beetle, *Tribolium castaneum* Herbst.	36	22
Confused flour beetle, *Tribolium confusum* J du V.	33	21
Khapra beetle, *Trogoderma granarium* Everts.	38	22
Rice weevil, *Sitophilus (Calandra) oryzae* L.	29–31	18
Lesser grain borer, *Rhyzopertha dominica* F.	34	21
Flat grain beetle, *Cryptolestes pusillus* (= *minutus*) Schönherr	32	19

expected up to about 15 per cent moisture content (Cotton, 1954). The major insect pests of stored seeds are listed in Table 2.2 which indicates optimal temperatures of insect activity and temperatures below which stored seed is safe from their activities. Less is known about the response of mites to environment but since they depend to varying extents on skin respiration they are sensitive to low relative humidities (Parkin, 1956). They cannot survive below 60 per cent relative humidity and they tend to die out between 60 and 70 per cent; multiplication is rapid above 75 per cent relative humidity (Oxley, 1948). Temperature optima are lower than for most storage insects (Cotton, 1954) and Burrell (1970) suggests that they are not active outside the temperature range of about 3–31°C. Some estimate of numbers of common mites (*Acarus siro*, *Glycyphagus* spp., and *Cheyletus* sp.) found in stored grain under various combinations of temperature and moisture content have been made by Burrell and Laundon (1967).

Although it is normally preferable to control insect and mite activity by the manipulation of the seed environment, it is possible to effect some control of these organisms chemically. Parkin (1963) has reviewed the use of fumigants and contact insecticides. One of the problems of chemical control is that the chemicals can have an adverse effect on seed viability or vigour and some of them are dangerous to handle. Nevertheless, fumigants which have been used successfully include methyl bromide, hydrogen cyanide, phosphine, ethylene dichloride and carbon tetrachloride in a 3:1 mixture, carbon disulphide and naphthalene. Contact insecticides used in seed storage have included DDT, lindane and malathion. There is also a possibility of using bacteria, particularly *Bacillus thuringiensis* for biological control of insect pests of stored grain (Burges, 1964), but although such an approach might have value in large-scale grain storage for food, it seems unlikely that such techniques will become important in seed storage.

The design of seed storage systems

It is not proposed here to deal with the problems of storage systems for seeds of those species which are harmed by low moisture contents mentioned at the beginning of this chapter: insufficient general principles have so far emerged to deal with these adequately. In this section then I shall continue to concentrate on the major category of seeds in which viability is prolonged by reduced temperature, moisture content and oxygen pressure. The advantage to be gained by deliberately excluding oxygen from the storage environment seems to be relatively small in relation to the extra handling problems which it would raise. For most applications it will be sufficient to concentrate on the control of temperature and moisture content.

By this time it will have become apparent that the concept of a 'critical moisture content' for seed storage, often mentioned in the literature, is misleading: the basic viability equations do not indicate any discontinuities in the relationship between temperature, moisture content and viability. The temperature and moisture content at which it is proposed to store seed will depend first on the extent of the loss of viability that will be sanctioned and the period for which it is intended to store the seeds, and secondly on the relative costs of producing the alternative combinations of temperature and moisture content that will achieve the desired result. The problem of deciding which particular combination of temperature and moisture content to choose, once the alternatives have been defined, will often be one of systems engineering.

The control of temperature is a straightforward refrigeration problem. However, two alternative strategies may be employed for the control of moisture content. The first approach is to dry the seeds to the required moisture content and hermetically seal them so that the seed cannot re-absorb water from the atmosphere (this system is often used in modern methods of packaging seeds). There is not room here to deal with the problems of drying technology in any detail. On a small scale chemical desiccants, vacuum drying, infra-red drying or sun drying may be used; more often, and particularly on a large scale, techniques depend on passing heated air through seeds. Obviously the temperature of the air must not be too high, but the lower the seed moisture content, the higher is the drying temperature that may be used with safety. The laws of safe seed drying have not yet been fully worked out in spite of the fact that the physics of the process, though complex, is fairly well understood. When drying with a given air temperature, the temperature and moisture content of the seeds change with time; the rate of change also changes and the dynamic process depends on the physics of heat exchange and evaporation (Kreyger, 1963). What remains to be done is to attempt an integration of the changes in temperature and moisture content of the seed, as determined by physical laws, with the biological effects of these parameters on seed deterioration as indicated by the basic viability equations. Until this is done it is necessary to be content with rough guides to safe air temperatures for drying which have been determined empirically. A number of these practical suggestions for maximum drying temperatures have been gathered together in Appendix 4.

The second approach to the control of seed moisture content during storage is to place the seeds in 'open' storage and control the relative humidity of the air; systems based on this principle depend on the fact that the moisture content of each type of seed has a definable relationship with the relative humidity of the atmosphere. Figure 2.5 shows that two curves are necessary to define the relationship between relative humidity and seed moisture content because of a hysteresis effect: i.e. the equilibrium relationship is not the same on the absorption of water by the seeds as it is on desorption. The curve for desorption may be obtained by beginning with seeds of high moisture content and placing them in atmospheres of different controlled relative humidities for several weeks until they have reached equilibrium; the curve for absorption may be obtained in the same way except that one begins with dry seeds. Alternatively a quicker but more complicated method depends on conditioning seed samples to different moisture contents, putting them in closed systems, and then measuring the relative humidity of the air (Lubatti and Bunday, 1960). But there are many pitfalls: see Chapter 3 (pp. 74–75).

The difference between the absorption and desorption curves usually varies between about 0.6 and 1.6 per cent moisture content over the range of relative humidities from 10–75 per cent; the maximum differences in equilibrium moisture content for wheat and maize have been shown to occur between 20 and 30 per cent relative humi-

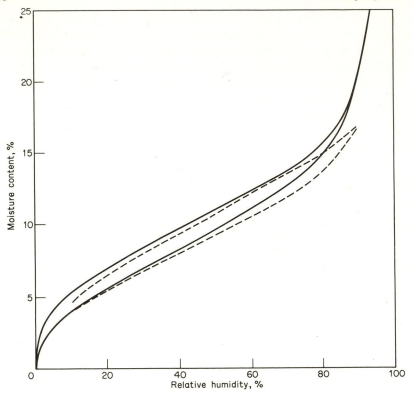

FIGURE 2.5 The hygroscopic equilibrium relationships of wheat at 35°C (———) and paddy (rice in the husk) at 25°C (-----). In both cases the upper curve represents the desorption relationship and the lower curve the absorption relationship. Data for wheat from Hubbard, Earle and Senti (1957) and for paddy from Breese (1955).

dity (Hubbard, Earle and Senti, 1957) and between 50 and 70 per cent relative humidity in rice (Breese, 1955).

Temperature has a small effect on the equilibrium relationship: the lower the temperature the greater the moisture content of the seeds for a given relative humidity. Becker and Sallans (1956) showed in wheat that over most of the relative humidity range a decrease in temperature from 50° to 25°C led to an increase in moisture content of about 3 per cent at a given relative humidity. And the data in Appendix 4 derived from Hubbard *et al.* (1957) show that a decrease in temperature from 35° to 25°C increases the moisture content of wheat by almost 1 per cent at 20 per cent relative humidity and by about 1.5 per cent at 75 per cent relative humidity. These observations agree with the theoretical hygroscopic equilibrium relationships developed by Henderson (1952) which would predict greater changes in equilibrium moisture contents for a given temperature change at higher relative humidities than at lower relative humidities. Henderson's work would also predict that a change of temperature at low temperatures would have a slightly greater effect than a similar change at high temperatures. But in contradiction to these expectations the experimental curves produced by Stermer (1968) for rice show that at 30 per cent relative humidity, for example, a reduction in temperature from 35° to 20°C increased the moisture content by almost 2 per cent but a further reduction to 5°C caused an additional increase of less than 1 per cent; on the other hand at 80–90 per cent relative humidity altering the temperature between 5° and 35°C had virtually no effect. From these observations it may be concluded that the effect of temperature varies with humidity and species but, in general, the increase in moisture content resulting from a reduction in temperature becomes less at lower temperatures.

As yet there is no single theory that will satisfactorily account for the hygroscopic relationship in seeds over the whole of the range of relative humidities, but Becker and Sallans (1956) working with wheat and Lubatti and Bunday (1960) working with peas have shown that Smith's equation fits the upper part of the curve while Henderson's equation fits the lower part of the curve. The equation proposed by Smith (1947) for the sorption of water by high polymers is:

$$w = w_b - w' . \ln(1 - p/p_0) \qquad (2.14)$$

where w is the water content (wet basis) expressed as a decimal fraction, w_b is the bound water, w' the amount of water condensed directly on the bound water; p/p_0 is the relative humidity calculated from the ratio of water vapour pressures, p_0 being the vapour pressure of saturated air and p the partial vapour pressure at equilibrium.

The equation suggested by Henderson (1952) is:

$$1 - rh = \exp(-k'M^n) \qquad (2.15)$$

where rh is the equilibrium relative humidity, expressed as a decimal; M is the equilibrium moisture content (per cent dry basis) and k' and n are constants.

As Lubatti and Bunday (1960) have pointed out, depending on which part of the curve it is in which one has most interest, either equation might be found useful in establishing the equilibrium curve in those cases where only a few data are available.

In fundamental studies on viability it is important to realise that it has been shown in wheat at least that the equilibrium hygroscopic relationship of the embryo is different from the endosperm: the

embryo has a lower moisture content than the embryo below 88 per
cent relative humidity, but a higher moisture content above this value
(Shelef and Mohsenin, 1966). The difference between the equilibria
of the embryo and the seed food reserves may be significant when
considering differences in loss of viability between different species
at the same overall seed moisture content.

Although there are many reports on the hygroscopic equilibrium
relationships of the seeds of many species, often only one curve has
been given and it is not always evident whether this represents the
absorption or desorption relationship. Another problem is that sometimes somewhat different results are reported for the same species by
different workers. Data for a number of species are given in Appendix
4. An examination of these figures shows that, broadly speaking, seeds
may be divided into two categories – the oily and non-oily. For a given
relative humidity the oily seeds show a lower equilibrium moisture
content: e.g. at 40 per cent relative humidity non-oily seeds like
cereals and beans show moisture contents of about 9 or 10 per cent,
whereas oily seeds like groundnuts or soya beans have moisture contents of about 6 or 7 per cent. Consequently, even if information is
not available for a particular species, it is possible to hazard an intelli-
gent guess as to the type of hygroscopic equilibrium to be expected.

Some of the biological factors to be taken into account in the design
of seed stores are shown in Fig. 2.6. An attempt has been made here
to collate the information on the environmental factors controlling
the activity of the organisms associated with seed in storage together
with information on the expected viability periods of five common
agricultural species predicted by the basic viability equations. There
is some difficulty in presenting all this material on one diagram since
the activity of storage organisms is related to relative humidity whereas the viability of seeds is related to their moisture content and, as
Appendix 4 shows, the relationship between relative humidity and
seed moisture content varies between species. However, the differences in the hygroscopic equilibria between the species included in
this diagram are not large and so, for convenience, the hygroscopic
relationship shown for wheat at 25°C (Hubbard *et al.*, 1957) has been
assumed for all species. Another difficulty is that although it is known
that the hygroscopic equilibrium varies with temperature it is difficult
to indicate this on a diagram in which the coordinates are temperature
and moisture content scales; however, the changes in the hygroscopic
equilibrium are not great and, as the diagram is presented, would only
tend to overestimate slightly the activity of micro-organisms at lower
temperatures. Finally the question arises as to which of the hygro-

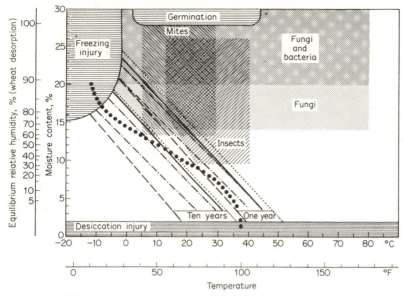

FIGURE 2.6   The relationship between seed moisture content and storage
problems at different temperatures. Diagonal lines represent those combinations
of moisture content and temperature which would lead to an expected drop of
viability to 95 per cent after one year and after ten years for the following five
species: – – – wheat; – · – · – broad beans; – · · – · · peas; —— barley; · · · · · rice.
The horizontally-hatched areas indicating the regions where freezing injury and
desiccation injury may take place are also related to the moisture content of the
seeds whereas the remaining information is, strictly speaking, related to relative
humidity. In this diagram it is assumed that the relationship between moisture
content and relative humidity is that for wheat at 25°C on desorption; however, as
a rough guide this may be taken as appropriate for a wide range of species (see
Appendix 4). The conditions under which various storage organisms may be
active are shown by different patterns of stippling. The line of solid circles
(● ● ● ●) indicates the rule-of-thumb suggested by James (1961) for long-term
storage conditions, i.e., the line indicates where value for per cent relative
humidity added to the number of degrees Fahrenheit totals 100.

scopic equilibrium relationships should be used – the absorption or
desorption curves. It is suggested that in designing an air-conditioned
store for 'open' seed storage it would be safest to assume that the seed
might not be properly dried before entering the store and consequently the desorption curve would be appropriate for indicating the
moisture content at which the seeds would equilibrate.

The decision as to what loss of viability will be tolerated in storage
does not rest solely on the ability to maintain a stock of seeds capable
of germination. Obviously this is important but if, for example, the

store is to be used for conserving genetic resources one has to consider the problem of genetic stability in storage. In Chapter 9 it is shown that loss of viability is correlated with an accumulation of genetic mutations in the surviving seeds. It is emphasised that the correlation is with percentage loss of viability and not with the chronological age of the seeds. The evidence suggests that surprisingly large amounts of genetic damage can occur in association with relatively small losses of viability and so it is suggested that, where the genetic integrity of the stock is important, it would be prudent to design storage systems which tolerate only small losses of viability. For this reason it is assumed in Fig. 2.6 that conditions are to be provided for periods of storage in which viability is not expected to fall below about 95 per cent; two storage periods have been indicated – 1 year and 10 years.

For a ten-year storage period of the species most susceptible to loss of viability it will be seen that a combination of 40 per cent relative humidity at 0°C or 10 per cent relative humidity at 10°C, for example, would be safe. It will also be seen that any conditions suitable for storing the most susceptible species for 10 years will automatically preclude any harmful activity of storage organisms. For some time now a useful guide to long-term storage has been that suggested by James (1961) which states that 'if percentage [relative] humidity and degrees Fahrenheit total 100 [or less], conditions are considered favourable for longevity.' A line indicating these conditions has been included in Fig. 2.6. In general it will be seen that this is still a fairly good rule-of-thumb, though it may be a bit optimistic at the lower relative humidities.

Sometimes it may be decided that greater tolerances in loss of viability are acceptable or longer storage periods may be required. In such cases the equilibrium hygroscopic relations shown in Appendix 4 used in conjunction with the nomographs in Appendix 3 may serve as useful guides. Further discussions based on first-hand experience of the practical problems involved in the design and operation of seed storage facilities are included in Appendices 1 and 2.

References to Chapter 2

ALLCOCK, H. J., JONES, J. R., and MICHEL, J. G. L., 1962. *The Nomogram.* 5th edit. Pitman, London.

ANON., 1954. *The preservation of viability and vigour in vegetable seed.* Asgrow Monograph No. 2. Asgrow Seed Co., New Haven, Conn.

ANTHANAKRISHNA RAO, P. N., 1959. A brief review of the results obtained at the Paddy Breeding Station, Mangalore. *Mysore agric. J.,* **34**, 42–48.

AUFHAMMER, G., and SIMON, U., 1957. Die Samen landwirtshaftlicher Kulturpflanzen im Grundstein des chamaligen Nürnberger Stadttheaters und ihre Keimfähigkeit. *Z. Acker-u PflBau.*, **103**, 454–72.

BACCHI, O., 1955. Secca da semente de cafe ao sol. *Bragantia*, **14**, 225–36.

BACCHI, O., 1956. Novos ensaios sôbre a seca da semente de cafe ao sol. *Bragantia*, **15**, 83–91.

BARTON, L. V., 1943. The storage of citrus seeds. *Contr. Boyce Thompson Inst.*, **13**, 47–55.

BARTON, L. V., 1961. *Seed Preservation and Longevity*. Leonard Hill, London.

BARTON, L. V., 1965. Viability of seeds of *Theobroma cacao* L. *Contrib. Boyce Thompson Inst.*, **23**, 109–22.

BARTON, L. V., 1966. Effects of temperature and moisture on viability of stored lettuce, onion, and tomato seeds. *Contr. Boyce Thompson Inst.*, **23**, 285–90.

BASS, L. N., 1965. Effect of maturity, drying rate and storage conditions on longevity of Kentucky bluegrass seed. *Proc. Ass. off. Seed Analysts N. Am.*, **55**, 43–46.

BECKER, H. A., and SALLANS, H. R., 1956. Studies on the desorption isotherms of wheat at 25°C and 50°C. *Cereal Chem.*, **33**, 79–91.

BOSWELL, V. R., TOOLE, E. H., TOOLE, V. K., and FISHER, D. F., 1940. *A study of rapid deterioration of vegetable seeds and methods for its prevention*. US Dept Agric. Tech. Bull. 708.

BOTTOMLEY, R. A., CHRISTENSEN, C. M., and GEDDES, W. F., 1950. Grain storage studies. IX. The influence of various temperatures, humidity, and oxygen concentrations on mould growth and biochemical changes in stored yellow corn. *Cereal Chem.*, **27**, 217–96.

BREESE, M. H., 1955. Hysteresis in the hygroscopic equilibria of rough rice at 25°C. *Cereal Chem.*, **32**, 481–87.

BRETT, C. C., ₁953. The influence of storage conditions upon the longevity of seeds, with special reference to root and vegetable crops. *Rep. 13th Inst. hort. Congr. 1952*, 1016–18.

BURGES, H. D., 1964. Control of insects with bacteria. *World Crops*, September, 1964, 229–43.

BURGES, H. D., and BURRELL, N. J., 1964. Cooling bulk grain in the British climate to control storage insects and to improve keeping quality. *J. Sci. Food Agric.*, **15**, 32–50.

BURGES, H. D., EDWARDS, D. M., BURRELL, N. J., and CAMMELL. M. E., 1963. Effects of storage temperature and moisture content on the germinative energy of malting barley, with particular reference to high temperature. *J. Sci. Food Agric.*, **14**, 580–83.

BURRELL, N. J., 1970. *Conditions for Safe Grain Storage*, Tech. Note 16. Home Grown Cereals Authority, London.

BURRELL, N. J., and LAUNDON, J. H. J., 1967. Grain cooling studies. I. Observations during a large scale refrigeration test on damp grain. *J. stored Prod. Res.*, **3**, 125–44.

CARLSON, G. E., and ATKINS, R. E., 1960. Effect of freezing temperature on seed viability and seedling vigour of grain sorghum. *Agron. J.*, **52**, 329–33.

CHING, T. M., PARKER, M. C., and HILL, D. D., 1959. Interaction of moisture and temperature on viability of forage seeds stored in hermetically sealed cans. *Agron. J.*, **51**, 680–84.

CHRISTENSEN, C. M., and LÓPEZ, F., 1963. Pathology of stored seeds. *Proc. int. Seed Test. Ass.*, **28**, 701–11.

COTTON, R. T., 1954. Insects. In *Storage of Cereal Grains and their Products*, eds. J. A. Anderson and A. W. Alcock, 152–220. Amer. Assoc. of Cereal Chemists, St Paul.

CURTIS, O. F., and CLARK, D. G., 1960. *An Introduction to Plant Physiology*, p. 569. McGraw-Hill, New York.

DAVIS, D. S., 1962. *Nomography and empirical equations*. 2nd edit. Reinhold, New York.

DORE, J., 1955. Dormancy and viability of padi seed. *Malayan agric. J.*, **38**, 163–73.

EVANS, G., 1957. The viability over a period of fifteen years of severely dried ryegrass seed. *J. Brit. Grassland Soc.*, **12**, 286–89.

FINNEY, D. J., 1962. *Probit Analysis*. 2nd edit. Cambridge University Press.

GANE, R., 1948. The effect of temperature, water content and composition of the atmosphere on the viability of carrot, onion and parsnip seeds in storage. *J. agric. Res.*, **38**, 84–89.

GANE, R., 1948. The effect of temperature, humidity, and atmosphere on the viability of Chewing's Fescue grass seed in storage. *J. agric. Sci.*, **38**, 90–92.

GHOSE, R. L. M., CHATGE, M. B., and SUBRAHMANYAN, V., 1956. *Rice in India*, 166. Indian Counc. Agric. Res., New Delhi.

GLASS, R. L., PONTE, J. G., CHRISTENSEN, C. M., and GEDDES, W. F., 1959. Grain storage studies. XXVIII. The influence of temperature and moisture level on the behaviour of wheat stored in air and nitrogen. *Cereal Chem.*, **36**, 341–56.

GROVES. J. F., 1917. Temperature and life duration of seeds. *Bot. Gaz.*, **63**, 169–89.

HABER, E. S., 1950. Longevity of the seed of sweet corn inbreds and hybrids. *Proc. Am. Soc. hort. Sci.*, **55**, 410–12.

HAFERKAMP, M. E., SMITH, L., and NOLAN, R. A., 1953. Studies on aged seeds. I. Relation of seed age to germination and longevity. *Agron. J.*, **45**, 434–37.

HARRISON, B. J., 1966. Seed deterioration in relation to storage conditions and its influence upon germination, chromosomal damage and plant performance. *J. nat. Inst. agric. Bot.*, **10**, 644–63.

HARRINGTON, J. F., 1963. Practical advice and instructions on seed storage. *Proc. int. Seed Test. Ass.*, **28**, 989–94.

HARRINGTON, G. T., and CROCKER, W., 1918. Resistance of seeds to desiccation. *J. agric. Res.*, **14**, 525–32.

HENDERSON, S. M., 1952. A basic concept of equilibrium moisture. *Agric. Engng*, **33**, 29–32.

HOLMES, G. D., and BUSZEWICZ, G., 1958. The storage of temperate forest tree species. *Forestry Abstr.*, **19**, Nos. 3 & 4.

HUBBARD, J. E., EARLE, F. R., and SENTI, F. R., 1957. Moisture relations in wheat and corn. *Cereal Chem.*, **34**, 422–33.

HUKILL, W. V., 1963. Storage of seeds. *Proc. int. Seed Test. Ass.*, **28**, 871–73.

HUTCHINSON, J. B., 1944. The drying of wheat. III. The effect of temperature on germination capacity. *J. Soc. chem. Ind.*, **63**, 104–7.

HUXLEY, P. A., 1964. Investigations on the maintenance of viability of robusta coffee seed in storage. *Proc. int. Seed Test. Ass.*, **29**, 423–44.

HYDE, M. B., and OXLEY, T. A., 1960. Experiments on the airtight storage of damp grain. I. Introduction, effect on the grain and the intergranular atmosphere. *Ann. appl. Biol.*, **48**, 687–710.

JAMES, E., 1961. In *Perpetuation and protection of germ plasm as seed*, ed. R. E. Hodgson. Publication No. 66, Amer. Assoc. Adv. Sci., Washington.

JAMES, E., 1967. Preservation of seed stocks. *Adv. Agron.*, **19**, 87–106.

JAMES, E., BASS, L. N., and CLARK, D. C., 1967. Varietal differences in longevity of vegetable seeds and their response to various storage conditions. *Amer. Soc. hort. Sci.*, **91**, 521–28.

JONES, J. W., 1926. Germination of rice seed as affected by temperature, fungicides, and age. *J. Amer. Soc. Agron.*, **18**, 576–92.

KALOYEREAS, S. A., MANN, W., and MILLER, J. C., 1961. Experiments in preserving and revitalizing pine, onion, and okra seeds. *Econ. Bot.*, **15**, 213–17.

KANTOR, D. J., and WEBSTER, O. T., 1967. Effects of freezing injury on viability of sorghum seed. *Crop Sci.*, **7**, 196–99.

KIESSELBACH, T. A., and RATCLIFFE, J. A., 1920. *Freezing injury of corn.* Univ. Nebraska Agric. Exp. Sta. Bull., No. 16, 1–96. [Data reproduced by W. J. Robbins, and K. F. Petch, 1932. Moisture content and high temperature in relation to the germination of corn and wheat grains. *Bot. Gaz.*, **93**, 85–92.]

KLINGMÜLLER, W., and LANE, G. R., 1960. Damaging effect of drying on *Vicia faba* seeds. *Nature, Lond.*, **185**, 699–700.

KREYGER, J., 1958. Recherches sur la conservation des orges de brasserie. *Le Petit Journal du Brasseur*, **66**, 811–16.

KREYGER, J., 1959. Recherches sur la conservation des orges de brasserie. *Le Petit Journal du Brasseur*, **67**, 7–10.

KREYGER, J., 1963. General considerations concerning the drying of seeds. *Proc. int. Seed Test. Ass.*, **28**, 753–84.

LEVAN, A., and LOFTY, T., 1951. Spontaneous chromosome fragmentation in seedlings of *Vicia faba*. *Hereditas*, **36**, 470–82.

LINDSTROM, E. W., 1942. Inheritance of seed longevity in maize inbreds and hybrids. *Genetics, Princeton*, **27**, 154.

LIPMAN, C. B., 1936. Normal viability of seeds and bacterial spores after exposure to temperatures near the absolute zero. *Pl. Physiol.*, **11**, 201–5.

LUBATTI, O. F., and BUNDAY, G., 1960. The water content of seeds. 1. The moisture relations of seed peas etc. *J. Sci. Food Agric.*, **12**, 685–90.

MACKAY, D. B., and FLOOD, R. J., 1968. Investigations in crop seed longevity. II. The viability of cereal seed stored in permeable and impermeable containers. *J. nat. Inst. agric. Bot.*, **11**, 378–403.

MACKAY, D. B., and TONKIN, J. H. B., 1967. Investigations in crop seed longevity. I. An analysis of long-term experiments, with special reference to the influence of species, cultivar, provenance and season. *J. nat. Inst. agric. Bot.*, **11**, 209–25.

MCFARLANE, V. H., HOGAN, J. T., and MCLEMONE, T. A., 1955. *Effects of heat treatment on the viability of rice.* Tech. Bull. No. 1129, US Dept Agric., Washington, DC.

MCROSTIE, G. P., 1939. The thermal death point of corn from low temperatures. *Scient. Agric.*, **19**, 687–99.

MILNER, M., and GEDDES, W. F., 1954. Respiration and heating. In *Storage of Cereal Grains and their Products*, eds. J. A. Anderson and A. W. Alcock, 152–220. Amer. Assoc. of Cereal Chemists, St Paul.

MITCHELL, F. S., and CALDWELL, F. Y. K., 1962. Influence of variations in harvesting and initial storage on wheat kept for several years. *J. agric. Engng Res.*, **7**, 27–41.

MUDALIER, C. R., and SUNDARARAJ, D. D., 1954. Dormancy and germination of a few crop seeds. *Madras agric. J.*, **38**, 163–73.

NAGAI, I., 1959. *Japonica Rice – its Breeding and Culture*, 706. Yokendo, Tokyo.

NUTILE, G. E., 1964. Effect of desiccation on viability of seeds. *Crop Sci.*, **4**, 325–28.

OWEN, E. B., 1956. *The Storage of Seeds for the Maintenance of Viability*. Commonwealth Agric. Bureaux, Farnham Royal, England.

OXLEY, T. A., 1948. *The Scientific principles of Grain Storage*. Northern Publishing Co., Liverpool.

OXLEY, T. A., and PIXTON, S. W., 1960. Determination of moisture content in cereals. II. Errors in the determination by oven drying of known changes in moisture content. *J. Sci. Food Agric.*, **11**, 315–19.

PARKIN, E. A., 1956. Stored product entomology. *Ann. Rev. Entom.*, **1**, 223–40.

PARKIN, E. A., 1963. The protection of stored seeds from insects and rodents. *Proc. int. Seed Test. Ass.*, **28**, 893–909.

PETERSON, A., SCHLEGEL, V., HUMMEL, B., CUENDET, L. S., GEDDES, W. F., and CHRISTENSEN, C. M., 1956. Grain storage studies. XXII. Influence of oxygen and carbon dioxide concentrations on mold growth and grain deterioration. *Cereal Chem.*, **33**, 53–66.

PORTER, R. H., 1949. Recent developments in seed technology. *Bot. Rev.*, **15**, 221–344.

RALL', JU. S., 1961. [The effect of constant reduced temperatures and winter conditions of storage on germination and biochemical properties of maize seeds.] (Russian, English summary.) *Izv. Timirjazevsk. S-H. Akad.*, **5**, 35–43. From *Field Crop Abstr.*, 1962, **15**, Abstr. No. 1330.

REES, A. R., 1963. A large-scale test of storage methods for oil palm seed. *J. West African Inst. Oil Palm Res.*, **4** (13), 46–51.

REES, A. R., 1970. Effect of heat-treatment for virus attenuation on tomato seed viability. *J. hort. Sci.*, **45**, 33–40.

RIEGER, R., and MICHAELIS, A., 1958. Cytologische und stoppwechselphysiologische Untersuchungen in aktiv-Meristem der Wurzelspitze von *Vicia faba* L. I. Der Einfluss der Unterwasser-Quellung der Samen auf die chromosomale Oberrationstrate. *Chromosoma*, **9**, 238–57.

ROBBINS, W. A., and PORTER, R. H., 1946. Germinability of sorghum and soybean exposed to low temperature. *J. Amer. Soc. Agron.*, **38**, 905–13.

ROBERTS, E. H., 1960. The viability of cereal seed in relation to temperature and moisture. *Ann. Bot.*, **24**, 12–31.

ROBERTS, E. H., 1961a. Viability of cereal seed for brief and extended periods. *Ann. Bot.*, **25**, 373–80.

ROBERTS, E. H., 1961b. The viability of rice seed in relation to temperature, moisture content, and gaseous environment. *Ann. Bot.*, **25**, 381–90.

ROBERTS, E. H., 1962. Dormancy in rice seed. III. The influence of temperature, moisture, and gaseous environment. *J. exp. Bot.*, **13**, 75–94.

ROBERTS, E. H., 1963. An investigation of inter-varietal differences in dormancy and viability of rice seed. *Ann. Bot.*, **27**, 365–69.

ROBERTS, E. H., and ABDALLA, F. H., 1968. The influence of temperature, moisture, and oxygen on period of seed viability in barley, broad beans, and peas. *Ann. Bot.*, **32**, 97–117.

ROBERTS, E. H., ABDALLA, F. H., and OWEN, R. J., 1967. Nuclear damage and the ageing of seeds with a model for seed survival curves. *Symp. Soc. exp. Biol.*, **21**,

65–100.

ROBERTS, H. M., 1959. The effect of storage conditions on the viability of grass seeds. *Proc. int. Seed Test. Assoc.*, **24**, 184–213.

ROBERTSON, D. W., LUTE, A. M., and GARDNER, R., 1939. Effects of relative humidity on viability, moisture content, and respiration of wheat, oats, and barley seeds in storage. *J. agric. Res.*, **59**, 281–91.

ROSENOW, D. T., CASADY, A. J., and HEYNE, E. G., 1962. Effects of freezing on germination of sorghum seeds. *Crop Sci.*, **2**, 99–102.

ROSSMAN, E. C., 1949. Freezing injury of inbred and hybrid maize seed. *Agron. J.*, **41**, 574–83.

SAHADEVAN, P. C., 1953. Studies on the loss of viability of rice seeds in storage. *Madras agric. J.*, **40**, 133–43.

SAYRE, J. D., 1947. Storage tests with seed corn. *Farm and Home Res.*, **32**, 149–54.

SEMENUIK, G., 1954. Microflora. In *Storage of Cereal Grains and their Products*, eds. J. A. Anderson and A. W. Alcock, 152–220. Amer. Assoc. of Cereal Chemists, St Paul.

SHANDS, H. L., JANISCH, D. C., and DICKSON, A. D., 1967. Germination response of barley following different harvesting conditions and storage treatments. *Crop Sci.*, **7**, 444–46.

SHELEF, L., and MOHENIN, N. N., 1966. Moisture relations in germ, endosperm, and whole corn kernel. *Cereal Chem.*, **43**, 347–53.

SIEGAL, S. M., 1953. Effects of exposure of seeds to various physical agents. II. Physiological and chemical aspects of heat injury in the red kidney bean embryo. *Bot. Gaz.*, **114**, 297–312.

SMITH, S. E., 1947. The sorption of water by high polymers. *J. Amer. chem. Soc.*, **69**, 646–51.

STERMER, R. A., 1968. Environmental conditions and stress cracks in milled rice. *Cereal Chem.*, **45**, 365–73.

TOOLE, E. H., and TOOLE, V. K., 1946. *U.S.D.A. Circ. 753.* Cited by Hukill, 1963.

TOOLE, E. H., and TOOLE, V. K., 1954. Relation of storage conditions to germination and to abnormal seedlings of beans. *Proc. int. Seed Test. Assoc.*, **18**, 123–29.

TOOLE, E. H., TOOLE, V. K., and GORMAN, E. K., 1948. *Vegetable-seed storage as affected by temperature and relative humidity.* US Dept Agric. Tech. Bull. 972.

TOUZARD, J., 1961. Influences de diverses conditions constantes de température et d'humidité sur la longévité des graines de quelques espèces cultivées. Adv. Hort. Sci. and their Applications. *Proc. 15th Internat. hort. Congr., Nice*, **1**, 339–47. Pergamon, Oxford.

VEGIS, A., 1963. Climatic control of germination, bud break, and dormancy. In *Environmental Control of Plant Growth*, ed. L. T. Evans, 265–87. Academic Press, New York.

VON SENBUSCH, 1955. Die Erhaltung der Keimfähigkeit von Samen bei tiefen Temperaturen. *Zuchter*, **25**, 168–69.

WATSON, E. L., 197c. Effect of heat treatment upon the germination of wheat. *Can. J. Plant Sci.*, **50**, 107–14.

WEBSTER, L. V., and DEXTER, S. T., 1961. Effects of physiological quality of seeds on total germination, rapidity of germination, and seedling vigour. *Agron. J.*, **53**, 297–99.

WEIBULL, G., 1953. The cold storage of vegetable seed and its significance for plant breeding and the seed trade. *Rep. 13th int. hort. Congr.*, *1952*.

WEIBULL, G., 1955. The cold storage of vegetable seeds, further studies. *Rep. 14th int. hort. Congr.*, *1954*, 647–67.

CHAPTER 3

Microflora and Seed Deterioration

C. M. Christensen

Many kinds of seeds harbour a great variety of microflora, especially fungi. This is particularly true of those kinds of seeds that are more or less exposed to contamination by airborne inoculum, as are the seeds of cereal grains, with the exception of maize. Seeds borne in pods, such as those of the legumes, or within fleshy fruits, such as tomatoes and melons, may at maturity be free of microflora if the fruits themselves are sound.

That some of the seed-borne microflora might reduce germinability of the seed when planted, or result in disease in the growing plant, has long been recognised. The importance accorded these seed-borne pathogens in the quality of seeds for planting is attested by the many papers devoted to them in the Proceedings of the International Seed Testing Association, such as those by de Tempe (1958, 1962, 1963), by Noble and Richardson (1968), by Malone and Muskett (1964), and by Christensen and López (1963). Grains and seeds that are sound when harvested may, during storage, be invaded by a variety of fungi that have been designated 'storage fungi'; these may result in various sorts of deterioration, ranging from decrease in germinability to complete spoilage. That these fungi might be involved in and responsible for deterioration of dormant seeds has been conclusively established only within the past few decades – before that few or none of those concerned with the problems of grain spoilage had any knowledge of fungi, and few or no mycologists or plant pathologists familiar with fungi had any acquaintance with problems of grain spoilage. Only since the early 1960s has it become generally known that some fungi growing in seeds either before or during harvest or in later storage can produce metabolites toxic to some kinds of animals – compounds known as mycotoxins or fungus toxins. Because of the possible importance of these mycotoxins in the health of humans and of our domestic animals, they are now undergoing fairly extensive and intensive investigation.

The present chapter will be devoted chiefly to deterioration of grains and seeds caused by storage fungi, but some of the more obvious deleterious changes caused by bacteria and by field fungi in seeds will also be discussed.

Bacteria

Relatively little work has been devoted to the relation of bacteria to deterioration of seeds. Muras (1964) stated that treatment, with a bactericide, of soya beans infected with *Xanthomonas phaseoli* var. *sojense* resulted in a 3–6.4 per cent increase in germination. Virgin (1940) reported that pea seeds soaked in a broth of bacteria isolated from pea seeds that germinated poorly were not reduced in germination percentage if the seed coats were intact; if the seed coats were chipped before the seeds were soaked in the bacterial broth, germination percentage was considerably reduced. It seems reasonable to suppose that even seed-borne pathogenic bacteria do not greatly reduce germination of the seeds they infect, if the infection has not progressed to the point of obvious decay; otherwise the problem would have been worked on more extensively. Bacteria are not likely to increase in stored seeds, since they require free water to grow, as do the seeds themselves – in other words, sufficient water for germination of the seeds. Seeds of some aquatic plants such as wild rice, *Zizania aquatica*, may be stored submerged in water, but seeds of most agricultural plants are stored at low moisture contents. In our experience, if seeds are kept at moisture contents that permit storage fungi to grow, the bacterial population of the seeds decreases very rapidly. Thermophilic bacteria may be involved in the final stages of biological heating of stored grains and seeds, at temperatures of 55–75°C, but by the time that temperature is reached the seeds are dead and decayed.

Fungi

From the standpoint of their ecology – the conditions that permit them to invade seeds – the fungi can be conveniently divided into two groups, (1) Field fungi, and (2) Storage fungi. Each group will be discussed separately.

FIELD FUNGI

These are fungi that invade seeds developing on the plants in the field, or after the seeds have matured and the plants are either still standing or are cut and swathed, awaiting threshing. The fungi require a moisture content in equilibrium with a relative humidity of at least 90–95 per cent to grow (Koehler, 1938). In the starchy cereal seeds this means a moisture content of 25–25 per cent on a wet weight basis, or about 30–33 per cent on a dry weight basis. Grains and seeds are not normally held at moisture contents this high for more than a short time after harvest. In some regions in some years a combination of wet weather during harvest, lack of drying facilities or lack of sufficient drying capacity, and lack of transport may result in much grain of high moisture content being piled on the ground. If such high-moisture content grain is of a temperature that allows microflora to grow, it will deteriorate rapidly. If the temperature is low throughout the pile, however, such grain may go through the winter with little or no damage. As an example, in the autumn of 1968 large quantities of corn and soya beans were piled on the ground in southwestern Minnesota, for lack of storage facilities. Numerous samples taken from these piles in February, 1969, with moisture contents of 15–20 per cent, still germinated 90 per cent, and had not undergone any detectable deterioration. Their temperature, from late October when they were harvested until February when the samples were taken, probably was continuously low.

Pepper (1960) lists approximately 180 species of filamentous fungi and about 20 species of yeasts that have been reported from barley kernels. Christensen and Kaufmann (1969) state that '. . . from a single gram of malting barley, about 25 kernels, we have isolated tens of thousands of colonies of filamentous fungi, hundreds of thousands of colonies of yeasts, and several million colonies of bacteria.' Malone and Muskett (1964) describe 77 species of seed-borne fungi, in 60 genera. Noble and Richardson (1968) list 20 species of fungi isolated from seeds of soya beans, 32 species from maize, 34 species from rice, 29 species from sorghum, and 28 species from seeds of wheat. The same publication lists *Alternaria* from seeds of more than 100 species of plants, and Fusarium from seeds of nearly 200 species. *Alternaria* must be an almost universal inhabitant of wheat kernels. Christensen and Kaufmann (1965) state 'Freshly harvested wheat from any of the many locations where we have obtained samples in the United States, Canada, Mexico, Colombia, South America and several countries in Europe have yielded *Alternaria* from nearly all of the surface-disinfected kernels cultured on a medium suitable for the growth of *Alternaria*. The growth of *Alternaria* from close to 100 per cent of surface-disinfected kernels of wheat indicates that the wheat probably was harvested recently, and that it was stored with a moisture content too low to permit invasion of the grain by storage fungi.' That is, *Alternaria* is present not only in stained or discoloured

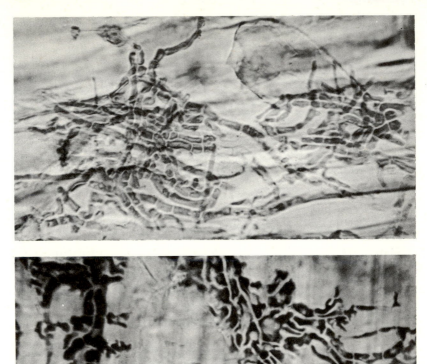

FIGURE 3.1 and 3.2   Fungus mycelium under the outer pericarp layers of wheat.

kernels of wheat, but also in clean, bright kernels, including those from irrigated and dry-land-farming areas, occurring commonly as mycelium under the outer pericarps of the kernels, as described by Hyde (1950) and by Christensen (1951). Figures 3.1 and 3.2 illustrate this mycelium. Concerning this subepidermal mycelium in general, Christensen (1951) stated, 'Mycelium was present beneath the pericarp of all the seeds examined. Most of this mycelium was dead. In the high-grade lots, most of the living mycelium beneath the pericarp was that of *Alternaria*, a fungus not known to cause deterioration of stored seeds. In the low-grade lots, most of the living mycelium beneath the pericarp was that of *Aspergillus* and *Penicillium* known to be involved in the deterioration of stored grain. No living mycelium was found in seeds more than eight years old.' Concerning the effects of field fungi in general upon seed quality, Christensen and Kaufmann (1969) state, 'Field fungi may discolor seeds, cause death of the ovules, shrivelling of the seeds or kernels, weakening or death of the embryos, and development of compounds toxic to man and to other animals. In grading of some grains, this discoloration is referred to as 'weathering' – a misnomer, since the discoloration is a product of growth of microflora.'

Reduced germination percentage of seed resulting from invasion of the seed by field fungi has been studied by several workers. Christensen and Stakman (1935) found a high correlation between increasing percentage of seed infected with *Helminthosporium* and *Fusarium* and decreasing germination percentage. Additional evidence that *Fusarium* and *Helminthosporium* can cause a reduction in germinability of barley seed was presented by Christensen (1936). Machacek and Greaney (1938) studied 'black point' or 'kernel smudge' of wheat, a dark discolouration usually of the lower or germ end of the kernel, resulting from invasion of the ripening kernel by fungi, especially *Alternaria*, *Helminthosporium* and *Fusarium*, although many other genera of the fungi as well as some bacteria, may be involved at times. Black point caused by *Alternaria* did not result in decreased germinability, but that caused by *Helminthosporium* did. Hanson and Christensen (1953) report the results of very extensive tests with different varieties of wheat grown in different locations over a period of 8 years. They found *Alternaria* to be the most prevalent fungus from kernels with black point (and also from sound, bright kernels), followed by *Helminthosporium*, then *Fusarium*. When only *Alternaria* was present, germination percentage of the seed was not affected, but infection by either *Fusarium* or *Helminthosporium* resulted in reduced germinability. Even if kernels whose pericarps are heavily discoloured as a result of invasion by *Alternaria*, the embryos normally are plump and bright – *Alternaria* does not usually invade the embryo. *Fusarium* may invade and kill the developing or mature embryo without causing any noticeable discolouration of either the pericarps or the embryo – the seeds appear sound, but actually are diseased or even dead.

STORAGE FUNGI

The storage fungi comprise chiefly several 'group species' of the genus *Aspergillus* and about an equal number of less well-defined species of *Penicillium*. The group species of *Aspergillus* that are involved – *A.*

restrictus, *A. glaucus*, *A. candidus*, *A. versicolor*, *A. ochraceus* and *A. flavus* (listed in order of increasing moisture required for growth) are relatively distinct from one another and relatively easy to identify with a high degree of certainty. The several different species of *Penicillium* found on deteriorating stored grains can be distinguished from one another only by a specialist in the group – and different specialists do not always agree as to the specific identity of a given isolate. This explains why in much of the literature dealing with deterioration of stored grains by fungi the species of *Aspergillus* are named, and those of *Penicillium* are not.

The fungi listed above are the major ones involved in deterioration of stored grains and seeds, but not the only ones. *Sporendonema sebi* sometimes is found in large numbers on occasional lots of wheat; it is able to grow at moisture contents about as low as those endured by *A. restrictus*; when inoculated on to wheat with moisture contents of 15–16 per cent and incubated for a time it has not, in my tests, produced any discolouration, has not made the grain unpalatable or toxic to experimental animals, and appeared to be restricted to the bran layers. Solomon *et al.* (1964) reported *S. sebi* to be antagonistic to the flour mite (*Acarus siro* L.); other than this the fungus has received little attention, and whether, from the practical standpoint, it deserves more attention seems doubtful. Under conditions that approach anaerobic in masses of stored grains with moisture contents of 18–22 per cent, one or more species of *Candida*, a filamentous yeast or budding fungus that in our experience is exceedingly difficult to distinguish from some isolates of the Imperfect fungus *Pullularia pullulans*, may predominate. Under conditions of optimum growth this fungus can raise the temperature of the material in which it is growing as much as 20°C in less than 24 hours. *Aspergillus halophilicus* is a minor storage fungus of considerable ecological but minor practical interest that will be described briefly. In the testing of thousands of samples of cereal grains – barley, wheat, oats, rice, sorghum and maize – from commercial bins, the two group species consistently associated with beginning or incipient deterioration have been *A. restrictus* and *A. glaucus*. The other common species of storage fungi – *A. candidus*, *A. ochraceus*, *A. versicolor*, *A. flavus* and *Penicillium* develop only later, after the growth of *A. restrictus* and *A. glaucus* has increased the moisture content of the grain mass, or of a portion of it, sufficiently to permit the growth of these higher-moisture-requiring or less xerophytic species. Species of *Aspergillus* such as *A. fumigatus*, *A. niger* and *A. terreus*, at times so abundant on decaying vegetation of various kinds, are not normally involved in deterioration of stored seeds, at least until the spoilage has progressed to the final stages and the moisture content of the decaying mass reaches that in equilibrium with a relative humidity above 90 per cent.

All of the species listed above as comprising the major storage fungi have the ability to grow in materials whose moisture contents are in equilibrium with relative humidities of 85 per cent or below. Each has its own rather sharply defined lower limit of moisture, below which it cannot grow. Competition may define almost as sharply the upper limit of moisture at which a given species will prevail or survive. If wheat from commercial bins, for example, bearing a mixture of field and storage fungi, is stored for some months in the laboratory with a moisture content of 13.2–13.5 per cent and a temperature of 22–25°C, *A. halophilicus* will develop on the germs and many of the seeds (Christensen, Papavizas and Benjamin, 1959). It has similarly been found by López and Christensen (1963) on sorghum seeds taken from commercial bins and stored for nearly a year at moisture contents of 13.3–13.8 per cent and 22–25°C. If the samples of wheat or sorghum are stored with a moisture content of 14.0–14.5 per cent, *A. restrictus* will predominate – in our repeated tests, at least, to the total exclusion of *A. halophilicus*. At moisture contents between 14.5 and 15.0 per cent, species of the *A. glaucus* group predominate, and in some storage tests with small quantities of wheat in the laboratory, where moisture content could be fairly rigidly controlled, a difference of 0.2 per cent in the range between 14.5 and 15.5 per cent, was critical for a given subspecies of the *A. glaucus* group. These relatively xerophytic fungi can sense moisture more precisely than we can measure it. It perhaps should be emphasised that at these relatively low moisture contents (low for biological activity) these fungi are growing in the embryos of the grain chiefly or almost exclusively. The embryos of these seeds contain much more oil than does the endosperm, and therefore at a given relative humidity will have a different, and lower equilibrium moisture content than does the endosperm. When we speak of a moisture content of 14.3 per cent in wheat or sorghum, we are speaking of the moisture content of the whole kernel – we do not know what the moisture content of the embryo is.

A. halophilicus is ecologically interesting because it has been detected only under the conditions described above – in seeds stored with a moisture content within a very narrow range, and for some months at moderate temperature. The source of inoculum of this fungus remains a mystery. Supposedly the fungus is present on the samples of wheat when we collect these from bins. We have repeatedly tried to isolate *A. halophilicus* from freshly harvested seed and from

TABLE 3.1 *Minimum relative humidity for the growth of common storage fungi at their optimum temperature for growth (27–30°C).*

Fungus	Minimum relative humidity, %
Aspergillus halophilicus	68
A. restrictus, Sporedonema	70
A. glaucus	73
A. candidus, A. ochraceus	80
A. flavus	85
Penicillium (depending on species)	80–90

seed recently collected from bins. In these attempts we have used nonsurface-disinfected kernels as well as surface-disinfected kernels, and have used agar media of high salt or sugar content on which only *A. halophilicus* or some members of the *A. restrictus* group can grow. None of these attempts to isolate *A. halophilicus* have been successful. We have been able to isolate it only from seeds stored for months in equilibrium with a relative humidity of about 68–70 per cent. The actual source of the fungus in nature remains a mystery. This fungus is described in some detail to emphasise the fact that some of the common storage fungi are, ecologically, highly specialised, and survive and thrive only in a very narrow ecological niche. The genus *Aspergillus* is treated extensively in the manual by Raper and Fennel (1965). The minimum relative humidities that permit growth of the various common storage fungi are summarised in Table 3.1, and the moisture contents of various common grains are summarised in Table 3.2.

TABLE 3.2 *Moisture content (%) of various grains and seeds in equilibrium with different relative humidities, at 25–30°C.*

Relative humidity, %	Wheat, corn sorghum	Rice		Soybeans	Sunflower	
		Rough	Polished		Seeds	Meats
65	12.5–13.5	12.5	14.0	12.5	8.5	5.0
70	13.5–14.5	13.5	15.0	13.0	9.5	6.0
75	14.5–15.5	14.5	15.5	14.0	10.5	7.0
80	15.5–16.5	15.0	16.5	16.0	11.5	8.0
85	18.0–18.5	16.5	17.5	18.0	13.5	9.0

The salient characteristics of each of the major storage fungi are described below.

Aspergillus restrictus

Lower limit of moisture for growth
Corn and wheat   13.5–14.5 per cent
Sorghum   14.0–14.5 per cent
Soya beans   12.0–12.5 per cent

Effects
Kills and discolours germs; causes 'sick' or germ-damaged wheat, blue eye in corn stored with moisture content of 14.0–14.5 per cent for some months; does not cause heating (because it grows too slowly).

Possible toxicity
Not known to produce compounds toxic to animals. We have grown a number of isolates of it in autoclaved moist grain and fed these to rats, also we have fed to rats samples of wheat very heavily invaded with different strains of *A. restrictus*; no illness and no lesions were observed in any of the rats so fed.

Other comments
A. restrictus is likely to be of significance only as a cause of germ damage and mustiness in grains stored for several months to a year or more with moisture contents as listed above; at higher moisture contents it is unable to compete with other storage fungi.

Aspergillus glaucus

Lower limit of moisture for growth
Corn and wheat   14.0–14.5 per cent
Sorghum   14.5–15.0 per cent
Soya beans   12.5–13.0 per cent

Effects
Kills and discolours germs, very slowly at moisture contents near lower limit for growth, more rapidly at higher moisture content. Causes blue eye in corn stored with 14.5–15.0 per cent moisture content; also causes mustiness and caking. Normally does not cause an appreciable rise in temperature, and so its increase is not detected by temperature sensing systems, but it may gradually increase the moisture content of the grain in which it is growing, and if the moisture increases to where *A. candidus* can grow rapidly, heating and spoilage may follow within a few days. That is, the increase in *A. glaucus* in itself may not be damaging, but such increase indicates that trouble might occur in the future.

Possible toxicity
Many isolates have been tested by us and others (our tests involved growing these isolates of *A. glaucus* in autoclaved corn-rice, and feeding this to rats) and none have been found to produce any symptoms or lesions in the test animals.

Other comments
Increase of *A. glaucus* in early stages, and up to when incipient spoilage may be under way, is not detectable by inspection with the unaided eye; microscopic examination or culturing of surface disinfected kernels, or both, are necessary to detect this. Many lots of wheat, corn, and sorghum that have been stored for several months will yield *A. glaucus* from 10–20 per cent of the surface disinfected kernels. If a lot yields *A. glaucus* from 20–50 per cent of the surface disinfected kernels it should be regarded as of questionable storability, especially if the moisture content is near or at the level where *A. glaucus* or *A. restrictus* can continue to grow. If the lot yields *A. glaucus* from 50–100 per cent of the surface disinfected kernels when cultured on agar, the lot is partly deteriorated, whether this is or is not visible by inspection. If percentage of surface disinfected kernels of grain in a given bin that yield *A. glaucus* increases from one sampling period to the next, incipient spoilage is under way, that may lead eventually to rapid heating and spoilage. Samples should be withdrawn from such lots at intervals of a few weeks and tested for number and kinds of fungi, moisture content, and damage. In this way any condition likely to result in serious damage can be detected before it becomes of any practical significance, and the grain can be either aerated, turned, dried, or processed before it can heat or spoil.

Aspergillus candidus

Lower limit of moisture for growth
Corn and wheat   15.0–15.5 per cent
Sorghum   16.0–16.5 per cent
Soya beans   14.5–15.0 per cent

Effects
Kills and discolours germs of seeds very rapidly. Causes heating up to 55°C, discolouration of entire kernels, total decay. In commercial storage bins *A. candidus* and *A. flavus* are the major causes of heating in all cases that we have investigated, in all kinds of grains and seeds. Once *A. candidus* has begun to grow in a given lot of grain, heating and spoilage are likely to follow within a few days to a few weeks; its presence in surface disinfected kernels is evidence of poor storage conditions in the past, and its increase is an indicator that an emergency exists NOW.

Possible toxicity
Some isolates of *A. candidus*, when grown under the right conditions in the laboratory, produce compounds toxic to experimental animals (most isolates do not), but this might be a laboratory phenomenon of little or no practical significance. In some of our tests, samples of severely heat damaged corn that had been decayed by a variety of fungi, including *A. candidus*, when fed to rats as their sole ration resulted in about the same weight gain as, and in some cases a slightly higher weight gain than sound, food-grade corn.

Other comments
Any increase in *A. candidus* in successive sampling periods of grain in a given bin

is cause for alarm; it means that some of the grain may be spoiling, and the location and size of that portion should be determined by removing samples by probe.

Aspergillus ochraceus

Lower limit of moisture for growth (same as for *A. candidus*)
Corn and wheat   15.0–15.5 per cent
Sorghum   16.0–16.5 per cent
Soya beans   14.5–15.0 per cent

Effects
A. ochraceus kills and discolours germs.

Possible toxicity
Some isolates of *A. ochraceus* produce a toxin, *ochratoxin*, similar to and just as toxic as *aflatoxin*. In tests of 164 samples of corn at the USDA Northern Regional Research Laboratory at Peoria *only one sample* was found to contain ochratoxin, and this one was of sample grade, damaged, and musty. It seems unlikely that *ochratoxin* is likely to be of much significance in the numerical grades of grain in the USA.

Other comments
A. ochraceus, in our experience, never predominates in a given lot of grain, even one that is undergoing or has undergone spoilage. We seldom have recovered it from more than 5 per cent of surface disinfected kernels of wheat or corn undergoing spoilage in the USA; in Mexico we have recovered *A. ochraceus* from 20–40 per cent of some lots of partially spoiled corn from commercial storage. If it were recovered from more than 5–10 per cent of surface disinfected kernels of a given lot, either that amount of partly deteriorated grain was added to or mixed with the lot at some time in the past, or deterioration is under way at present.

Aspergillus flavus

Lower limit of moisture for growth
Corn and wheat   18.0–18.5 per cent
Sorghum   19.0–19.5 per cent
Soya beans   17.0–17.5 per cent

Effects
Kills and discolours germs and decays and discolours whole kernels, causes rapid heating up to 55°C. *A. flavus* and *A. candidus* are the chief causes of heating of stored grains up to 55°C.

Possible toxicity
Some isolates, under some conditions of growth, produce aflatoxins. According to the evidence of the USDA workers at the Northern Regional Research Laboratory at Peoria, Ill., aflatoxins are not likely to be present in significant amounts in grains such as wheat, corn, and sorghum, or in soya beans; peanuts, peanut meal,

cottonseed and cottonseed meal, and fishmeal are more likely to contain aflatoxin than the cereal grains or soya beans.

Other comments
Presence of *A. flavus* in surface-disinfected kernels is evidence of poor storage in the past or of spoilage under way at present in the bin from which samples were taken. If *A. flavus* increases between sampling periods from a given bin, it is evidence of spoilage somewhere in the bin, with heating and more spoilage likely to develop rapidly. As with *A. candidus*, any increase in *A. flavus* in the grain in a given bin is cause for immediate action – aeration, turning, drying, or utilisation of the grain. If the moisture content of the grain is high enough to permit growth of *A. flavus*, aeration with air of low temperature may be required to prevent aflatoxin production.

Penicillium

Lower limit of moisture for growth
Corn and wheat   16.5–19.0 per cent
Sorghum   17.0–19.5 per cent
Soya beans   16.0–18.5 per cent

Effects
Kills and discolours germs and whole kernels or seeds, causes mustiness and caking, may be involved in early stages of heating, but does not cause such rapid heating or to so high a temperature as *A. candidus* and *A. flavus*. Causes blue-eye in corn stored with a moisture content above 18.5 per cent and at low temperature (*A. restrictus* and *A. glaucus* causes blue-eye in corn stored with moisture contents of 14.0–15.5 per cent).

Possible toxicity
Isolates of several different species of *Penicillium* when grown as pure cultures in the laboratory produce compounds toxic to various kinds of animals; how much of this is a laboratory phenomenon only is not known at present. The hemorrhagic syndrome in chicks, and an occasionally severe disease of turkey poults, involving liver lesions, are suspected to be due to consumption of feeds heavily invaded by certain species of *Penicillium*, but this has not been proven beyond question. Feeds invaded by *Penicillium* and by some other fungi may result in less efficient weight gains than sound feeds.

Other comments
Some common seed-invading species of *Penicillium* are able to grow at a temperature of 2–5°C, and some can grow slowly at a temperature below freezing. In spite of this, low temperature often is a very effective preservative of quality of stored grains of high moisture content.

WHEN STORAGE FUNGI INVADE SEEDS

It has been supposed by many of those in charge of and responsible for maintenance of quality in stored grains and seeds that, if loss in quality caused by storage fungi occurred in the seeds under their care, it must have resulted from damaging invasion of the seeds by storage fungi before harvest. They insist, often with vehemence, that the condition of the grain in question when under their care was such that storage fungi could not possibly have grown in it. This point is an important one, and for this reason considerable research has been devoted to determining whether seeds are invaded by storage fungi before harvest or after. The evidence from all of this is consistent: there is no significant invasion of seeds by storage fungi before harvest. The kinds of seeds studied have included wheat, barley, oats, rice, sorghum, maize, common beans, soya beans, seeds of various kinds of vegetables, and the regions where this work has been done, or from where seeds were collected, included North and South America, Europe, Africa and Southeast Asia. Many thousands of samples have been tested, over a period of more than 20 years. It is of course theoretically possible that some kinds of seeds under some circumstance may be heavily invaded by storage fungi before harvest – all we can say is that up to now, no such lots have been found. Some of the work concerning this is worth looking at more closely.

Tuite and Christensen (1957) collected ripe heads from plants of different varieties and classes of wheat in several states over a period of three harvest seasons. Some of the plants had been left standing in the field for as long as a month after normal harvest time, or were in shocks or in windrows, at times exposed to frequent rains. From 50 to 100 kernels of each sample were surface disinfected and put on an agar medium favourable to the growth of storage fungi. Storage fungi grew from less than 5 per cent of the kernels. About 3,000 kernels of wheat were collected from the field at harvest time and from combines and, *without being subjected to any surface disinfection*, were similarly put on an agar medium favourable to the growth of storage fungi. *Aspergillus glaucus* grew from about 5 per cent of these and other storage fungi grew from less than slightly less than 7 per cent – and probably most of these were contaminants. Tuite (1959) cultured 73,200 surface-disinfected kernels of soft red winter wheat, from 732 samples collected in Indiana, some from fields that remained unharvested for several weeks after the grain had ripened, because of heavy rains. Storage fungi grew from only 25 kernels, or 1 in 300, or 0.33 per cent. Tuite and Christensen (1955) collected many samples of mature barley kernels from plants in fields in Minnesota during moist weather with frequent showers, when relative humidities above 75 per cent sometimes prevailed for several days at a time. Some of the plants from which the samples were taken were lodged, and many kernels were severely discoloured by field fungi. Kernels were lightly surface disinfected and put on malt-salt agar for the detection of storage fungi. Some colonies of *Aspergillus flavus* developed in one of the dishes, almost certainly from contamination by airborne spores during the culturing process, and not from the grain itself. No storage fungi, other than those contaminents grew from the seeds. Both Tuite (1961) and Qasem and Christensen (1958) reported that very few surface-disinfected kernels of corn from plants in the field yielded storage fungi, even when the plants from which the kernels were collected had stood for some time in the field after normal harvest time and had been exposed to frequent rains. Kaufmann (1959) tested many hundreds of samples of wheat from lots arriving at terminal elevators in different sections of the United States over a period of 3 years. *Alternaria*, a common field fungus in wheat and in other small grains, grew from 48 to 83 per cent of the surface-disinfected kernels, *A. glaucus* from 1.4 to 10.9 per cent, *A. flavus* from 0 to 3.7 per cent, *A. candidus* from 0 to 0.5 per cent, and *Penicillium* from 0 to 0.8 per cent. Some of these lots from which the samples were taken had passed through country elevators on their way to terminals, and it is highly probable that they contained at least some grain from a previous year's crop, since such 'mixing off' is a common practice. It is possible that the different amounts of infection by storage fungi in these samples reflected the different amounts of such mixing in of old grain in the lots from which the samples came. In any case, very few of the thousands of samples examined were very extensively invaded by storage fungi – and it is highly probable that those few had been in storage for some time before they arrived at the terminals. Thus in many thousands of samples of different kinds of grains taken from plants at or immediately after harvest and over a period of 20 years we have yet to find a single sample with more than a very light invasion of storage fungi in more than a small percentage of the seeds.

Under the conditions favourable to the growth of storage fungi, however, such invasion can occur within a few days. As an example, some years ago country elevator men in several locations in the winter wheat growing area of the United States collected maturing wheat heads and sent them to us. We removed the kernels and put them on agar to determine whether they were invaded by storage fungi. The results of this study are included in the data given above. No storage fungi ever were isolated from these samples, with one exception. In that one exception the green heads of wheat had been wrapped in waxed paper before they were mailed to us. The waxed paper served as a moist chamber, and in the 3 days that elapsed between the time the heads were collected in the field and the time they arrived at the laboratory, 100 per cent of the surface-disinfected kernels that were put on agar yielded *Aspergillus glaucus*; when the heads were removed from the envelope in which they were sent, in fact, *A. glaucus* could be seen sporulating heavily on the glumes that covered the kernels. Some of the data from work on this problem are summarised in Tables 3.3 and 3.4.

CONDITIONS FOR GROWTH

The basic requirements for the growth of storage fungi are similar to those for the growth of other living things – food, water, a favourable temperature, a suitable atmosphere, and time. The rate at which they will develop on any given lot of seeds or grains will depend greatly on the history and condition of the grain – the degree to which it already has been invaded by these fungi, the amount of cracked and broken seeds, the amount, nature, and distribution of the debris, whether the embryos are alive or dead, the presence, numbers, and activities of insects and mites, to name what appear to be the main ones. All of these factors are interrelated and interacting, and while here they are discussed separately, in practice they must be considered together.

TABLE 3.3 *Storage fungi from wheat collected at various places from field to terminal, cultured without surface-disinfection on malt-salt agar. (From Tuite and Christensen, 1955.)*

Sources of samples	No. of samples	No. of kernels	Percentage of kernels yielding fungi	
			A. Glaucus	Other storage fungi
Heads of standing, shocked and windrowed wheat	27	2050	4.8	2.7
Combines	7	1000	4.8	6.8
Country elevators, new crop	16	800	50.0	8.0
Trucks from country elevators unloading at terminal	12	575	64.0	6.0

TABLE 3.4 *Percentage of surface-disinfected kernels of wheat yielding* Alternaria *and storage fungi in Crop years 1953–54, 1954–55, and 1955–56 as the grain arrived at terminal elevators. (From Kaufman, 1959.)*

Region of United States	Storage fungi				
	Alternaria	*A. Glaucus*	*A. Flavus*	*A. Candidus*	*Penicillium*
East	71.2	6.6	0.8	0.3	0.8
Southeast	66.6	8.0	3.7	0.5	0.4
Northwest	60.5	10.9	1.1	0.3	0.3
Central	83.4	10.0	1.6	0.1	0.3
South	58.2	9.2	1.8	0.4	0.3
Southwest	67.8	7.5	0.7	0.2	0.1
Pacific Northwest*	48.4	1.4	0.0	0.0	0.0

*One year only.

MOISTURE

The minimum relative humidities that permit the various common storage fungi to grow are summarised in Table 3.1. The moisture contents of representative grains and seeds in equilibrium with those relative humidities are summarised in Table 3.2 (see also Appendix 4), and the lower moisture contents that will permit invasion by the individual fungi are given in the previous section in which the individual species are described. Actually the specification of very precise limits of moisture for the growth of a given fungus in a given kind of seed oversimplifies the matter, because the moisture content of a given kind of grain in equilibrium with a given relative humidity will vary to some extent with the treatment of the grain since harvest; it is especially dependent on whether the grain has been artificially dried and, if so, at what temperature and for how long. Fairbrother (1929) mixed wheats of different moisture content and allowed the mixture to stand until equilibrium was reached; the originally moister lot retained a moisture content from 1 to 2 per cent above that of the originally dried lot. The same has been shown in corn by Hubbard et al. (1957) and by Tuite and Foster (1963), and in rice by Schroeder and Sorenson (1961). Sound sorghum seed dried for 18 hours at 70°C (158°F), then exposed to a relative humidity of 75 per cent, attained a moisture content of 14.3 per cent (my own unpublished work). Seed of the same lot, moistened to 20 per cent and then exposed to 75 per cent relative humidity in the same chamber attained an equilibrium moisture content of 15.2 per cent. Yet after storage for some months the sample of the *lower* moisture content was more heavily invaded by storage fungi (chiefly species of the *Aspergillus restrictus* group) than was the sample at the higher moisture content. In similar and also unpublished work with various samples of maize treated in different ways and then exposed to a uniform relative humidity, sometimes the samples originally of lower moisture content have had, at equilibrium, a moisture content higher than the samples that originally were moister. Even the complications are complicated.

Most maize in commercial channels in the United States has been artificially dried, since most of it is harvested with a moisture content too high for safe storage. A common practice is to dry a portion of a given lot to a moisture content well below the limit of 15.5 per cent specified for Grade No. 2 corn, then mix this with another portion of higher moisture content to achieve an average of 15.5 per cent. If it is to be stored for some time, the moisture content aimed for in the mixture may be 14.5 or 15.0 per cent. Even if an average of 14.5 per cent moisture is achieved in the mix, a portion of the grain may have a moisture content of 13.5 per cent and a portion of 15.5 per cent or more. Depending on how long and at what temperature these lots were stored before they were dried and mixed, one or both may have been invaded to a moderate extent by storage fungi and may be of very high storage risk. That is, among a hundred lots of maize, all of which are Grade No. 2 according to the characteristics specified in the Official Grain Standards of the United States, there may be a tremendous range in storability and in deterioration risk. This can be detected by laboratory tests that determine the number and kinds of fungi present.

The Official Grain Standards of the United States (USDA, 1964) specifies the maximum moisture content permitted in each of the several numerical grades of grains and seeds. As emphasised by Christensen and Kaufmann (1969), these specifications were developed to promote orderly marketing, and have very little to do with storage. The upper moisture limit for Grade No. 2 or 3 corn (15.5 and 17.5 per cent, respectively) or for Grades No. 2 and 3 soya beans (14 and 16 per cent, respectively) are too high for safe storage for more than a very short time unless the temperature is so low that storage fungi can now grow.

Methods of sampling of grain in trucks or railway cars, or being loaded into or out of bins or ships, have been developed to assure that the samples are representative of the bulks from which they were taken. These tell nothing about the range in characteristics from place to place in the bulk, and in judging storability this range often is critical. If we want to know the condition of the grain in a specific portion of the bulk, we must take a sample from that portion and examine it individually.

Methods of measuring moisture content also are specified (USDA, 1959), but in practice moisture content of grains and seeds is measured by means of an electric moisture meter. The meters now in general use give results in close agreement with those obtained by oven-drying samples at 103°C for 3 days, so long as the meter and the operator are functioning properly. Sometimes, however, and especially with grain whose moisture content is in the range of 14.5–16.5 per cent, there may be a difference of 1 per cent in the moisture content as determined by a given meter and by oven drying. The reason for this occasional and unpredictable discrepancy is not known. This can be of some practical importance. If, for example, the meter indicates that a given sample taken from a bulk of corn has a moisture content of 14.5 per cent, the lot is, in this respect, well within the limits for No. 2 corn. If the meter happens to read 1 per cent low, then the actual moisture content of the sample is 15.5 per cent, and the bulk from which it was taken probably had, in some portions, a moisture content of at least 16.5–17 per cent, and might be so heavily invaded by fungi as to be on the verge of spoilage. And if the sample represents or indicates the condition of a bulk of 50,000 bushels, most of which later spoils from 'mysterious' causes, the discrepancy can indeed be serious.

In grains and seeds stored in bulk there almost inevitably are differences in temperature between different portions of the bulk. If the grain goes into storage with a relatively high temperature, as often is the case in the north temperate zone, and later is subjected to much lower temperatures, the top and outer portions of the grain acquire a much lower temperature than the inner portions. Slow air currents are set up, moving upward through the centre of the mass and downward on the outside. As the air moves upwards through the warmer portion of the grain it of course comes to equilibrium with the moisture content of the grain there. As it reaches the upper and cooler portions the relative humidity of the air is increased and the grain absorbs moisture from it. Johnson (1957) calculated that in a large bin of corn with 14.5 per cent moisture, a temperature differential of about 22°C between the interior and the surface of the bulk would result, in 20 days, in sufficient transfer of moisture to raise the moisture content of a layer 6 inches deep on the surface of the bulk to 20 per cent. Holman (1950) reported that soya beans stored in November

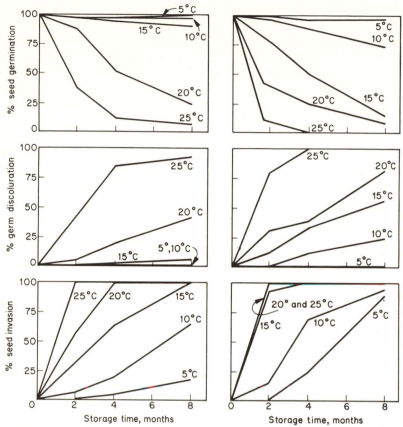

FIGURE 3.3   Percentage of seeds invaded, germs discoloured, and germination of yellow dent corn inoculated with a mixture of fungi and stored 8 months at 5–25°C. Left, 16 per cent moisture content; right, 18 per cent moisture content. (From Qasem and Christensen, 1958.)

1942, with an average moisture content of 12–13 per cent, had, by February 1943, a moisture content of 16–17 per cent in the centre of the upper surface of the bulk. By February 1944, after 15 months of storage, the moisture content of the beans in the centre of the upper surface of the bulk ranged from 20 to 24 per cent. Christensen (1970) stored sorghum seeds of 14.3 per cent moisture in a gallon glass jug and maintained a temperature 10–15°C higher in the grain near one side than in the grain near the opposite side. Samples were removed periodically from each side and tested for moisture content. After only 3 days the moisture content of the grain on the cool side was 1.4 per cent higher than that of the grain on the warm side, and after 6 days the difference was 2 per cent. Additional tests with other kinds of grains and seeds have confirmed the fact that relatively small differences in temperature, if consistently maintained, will result in relatively rapid transfer of moisture from the warm to the cool portion of the bulk. The higher the average moisture content of the grain and the greater the temperature differential between different portions, the greater the moisture transfer. One of the functions of aeration is to maintain a uniform temperature throughout a given bulk and reduce such tranfer to a minimum.

When all of the above are combined – mixtures of lots of different moisture content that never come to the same equilibrium, methods of sampling specifically designed to prevent one from knowing the condition of the grain in a given portion of a bulk, error in measurement of moisture content of a given sample, slow or rapid moisture transfer from one portion of a bulk to another – it is not surprising that the warehouseman frequently does not know within 2–5 per cent or more the moisture content of the grain under his charge. If his moisture meter is functioning well he may know rather precisely the moisture content of a single sample that may or may not be represen-

TABLE 3.5   *Germination, germ discolouration, and fungus invasion of yellow dent corn inoculated with* Aspergillus candidus *and stored at 18 per cent moisture content and 5, 15, and 25°C for 1, 3, and 6 months.*

Months stored	Temperature, °C	Germination, %	Germs discoloured, %		Seeds invaded[a], %
			Ochre	Brown	
1	5	100	0	0	0
	15	96	4	0	42
	25	76	12	16	100
3	5	98	0	0	0
	15	72	20	8	100
	25	2	24	32	100
6	5	98	0	0	0
	15	53	16	26	100
4.5	25	0	0	100	100
Uninoculated controls					
6	5	98	0	0	0
6	15	97	0	0	6[b]
4.5	25	94	2	0	8[b]

[a] Seeds shaken 1 minute in 1 per cent sodium hypochlorite, rinsed twice in sterile water, and cultured on malt-salt agar.
[b] *Aspergillus repens.*

tative of the bulk. This gives him no information on the range of moisture within the lot or bulk in question, and it is this which determines its storability. The range in moisture content within a given lot can be determined only by withdrawing samples from different portions and testing them separately. This should be an integral part of good storage practices.

TEMPERATURE

Figure 3.3 and Table 3.5 from Qasem and Christensen (1958) show the influence of temperature upon invasion of stored corn (maize) by storage fungi and on reduction in germinability of the seed. We stated that 'low temperature was as effective as low moisture content in preventing damage by the fungi tested.' Data of Papavizas and Chris-

TABLE 3.6 *Fungus invasion, seed viability, and development of brown germs of white wheat, not inoculated and inoculated with members of the* Aspergillus glaucus *group. Kept for 8 days at 25° ± 2°C and stored for 12 months at 15.0–15.5 per cent moisture content and at 5° and 10°C.*[a]

Inoculum	Final moisture content, % (original, 15.5 per cent)	Seed viability, %	Brown germs, %	Fungi invading seeds[b], %			
				A. amstelodami	*A. repens*	*A. ruber*	*A. restrictus*
	Stored at 5°C						
A. amstelodami	15.3	99	0	0	0	0	0
A. repens	15.3	95	0	0	0	0	0
A. ruber	15.1	98	0	0	0	0	0
A. restrictus	15.2	99	0	0	0	0	0
Not inoculated	15.4	100	0	0	0	0	0
	Stored at 10°C						
A. amstelodami	14.9	96	0	84	0	0	0
A. repens	15.1	99	0	0	0	0	0
A. ruber	15.2	98	0	0	0	0	0
A. restrictus	15.0	99	0	0	0	0	0
Not inoculated	15.1	99	0	0	0	0	0

[a] Each figure is an average of four replicates.
[b] As determined by culturing 100 surface-disinfected seeds on malt-salt agar.

tensen (1958) are summarised in Tables 3.6 and 3.7. In white wheat stored at moisture contents of 15–15.5 per cent and at 5° and 10°C there was no invasion by storage fungi and no reduction in germinability in 12 months (Table 3.6) but with moisture contents of 16–16.5 per cent and at the same temperatures the reduction in germinability was considerable in 12 months (Table 3.7).

TABLE 3.7 *Fungus invasion, seed viability, and development of brown germs of white wheat, not inoculated and inoculated with members of the* Aspergillus glaucus *group, and stored for 12 months at moisture contents of 16.0–16.5 per cent and at 5° and 10°C.*[a]

Inoculum	Final moisture content % (original 16.5 per cent)	Seed viability, %	Brown germs, %	Fungi invading seeds[b], %			
				A. amstelodami	A. repens	A. ruber	A. restrictus
Stored at 5°C							
A. amstelodami	16.2	53	22	97	0	0	0
A. repens	16.1	62	21	0	66	0	0
A. ruber	16.4	65	27	0	0	77	0
A. restrictus	16.0	52	41	62[c]	12[c]	0	5
Not inoculated	16.4	94	0	2[c]	0	0	0
Stored at 10°C							
A. amstelodami	16.2	27	58	94	0	0	0
A. repens	16.0	23	47	0	89	0	0
A. ruber	16.0	18	28	0	0	95	0
A. restrictus	16.2	12	57	90[c]	0	0	15
Not inoculated	16.3	78	4	12[c]	6[c]	0	0

[a] Each figure in the table is an average of four replicates.
[b] As determined by culturing 100 surface-disinfected seeds on malt-salt agar.
[c] Invasion apparently resulting from inoculum originally present in the seed and not eliminated by the treatment with sodium hypochlorite solution at the beginning of the test.

Approximate minimum, optimum and maximum temperatures for growth of common storage fungi are summarised in Table 3.8. These are compiled from tests and observations in the laboratory and in commercial bins of grain and refer to the growth of the fungi on grains with moisture contents near the optimum for each of the several species. The cardinal temperatures for growth on agar may

TABLE 3.8   *Approximate minimum, optimum, and maximum temperatures for growth of common storage fungi on grains.*

Fungus	Temperature for growth, °C		
	Minimum	Optimum	Maximum
Aspergillus restrictus	5–10	30–35	40–45
A. glaucus	0–5	30–35	40–45
A. candidus	10–15	45–50	50–55
A. flavus	10–15	40–45	45–50
Penicillium	– 5–0	20–25	35–40

be somewhat different. Also individual strains of a given group species of *Aspergillus*, and different species of the genus *Penicillium* may differ from the figures given.

OXYGEN

So far as is known, there are no obligate anaerobes among the fungi, and certainly there are none among the storage fungi. One can find statements in the literature to the effect that all fungi are strictly aerobic – the statement, in fact, has been repeated so often and for so long that it has become part of the credo of even professional mycologists. Generalities such as this are likely to be only partial truths. Tabak and Cooke (1968) incubated a number of fungi, isolated from sewage sludges, in microaerophilic and anaerobic conditions in the laboratory. All of the fungi they tested grew to some extent under conditions as anaerobic as could be maintained in the laboratory. As judged by the weight of cell mass produced, the rate of growth under strict anaerobic conditions ranged from 20 to 50 per cent as much as that under aerobic conditions. Yeasts, including *Candida* (or *Pullularia*) are common on maize silage and on some lots of moist grain in which near-anaerobic conditions have prevailed. It is probable that the lack of obvious development of filamentous fungi on corn (maize) silage is due to competition with the acid-forming bacteria that usually predominate, rather than to their inability to grow under conditions of high carbon dioxide and low oxygen tension. Sealed storage of high moisture grains has been tested as a method of preserving quality, as will be mentioned later; certainly there will be no significant development of storage fungi under such conditions. At lower moisture contents, where storage fungi ordinarily prevail, the case may be otherwise.

TABLE 3.9   *Germination, germ discolouration, and fungus invasion of 3 commercial samples of grade No. 2 yellow dent corn and 1 sample of seed-grade yellow dent corn stored 2 and 4 months at 16 per cent moisture content and 25°C.*

Sample	Months stored	Germina-tion, %	Germs dis-coloured, %		Surface-disinfected kernels yielding *Aspergillus glaucus*, %	Mould count, thousands
			Brown	Ochre		
Commcl. A	0	50	1	16	40	7.5
	2	12	2	14	80	60
	4	0	16	26	100	442
Commcl. B	0	62	2	11	55	11.5
	2	1	6	28	100	350
	4	0	18	46	100	960
Commcl. C	0	48	5	22	46	81
	2	6	10	36	100	700
	4	0	28	58	100	1200
Seed Grade	0	98	0	0	4	4
	2	98	0	2	48	48
	4	69	4	14	100	92

DEGREE OF INVASION BY STORAGE FUNGI

Seeds that are already invaded by storage fungi by the time they reach a given storage site are already partly deteriorated. That is, once invasion has begun there is no sharp line dividing sound from deteriorated seeds; it becomes a matter of degree and, from a practical standpoint, of the use to which the lot in question is to be put. A degree of invasion by storage fungi sufficient to reduce germinability to 80 per cent may make the seed unacceptable for malting or for planting, but the grain may be perfectly suitable for milling, chemical processing, or feed. Seeds already invaded to some extent by storage fungi will, if kept under conditions that permit the fungus to increase, be invaded further, and will lose quality faster, than sound seeds similarly stored. An example of this is given in Table 3.9, from Qasem and Christensen (1960).

DEBRIS AND FOREIGN MATERIAL

Other things being equal, grain containing appreciable amounts of foreign material and debris is more subject to damage by storage fungi than clean grain. In part this is because such debris, or a portion of it, may be inherently more susceptible to attack by these fungi than is sound grain. Second, when bulk grain is loaded into a bin, tank, shiphold, or other container, the fine material is likely to accumulate in the grain directly below the loading spout, forming a portion called by practical men the 'spout line'. When maize containing only 2–3 per cent of foreign material is loaded into a bin from a central over-head spout, the grain in the central core of the bin may contain 50 per cent foreign material. If air is forced through a bin of grain such as that, as often is done to reduce the temperature and make it uniform throughout the bin, the air will channel around this tightly packed material. One practical means of reducing the hazard posed by such an accumulation of foreign material in a central core in the bulk, at least in bins with a central unloading hopper, is as follows. As soon as the bin has been loaded, open the hopper and let the central core, where the foreign material is concentrated, run out; it then can be screened to remove the foreign material, and the sound grain loaded back into the bin.

INSECTS AND MITES

Some kinds of grain-infesting insects and mites are regularly asso-ciated with storage fungi. This is particularly true of those kinds of insects whose larvae develop within the kernels of stored grains, such as the weevils. Weevils may carry large numbers of spores of storage fungi, especially of *Aspergillus restrictus* and *A. glaucus*, into the grain when they invade it, as shown by Agrawal *et al.* (1957), and may provide the conditions, especially the moisture, to promote rapid growth of storage fungi, as shown by Agrawal *et al.* (1958) and by Christensen and Hodson (1960). According to Griffiths *et al.* (1959) the same may be true of the grain-infesting mites. Much information on the interrelation of grain infesting insects and fungi is summarised by Sikorowski (1964). Fumigation may eliminate the insects from the grain, but if the fungi have developed to the point where further deterioration is a self-perpetuating process, spoilage will continue, since the fumigants used, in the dosages used, ordinarily have little or no effect on the fungi growing within the seeds.

Effects of storage fungi upon grains and seeds

The major effects of storage fungi upon seeds are, in the usual, but not inevitable, order of their appearance: (1) Decrease in germin-ability; (2) discolouration; (3) production of mycotoxins; (4) heating; (5) development of mustiness and caking; (6) total decay.

DECREASE IN GERMINABILITY

When storage fungi were first found to be commonly associated with deterioration in stored grains and seeds it was maintained by some that the fungi did not precede the death and discolouration of the seeds, but *followed it*. There was no experimental or even observational evidence to support this position; the view was maintained, evidently, that because no such causal relationship has been established by investigators in the past, no such relationship could exist. Several investigators, it is true, in the study of respiration of moist stored seeds, had treated the seeds with one compound or another that was supposed to be fungicidal, and since respiration was not inhibited they concluded that fungi could not be responsible. Actually the compounds they used, under the conditions of use, were not at all

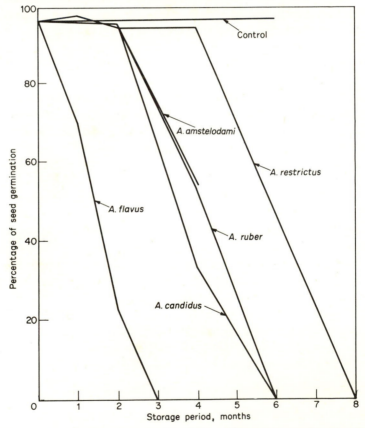

FIGURE 3.4   Effects of various species of *Aspergillus* upon germination of pea seeds stored at 30°C and 85 per cent relative humidity. (From Fields and King, 1962.)

fungicidal to the fungi concerned, and thus this evidence was invalid.

As emphasised by Christensen and Kaufmann (1969) the fact that invasion of seeds by storage fungi is a direct cause of loss of germinability of the seed has been difficult to establish. An essential requirement was that the seeds with which the work was to be done should be known by actual test to be free of storage fungi, and that those to be kept free of storage fungi should actually be kept so during the course of the tests. It eventually was found that some kinds of seeds, such as those of legumes, borne in pods, could easily be obtained free of storage fungi. Also it was found that if cobs of corn free of any obvious fungus infection were selected at maturity and the kernels removed by hand in a clean place and dried immediately, they would be free of invasion by storage fungi. This is now well known, and seems obvious; but it was not so when work first got under way on this problem in the 1950s. It was found difficult to obtain lots of wheat or barley of which even small samples could be kept free of storage fungi for some months when stored at moisture contents and temperatures that permitted storage fungi to grow. Seed treatment fungicides proved to be relatively worthless for the purpose. Also, the temperature-moisture-content-time combinations that result in invasion of seeds by storage fungi are those which result in increasing respiration of the seeds themselves. Eventually, however, evidence was accumulated to show without question that storage fungi are indeed capable of killing seeds. Some of this evidence, from Christensen and López (1963) is summarised in Table 3.10. The data of Fields and King (1962) from their work with pea seeds (Fig. 3.4) shows very strikingly the loss of germinability resulting from invasion by storage fungi. Figure 3.5 shows the effect of *Aspergillus candidus* on germin-

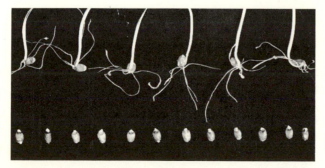

FIGURE 3.5 Reduction in germinability of wheat caused by *Aspergillus candidus*. Top, controls, free of fungi, germination 100 per cent; bottom, seeds inoculated with *A. candidus*, germination 0 per cent. Both stored for 25 days at 15.4 per cent moisture and at 25°C. (From Christensen and Kaufmann, 1969).

TABLE 3.10   *Reduction in germination percentage of seeds, caused by storage fungi.*

Kind of seed	Conditions of storage			Treatment	Germina-tion, %	Authority
	Moisture content, %	Temp., °C	Time			
Wheat	15.5–15.7	20–25	6 wks.	Free of storage fungi	95	Christensen (un-published data)
				Inoculation with		
				A. restrictus	5	
	16.0–16.4	25	2 mos.	Free of storage fungi	90	Papavizas and Christensen
				Inoculation with		
				A. candidus	25	
				A. amstelodami	38	
				A. restrictus	40	
	17.0–17.2	25	1 mo.	Free of storage fungi	93	Papavizas and Christensen
				Inoculation with	20	
				A. candidus	20	
				A. ruber	32	
				A. restrictus	34	
Maize	17.0–18.0	15	2 yrs.	Free of storage fungi	96	Qasem and Christensen (1960)
				Inoculation with storage fungi	0	
	18.5	20	3 mos.	Free of storage fungi	96	Qasem and Christensen (1958)
				Inoculation 4 spp.		
				A. glaucus-avg. of all	56	
Sorghum	15.8	28	40 days	Free of storage fungi	95	Christensen and López (un-published data)
				Inoculation with storage fungi	35	

ability of wheat seed. Under the conditions that usually prevail when storage fungi invade seeds, assuming that the seeds were of high viability when stored, it seems likely that the storage fungi may contribute significantly to reduced germinability. It is not implied that this is true of all kinds of seeds, since there are many conditions that may reduce seed viability. Barton (1961), for example, mentions a number of kinds of seeds whose viability is reduced, sometimes very rapidly, by desiccation.

DISCOLOURATION

As mentioned earlier, field fungi may cause discolouration of seeds before harvest. In the small cereal grains this discolouration is limited to or more prominent in the glumes or pericarps than in the embryo or endosperm of the caryopsis. Also the discolouration by field fungi does not continue to develop in seeds in storage; there are exceptions to this, of course, as in corn (maize) stored on the cobs in cribs, with a high moisture content, or in seeds of various kinds on plants that remain in the field over winter in climates where they are subjected to rain and snow.

The storage fungi, developing at lower moisture contents and over a relatively long period of time can cause discolouration of the embryo or of the whole seed or kernel. 'Sick wheat' or 'germ damaged wheat' is wheat with brown to black embryos. In the laboratory, discolouration of this sort has been caused by various conditions. In commercial storage, however, it seems highly probable that storage fungi are a major cause of this degrading (in a quality sense) discolouration. Christensen and Kaufmann (1969) state, 'however, neither in the laboratory of the Department of Plant Pathology of the University of Minnesota, where since about 1948 thousands of samples of wheat have been examined, including many hundreds with different amounts of germ damage, nor at the Grain Research Laboratory at Cargill, Inc., where since 1952 many thousands of samples of wheat from commercial storage throughout the United States have been examined, have we ever encountered a single case of germ damaged or sick wheat in which storage fungi were not involved.' Seeds that are discoloured brown to black throughout are, in the United States, referred to as 'bin burned' or 'heat damaged'. Usually this is an advanced stage of spoilage in which there has been some microbiologically engendered heating, as described below. The dark colour, however, can develop without heating. Schroeder (1963) stated that what is called heat damage in rice may develop without any detectable rise in temperature, the discolouration presumably being caused by storage fungi.

HEATING

That fungi and bacteria growing in plant materials would cause heating was shown 80 years ago by Cohn (1890). Darsie *et al.* (1914) reported that germinating seeds in Dewar bottles, with moisture contents and respiration rates much higher than those of seeds in storage, raised the temperature only 1–3°C, whereas the lots of seeds overgrown with fungi increased 10°C in temperature in 4 or 5 days, at which time the tests were terminated. Seed respiration could not possibly raise the temperature higher than the germinating seeds themselves could endure, which probably would not be above 30°C; this obvious biological fact evidently was disregarded by those who maintained that respiration of the seeds themselves was responsible for heating up to 50–60°C. Dead seeds do not respire. Ramstad and Geddes (1943) implicated fungi in the heating of moist soya beans, and Milner and Geddes (1945, 1946) established beyond question the course of heating in moist stored soya beans. Briefly, fungi, primarily *Aspergillus candidus* and *A. flavus* will when growing rapidly raise the temperature up to about the maximum they can endure, 50°C for *A. flavus* and 55°C for *A. candidus*. During this time metabolic water is produced by the fungi, which may raise the moisture content high enough for thermophilic bacteria to take over. These may raise the temperature to 70–75°C. With the right combination of cirumstances, nonmicrobiological heating may then raise the temperature still further, sometimes to the point of ignition. Carlyle and Norman (1941), Carter (1950) and Christensen and Gordon (1948) also contributed evidence that microflora were principally involved in the heating of moist grains and other plant materials. Walker (1967) revealed the vital role of water in spontaneous combustion of various plant and animal materials. There is no evidence whatever that respiration of the seeds themselves, or seed enzymes, are at all concerned in the heating of stored grains. Hummel *et al.* (1954) found that respiration of wheat free of storage fungi, with moisture contents of 14–18 per cent, and kept at 35°C, was so low that it was not detectable. No one, so far as I am aware, ever has measured respiration of grains, either free of or invaded by storage fungi, at moisture contents in equilibrium with relative humidities of 70–80 per cent, the moisture contents at which, in actual storage, deterioration gets under way. By the time that the fungi have grown enough to have raised the moisture content high enough so that seed respiration might become measureable, the seeds are dead and decayed.

MYCOTOXINS

It has long been known that some common fungi when growing in plant materials might produce metabolites toxic to other organisms, including domestic animals and man. The great impetus to the study of mycotoxins, however, came from the discovery of aflatoxin in poultry feed in England in the early 1960s. Mycotoxins are of no importance in seed used for planting, but may be of great importance in seeds used for food and feed. The study of mycotoxins now is in approximately the same stage of development as was the study of bacteriology about a century ago. A large number of common fungi, when grown under suitable conditions in the laboratory, produce compounds toxic to one or another kind of animals. With the exception of a few such as *Aspergillus flavus*, which produces aflatoxin, and *Fusarium roseum*, which produces the œstrogenic compound designated F–2, the extent to which they do so in nature is not known. The problem deserves, and is receiving, considerable attention. The book edited by Goldblatt (1969) is an excellent summary of most of the important aspects of aflatoxins and contains some information on other mycotoxins.

MUSTINESS, CAKING AND TOTAL DECAY

These are the final stages of spoilage caused by fungi when the organisms involved become detectable by the unaided eye, and often also by the unaided nose. To many warehousemen and seedsmen these are thought to be the first stages of decay by fungi – a totally erroneous idea.

Control of storage fungi

It often is assumed that because many plant diseases as well as decay and rot of many kinds of materials can be prevented with suitable fungicides, the same should apply to storage fungi on seeds. The storage fungi grow in seeds and other materials with moisture contents in equilibrium with relative humidities of 70–90 per cent. No free water is available. Fungicides whose effectiveness depends on their being dissolved in water may, under those conditions, exert no fungicidal action whatever. Milner *et al.* (1947) tested more than 100 supposedly fungicidal compounds for control of storage fungi on wheat; none of them greatly inhibited the fungi without also killing the seed. It is possible that fungicides for this purpose will be developed, but so far they have not been.

Air-tight storage has been reported by Hyde and Oxley (1960) to preserve the quality of damp grain for use as feed, but probably not for use as food, because of the fermentation odours that develop. This appears to be in at least fairly common use as a means of storing shelled corn (maize) of high moisture content. One drawback of the method is that the grain must be used very soon after being removed from storage, because it is prone to rapid spoilage.

Refrigeration also is being tested as a means to preserve quality in stored grains and seeds of high moisture content. In regions of low winter temperature, low-volume aeration suffices to lower the temperature throughout the grain mass, and grain of high moisture content can be very effectively preserved in this way through the winter and into early spring. Artificial refrigeration is being used to a slight extent in some regions to preserve grain quality; the effectiveness of this method evidently is not questioned, but where and under what circumstances it will be economically feasible can be determined only by experience.

By far the most generally used method of preserving quality in stored grains and seeds is storage at moisture contents too low for fungi to grow. As should be evident by now, either a low temperature or a low moisture content is effective in prolonging viability and maintaining quality in most agricultural seeds. A combination of the two is even better. A low and uniform moisture content and a low and uniform temperature also reduce the possibility of moisture transfer within the bulk or package, and this adds greatly to the storage life of the grain.

References to Chapter 3

AGRAWAL, N. S., CHRISTENSEN, C. M., and HODSON, A. C., 1957. Grain storage fungi associated with the granary weevil. *J. Econ. Entomology*, **50**, 659–63.

AGRAWAL, N. S., HODSON, A. C., and CHRISTENSEN, C. M., 1958. Development of granary weevils and fungi in columns of wheat. *J. Econ. Entomology*, **51**, 701–2.

BARTON, L. V., 1961. *Seed Preservation and Longevity*. Interscience Publishers, New York.

CARLYLE, R. E., and NORMAN, A. G., 1941. Microbial thermogenesis in the decomposition of plant materials. *J. Bacteriology*, **41**, 699–724.

CARTER, E. P., 1950. *Role of fungi in the heating of moist wheat*. USDA Circular, 838.

CHRISTENSEN, C. M., 1970. Moisture content, moisture transfer, and invasion of stored sorghum seeds by fungi. *Phytopath.*, **60**, 280–83.

CHRISTENSEN, C. M., 1964. Effect of moisture content and length of storage period upon germination percentage of seeds of corn, wheat and barley free of storage fungi. *Phytopath.*, **54**, 1464–66.

CHRISTENSEN, C. M., 1962. Invasion of stored wheat by *Aspergillus ochraceus*. *Cereal Chem.*, **39**, 100–6.

CHRISTENSEN, C. M., 1951. Fungi on and in wheat seed. *Cereal Chem.*, **28**, 408–15.

CHRISTENSEN, C. M., and GORDON, D. R., 1948. The mould flora of stored wheat and corn and its relation to heating of moist grain. *Cereal Chem.*, **25**, 42–51.

CHRISTENSEN, C. M., and HODSON, A. C., 1960. Development of granary weevils and storage fungi in colums of wheat. II. *J. Econ. Entomology*, **53**, 375–80.

CHRISTENSEN, C. M., and KAUFMANN, H. H., 1969. *Grain Storage – the Role of Fungi in Quality Loss*. 153 pp. University of Minnesota Press, Minneapolis.

CHRISTENSEN, C. M., and LÓPEZ, L. C., 1963. Pathology of stored seeds. *Proc. int. Seed Test. Ass.*, **28**, 701–11.

CHRISTENSEN, C. M., PAPAVIZAS, G. C., and BENJAMIN, C. R., 1959. A new halophilic species of Eurotium. *Mycologia*, **51**, 636–40.

CHRISTENSEN, J. J., 1936. Association of micro-organisms in relation to seedling injury arising from infected seed. *Phytopath.*, **26**, 1091–105.

CHRISTENSEN, J. J., and STAKMAN, E. C., 1935. Relation of *Fusarium* and *Helmin-*

thosporium in barley seed to seedling blight and yield. *Phytopath.*, **25**, 309–27.

COHN, F., 1890. Ueber Warme Erzeugung durch Schimmelpilze und Bakterien. Jahresberichte Schles. *Gesellschaft (Breslau)*, **68**, 23–29.

DARSIE, M. L., ELLIOTT, C., and PEIRCE, G. J., 1914. A study of the germinating power of seeds. *Bot. Gaz.*, **58**, 101–36.

FAIRBROTHER, T. H., 1929. The influence of environment on the moisture content of wheat and flour. *Cereal Chem.*, **6**, 379–95.

FIELDS, R. W., and KING, T. H., 1962. Influence of storage fungi on deterioration of stored pea seed. *Phytopath.*, **52**, 336–39.

GOLDBLATT, L. A. (ed.), 1969. *Aflatoxin, Scientific Background, Control, and Implications*. Academic Press, New York and London.

GRIFFITHS, D. A., HODSON, A. C., and CHRISTENSEN, C. M., 1959. Grain storage fungi associated with mites. *J. Econ. Entomology*, **52**, 514–18.

HANSON, E. W., and CHRISTENSEN, J. J., 1953. *The black point disease of wheat in the United States*. University of Minnesota Agricultural Experiment Station Technical Bulletin 206.

HOLMAN, L. H., 1950. Handling and storage of soybeans. In *Soybeans and Soybean Products*, Vol. 1, eds. K. S. Markley, 455–82. Interscience Publishers, Inc., New York.

HUBBARD, J. E., EARLE, F. R., and SENTI, F. R., 1957. Moisture relations in wheat and corn. *Cereal Chem.*, **34**, 422–33.

HUMMEL, B. C. W., CUENDET, L. S., CHRISTENSEN, C. M., and GEDDES, W. F., 1954. Grain Storage Studies 13: Comparative changes in respiration, viability, and chemical composition of mould-free and mould-contaminated wheat upon storage. *Cereal Chem.*, **31**, 143–50.

HYDE, M. B., 1950. The subepidermal fungi of cereal grains. 1. A survey of the world distribution of fungal mycelium in wheat. *Ann. appl. Biol.*, **37**, 179–86.

HYDE, M. B., and OXLEY, T. A., 1960. Experiments on the airtight storage of damp grain. 1. Introduction, effect on the grain and the intergranular atmosphere. *Ann. appl. Biol.*, **48**, 687–710.

JOHNSON, H. E., 1957. Cooling stored grain by aeration. *Agric. Engng*, **38**, 597–601.

KAUFMANN, H. H., 1959. Fungus infection of grain upon arrival at terminal elevators. *Cereal Science Today*, **4**, 13–15.

KOEHLER, B., 1938. Fungus growth in shelled corn as affected by moisture. *J. agric. Res.*, **56**, 291–307.

LÓPEZ, L. C., and CHRISTENSEN, C. M., 1963. Factors influencing invasion of sorghum seed by storage fungi. *Pl. Disease Reporter*, **47**, 597–601.

MACHACEK, J. E., and GREANEY, F. J., 1938. The 'black-point' or 'kernel smudge' disease of cereals. *Can. J. Res.*, *C*, **16**, 84–113.

MALONE, J. P., and MUSKETT, A. E., 1964. Seed-borne fungi – description of 77 fungus species. *Proc. int. Seed Test. Ass.*, **29**, 179–384.

MILNER, M., and GEDDES, W. F., 1945. Grain Storage Studies 2: The effect of aeration, temperature, and time on the respiration of soybeans containing excessive moisture. *Cereal Chem.*, **22**, 484–501.

MILNER, M., and GEDDES, W. F., 1946. Grain Storage Studies 3: The relation between moisture content, mould growth, and respiration of soybeans. *Cereal Chem.*, **23**, 225–47.

MILNER, M., CHRISTENSEN, C. M., and GEDDES, W. F., 1947. Grain Storage Studies

7: Influence of mould inhibitors on respiration of moist wheat. *Cereal Chem.*, **24**, 507–17.

MURAS, V. A., 1964. [Bacterial diseases of soybean and their causal agents.] *Rev. appl. Mycol.*, **44**, p. 244, entry 1326 (Russian).

NOBLE, M., and RICHARDSON, M. J., 1968. *An annotated list of seed-borne diseases.* Int. Seed Test. Ass. Handbook on Seed Health Testing, Series 1, and Commonwealth Mycological Institute Phytopathological Papers, No. 8.

PAPAVIZAS, G. C., and CHRISTENSEN, C. M., 1958. Grain Storage Studies 26: Fungus invasion and deterioration of wheats stored at lower temperatures and moisture contents of 15–18 per cent. *Cereal Chem.*, **35**, 27–34.

PEPPER, E. H., 1960. *The microflora of barley kernels; their isolation, characterization, etiology, and effects on barley, malt, and malt products.* Ph.D. thesei, Department of Botany and Plant Pathology, Michigan State University, East Lansing (unpublished).

QASEM, S. A., and CHRISTENSEN, C. M., 1960. Influence of various factors on the deterioration of stored corn by fungi. *Phytopath.*, **50**, 703–9.

QASEM, S. A., and CHRISTENSEN, C. M., 1958. Influence of moisture content, temperature, and time on the deterioration of stored corn by fungi. *Phytopath.* **48**, 544–9.

RAMSTAD, P. E., and GEDDES, W. F., *The respiration and storage behaviour of soybeans.* Minnesota Agricultural Experiment Station Technical Bulletin 156.

RAPER, K. B., and FENNELL, D. I., 1965. *The Genus Aspergillus.* The Williams & Wilkins Company, Baltimore, USA.

SCHROEDER, H. W., 1963. The relation between storage moulds and damage in high moisture rice in aerated storage. *Phytopath.*, **53**, 804–8.

SCHROEDER, H. W., and SORENSON, J. W., Jr., 1961. Mould development of rough rice as affected by aeration during storage. *Rice J.*, **64**, 8–10, 12, 21–23.

SIKOROWSKI, P., 1964. *Interrelation of fungi and insects to deterioration of stored grains.* Washington State University, Institute of Agricultural Sciences, Technical Bulletin 42.

SOLOMON, M. E., HILL, S. T., CUNNINGTON, A. M., and AYERST, G., 1964. Storage fungi antagonistic to the flour mite (*Acarussiro L.*). *J. appl. Ecol.*, **1**, 119–25.

TABAK, H. A., and COOKE, W. B., 1968. Growth and metabolism of fungi in an atmosphere of nitrogen. *Mycologia*, **60**, 115–40.

TEMPE, J. de, 1958. Three years of field experiments on seed borne diseases and seed treatment of cereals. *Proc. int. Seed Test. Ass.*, **23**, 38–67.

TEMPE, J. de, 1962. Comparison of methods for seed health testing. *Proc. int. Seed Test. Ass.*, **27**, 819–28.

TEMPE, J. de, 1963. On methods of seed health testing; principles and practice. *Proc. int. Seed Test. Ass.*, **28**, 97–105.

TUITE, J. F., 1959. Low incidence of storage moulds in freshly harvested seed of soft red winter wheat. *Pl. Disease Reporter*, **43**, 470.

TUITE, J. F., 1961. Fungi isolated from unstored corn seed in Indiana in 1956–58. *Pl. Disease Reporter*, **45**, 212–15.

TUITE, J. F., and CHRISTENSEN, C. M., 1955. Grain Storage Studies 16: Influence of storage conditions upon the fungus flora of barley seed. *Cereal Chem.*, **32**, 1–11.

TUITE, J. F., and CHRISTENSEN, C. M., 1957. Grain Storage Studies 23: Time of invasion of wheat seed by various species of *Aspergillus* responsible for deterioration of stored grain, and source of inoculum of these fungi. *Phytopath.*, **47**, 265–68.

TUITE, J., and FOSTER, G. H., 1963. Effect of artificial drying on the hygroscopic properties of corn. *Cereal Chem.*, **40**, 630–37.

USDA, Agricultural Marketing Service, Grain Division, 1964. *Official Grain Standards of the United States. Revised.* US Government Printing Office, Washington, DC.

VIRGIN, W. J., 1940. Low germination of peas associated with the presence of bacteria in the seed. *Phytopath.*, **30**, 790–91. Abstr.

WALKER, I. K., 1967. The role of water in spontaneous combustion of solids. *Fire Research Abstracts and Reviews*, **9**, 5–22.

CHAPTER 4

Effects of Mechanical Injuries on Viability

R. P. Moore

Within mechanised seed production programmes, mechanical injuries are major destructive forces in reducing seed soundness and viability. These injuries cannot entirely be avoided but their extent and seriousness can be greatly reduced. A knowledge of seed structures and of the nature of the injuries to these structures can help in preventing excessive damage and early loss of viability.

DEFINITIONS

Viability

The term viability is used in this chapter with a broad meaning. Basically it refers to the capability of a seed to develop into an acceptable seedling, even under conditions which may not be entirely ideal, such as commonly occur in the field. According to this definition the protrusion of a radicle is insufficient evidence that a seed is viable. Conversely a non-viable seed may not be wholly dead but only partly dead or fractured. The term non-viable refers not only to dead or diseased seeds but also to those that produce, or are expected to produce, abnormal or diseased seedlings that are unacceptable for inclusion in the germination figures quoted by seed-testing laboratories (see Chapter 7).

Mechanical injuries

The concept of mechanical injury is restricted to detectable disturbances resulting from destructive forces encountered in harvesting, elevating, handling, etc. Mechanical injury is often also referred to as mechanical disturbance or damage. Since the effects of injuries become more severe with time, the concept of mechanical injury includes various stages of progressive damage initiated by mechanical injury.

Water damage

Water damage, which is often mistaken for mechanical injury, denotes numerous types of distrubance associated with rapid and uneven water uptake and/or water loss. It is especially prevalent in large seeded legumes.

Seed morphology and the nature of mechanical injuries

Most mechanical injuries are not readily detected. Commonly used tests for mechanical injuries include observations of fractured seed coats or of seedling structures in growth tests. Less obvious seed coat injuries have been made conspicuous by use of iodine, fast green, methylene blue, or other stains. For alfalfa seed (*Medicago sativa*) Cobb and Jones (1960) suggested that each seed should be examined carefully on all sides at 10 × magnification under a bright light. Under many conditions, the method suggested by Cobb and Jones for freshly harvested alfalfa seeds provides only a rough estimate of internal damage under other conditions. Munn (1928), for example, stated that it is not possible to determine the amount of broken growth in red clover seed germination by an arbitrary scheme of classifying seeds on the basis of extent, nature, or character of seed coat injuries. He found that seeds which showed no perceptible injuries under a 3 × hand lens gave half as many broken sprouts as did those upon which injuries were plainly evident. Since the red clover seed samples studied by Munn were probably produced in a more humid section of the United States than were the alfalfa seeds studied by Cobb and Jones, internal bruising could possibly account for some of the observations noted by Munn.

A study of fragmented embryos and seedlings, infections, misshapen seedlings, scar tissue, etc., in growth tests provides a more realistic evaluation of mechanical damage than studies involving observations of embryos through seedcoats. Infections and healing in growth tests, however, often conceal a considerable amount of mechanical damage. Symptoms of mechanical damage in standard growth tests are variable. They include detached seed structures, breaks within structures, abnormally shaped structures, scar tissues, infections, restricted growth, uneven placement of cotyledons, unnatural shrinkage of cotyledons, and splits or otherwise abnormally developed hypocotyls and primary roots (Fig. 4.4–C, E, F). Injured roots often appear dwarfed and twisted and the tips are often blunt and dull in appearance. Injuries also tend to be reflected by reduced viability, and by irregularities in germination and early seedling development. Bulat (1969) discusses these and other abnormalities in light of her observations from tetrazolium tests.

The X-ray method as used by Kamra (1967) provides another useful method for revealing internal injuries associated with immediate or premature loss of viability. By revealing injuries on individual embryo structures, this method compares somewhat favourably with the tetrazolium method. The view presented from only one angle lacks the precision needed to separate seeds with injuries of a borderline nature into viable or non-viable classes; however, this extra degree of precision would often not be required in commercial practices.

The tetrazolium test has been found to be highly useful in revealing the presence and nature of mechanical injuries. The red stain which develops in the test makes obvious many embryo conditions that are commonly invisible in non-stained seeds. When stained by tetrazolium, sound embryo tissues develop a normal carmine red tint; whereas recently injured but living tissues develop a deep red stain. Critically injured tissues remain white or produce an abnormal, dark brownish red, granular stain. Mechanically bruised tissues often appear flaccid and waterlogged. The nature of staining patterns provides many useful clues for establishing the presence, extent and seriousness of damage from mechanical and other sources (Fig. 4.3–A, D, E, G, H, I; Fig. 4.4–A, G, H).

Often the embryos in tetrazolium tests stain normally, but the seeds fail to germinate because of critically located breaks (Fig. 4.3–I) in one or more embryonic structures. Further aspects of tetrazolium testing are discussed by Delouche, Still, Raspet and Lienhard (1962). The tetrazolium test deserves consideration as a special method for revealing internal mechanical injuries that influence both vigour and viability. The knowledge revealed by tetrazolium tests can be very useful in interpreting causes for the performance of seeds in other tests (Moore, 1969).

Viability cannot be distinctly separated from vigour or embryo soundness. Loss of vigour or vitality commonly reveals different levels in the progress of the same deterioration processes. Because of this relationship, vigour tests are useful in forewarning the approach of loss of viability from mechanical injuries and other causes (see Chapter 8). Vigour tests commonly involve adverse environmental conditions that hasten loss of viability. The nature of most vigour tests is such that it is not possible to separate the extent or influences of mechanical injuries from those of other forms of deterioration.

The location, shape, size and nature of individual seed structures account for wide differences in the frequency and seriousness of impact injuries. The diagrams of a snap bean seed (Fig. 4.1) and of a maize kernel (Fig. 4.2) show the basic structures present in most seeds.

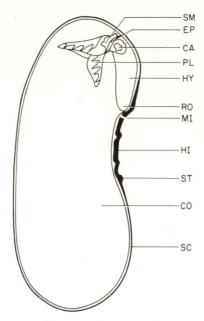

FIGURE 4.1 Snap bean seed (*Phaseolus vulgaris*) showing structures commonly present on legume and other dicotyledonous seeds. SM – shoot meristem, EP – epicotyl, CA – cotyledon attachment, PL – plumules, HY – hypocotyl, RO – root, MI – micropyle, HI – hilum, ST – strophiole, CO – cotyledon, and SC – seed coat.

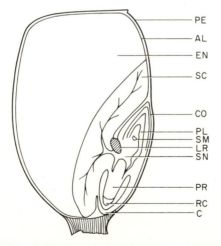

FIGURE 4.2 Maize kernel (*Zea mays*) showing structures commonly present in grass seeds. PE – pericarp, AL – aleurone layer, EN – endosperm, SC – scutellum (cotyledon), CO – coleoptile, PL – plumules, SM – shoot meristem, LR – lateral root, SN – scutellar node with mesocotyl immediately above, PR – primary root or radicle, RC – root cap, and C – coleorhiza.

With regard to the possibility of impact damage, special attention should be focused upon the location of the radicle of the snap bean seed and the plumules of maize in relation to the seed surface. Injuries to radicles, or to radicle-cotyledon attachments in legumes, are frequently responsible for loss of embryo soundness and viability (Fig. 4.4–E). When seeds are excessively dry, these structures are highly susceptible to fracturing.

Fractures of radicles sometimes extend only partially across the diameter. These intermediate injuries often heal sufficiently to produce countable seedlings. Radicle segments showing complete breaks have a tendency to separate sufficiently to prevent healing during germination. A seedcoat break above a radicle fracture tends to promote separation of the fractured parts. Injuries to radicles in vetch (*Vicia sativa*) and a few other crops that have good root regeneration capacity, can inactivate approximately two thirds of the tissue from the tip without causing loss of viability.

The uppermost part of a legume radicle (Fig. 4.3–G) frequently receives injuries that cause immediate or early loss of viability. Multiple fractures and/or deep-seated bruises commonly occur in roughly-handled seed lots. Either or both cotyledon attachments oftentimes become broken or bruised.

In moist seeds, radicle or root bruises tend to be the predominant type of injury associated with loss of viability. The depth of most bruised areas is often more critical than their location, but both aspects are important considerations in the loss of viability. Bruises that penetrate only the cortex usually do not result in the immediate loss of viability. Unless treated before germination, seeds with such injuries, may become infected and fail to produce normal, countable seedlings.

Localised bruises that involve the cortex and more than about one quarter of the diameter of the stele tend to prevent production of countable seedlings even under favourable germination conditions. In growth tests the radicles sometimes tend to curve away from the cotyledons and twist spirally, as reported by Kisser and Stasser (1930). These investigators report that the dry decoating of seeds of *Pisum sativum, Phaseolus vulgaris, Vicia villosa* and *Lens culinaris* causes radicle curvature which could be avoided by moistening seeds for six hours before decoating. They assumed that when dry seed coats adhere strongly to the radicle, the surface cells were overstretched and preconditioned for early death during seed coat removal. These studies provide an insight as to the kind of possible damage that might result when coats become loosened during harvesting and processing. Further studies are needed to confirm and clarify the nature of such injuries. I believe that rapid uptake of water by decoated embryos may have caused some of the troubles reported.

Shallow bruises on hypocotyls tend to be reflected in growth tests as diseased seeds or seedlings or as streaks of scar tissue. When seeds are treated with a fungicide, the outer surface of the scar tissue tends to appear granular and white. Without fungicide treatment and severe infection, the ridges of scar tissue appear brownish, thus indicating minor infections similar to those shown in Fig. 4.3–B. I have found that cleanly cut surfaces are much less damaging than bruised tissues and are less likely to promote infection and accelerated deterioration.

Seeds respond differently to location of injuries. Certain species have greater natural recovery power than other species. La Rue (1935), in studying root recovery in seedlings of wheat, barley, maize, pop-corn, oats, sorghum, canary grass and Sudan grass, cut off the seminal roots of seedlings just below the seed. Abundance of adventitious roots was produced in all crops. Only in sorghum was any regeneration of seminal roots observed. In the other crops tested, new roots arose from the scutellar node, the first internode, the coleoptilar node, or less frequently, from the third node. Under common rules for testing, a seed with an inactive seminal root, as in a crop like sorghum without lateral root meristems, is considered abnormal. Many species of small-seeded grasses likewise require a functional primary root for acceptance as countable seedlings.

The nature of mechanical damage varies widely. The most intensive injuries reduce viability immediately. The small injuries often do not cause an immediate loss in viability, but become increasingly critical as ageing occurs. In seeds that are extremely dry and brittle, fracturing is the predominant type of injury. Bruising is usually prevalent in seeds with sufficient moisture for toughness. The intensity of bruising is usually related to seed moisture and softness of tissues, as well as to kinds, adjustments and operation of equipment. Toughness of such supporting tissues as endosperms, seedcoats, cotyledons, etc., tends to protect embryo structures against excessive injuries that reduce viability.

Tatum and Zuber (1943) report that corn processed at 14 per cent moisture showed 3–4 per cent damage and at 8 per cent moisture, 70–80 per cent damage. These evaluations are likely to reflect only the obvious external breaks. We know from experience that corn, soya beans, small grains, peanuts and many other crops tend to show the least critical damage when harvested at approximately 16–18 per cent moisture. This moisture level, however, is not suitable for safe

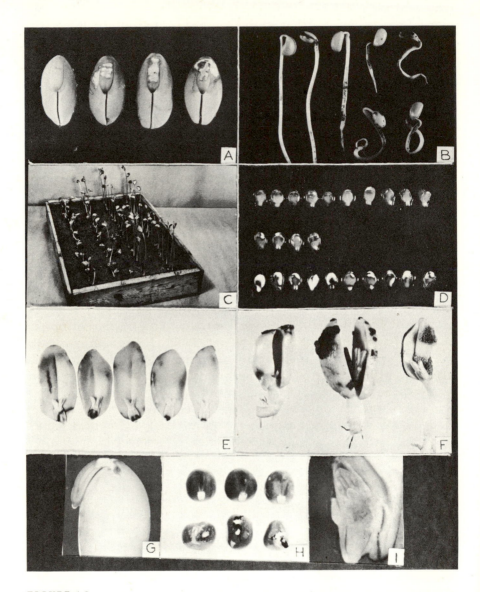

FIGURE 4.3

A. Soyabean embryos. Tetrazolium stained. Viable but weakened as a result of surface necroses (white areas) from water damage plus ageing of initially injured and nearby tissues.

B. Soyabean seedlings. Scar tissues on hypocotyls and roots and abnormalities in shape resulting from the type of water damage shown in A. The four seedlings on right were considered abnormal.

C. Soyabean seedlings. Rows of seedlings from left to right illustrate responses from mechanically caused extensive seed coat breakage, slight seedcoat storage (see Chapter 2). Bainer and Borthwick (1934) have presented extensive data and illustrations from studies with beans concerning the amount and nature of damage occurring at different seed moisture levels and different seed velocities at time of impact.

The nature of injuries varies with crops and embryo structures. Justice (1950), for example, states that the most frequently occurring type of abnormal seedlings (i.e. non-viable as defined here) is that type without radicles or with only stunted radicles as commonly found in timothy (*Phleum pratense*). He also called attention to abnormal seedlings in onions (*Allium cepa*) that usually have little or no root development. With reference to this condition in onions, I have myself noticed that the root-tips extend beyond the general outline of the seed and are easily injured during harvest.

In describing questionable seedlings of several small-seeded legumes, Andersen (1957) mentioned that lesions on hypocotyls consist of splits, cracks, or of granular tissues beneath epicotyls. My own tetrazolium studies have shown that small injuries are especially common on embryo hypocotyls. When the hypocotyls elongate by cell enlargement during seedling development, these injuries become streaks of scar tissue of granular appearance.

Kinds of seed vary widely with reference to both extent and intensity of damage and for different reasons. Because of weight and size, the large-seeded legumes in particular tend to be especially susceptible to injuries that reduce viability. Such seeds are often preconditioned for early loss of viability by exposure to alternate wetting and drying prior to harvest.

Seeds of small-seeded crops tend to escape serious injuries during

breakage, seedcoat pitting from water damage, and seeds with sound coats.

D. Sesame embryos. Tetrazolium stained. Top row-sound and viable; middle row, viable, but mechanically injured on cotyledon tips; bottom row, non-viable because of extent and location of injuries.

E. Peanut embryos. Tetrazolium stained. All embryos show mechanical injuries on roots plus progressive damage (dark area) initiated by mechanical injury to moist seed. Three embryos on left are non-viable; fourth embryo almost non-viable; fifth embryo viable, but weakened.

F. Peanut seedlings. Saprophytic infection within mechanically injured tissues.

G. Soyabean embryo. Tetrazolium stained. Mechanical injury (white area) causes seed to be non-viable.

H. Garden Peas. Tetrazolium stained. Upper row, viable and sound; lower row, viable with broken seedcoats and accompanying internal cotyledon injuries which accelerate deterioration and loss of viability.

I. Wheat embryo. Tetrazolium stained. Non-viable as a result of a critically located mechanical split within embryo near base of germ.

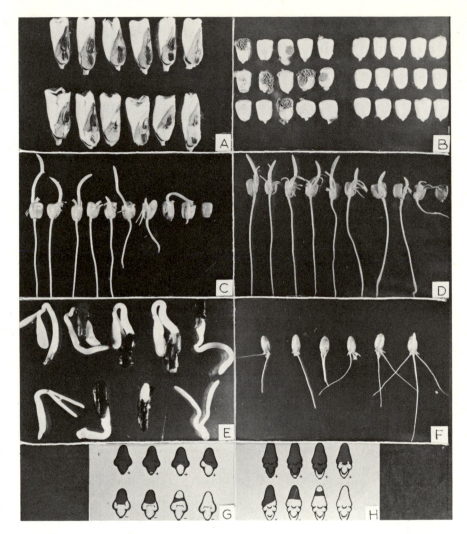

FIGURE 4.4

A. Maize. Tetrazolium stained. Internal injuries resulting from combine harvesting. First two seeds on upper left are viable. All other seeds are non-viable as a result of internal mechanical injuries (white areas or fractures on embryos).

B. Maize. Seeds of left sample were combined at 23 per cent moisture and then dried. Seeds on right were picked at 23 per cent moisture and dried on ear before shelling. Dry seeds of both samples were placed in a humidity chamber for 5 days to test for susceptibility to saprophytic fungi as shown on the combine-harvested seed.

C. Maize. Seedlings from the initial lot of dried seed from combine harvest as shown in B (left).

harvest. The mass and nature of vegetation harvested along with the seed provide valuable protection against impacts. Hard seededness in legumes, as well as the presence of a dry, firm endosperm, lemma and palea of many kinds of grass seeds also provide extra protection.

Flat seeds, such as sesame (*Sesamum indicum*) with very thin, flexible seedcoats, are extremely susceptible to critical mechanical injuries. Flax seeds (*Linum usitatissium*) with a firm brittle seedcoat are less susceptible than sesame, but critical injuries are of major concern in both crops.

Seeds that are spherical in shape are usually better protected against critical injuries than are seeds that are elongated, or irregularly shaped. In seeds of *Brassica* spp., for example, the spherical shape and the folded cotyledons tend to provide good protection for vital structures.

In sorghum seeds (*Sorghum vulgare*) the lower portion of the germ extends beyond the general outline of the endosperm. As a result of seed shape, radicles are often damaged to the extent that seedlings fail to produce primary roots. Such seeds are commonly considered non-viable. The natural protrusion of the tip of the radicle in onion and peanut seeds (*Arachis hypogaea*) likewise promotes root-tip injuries, which lead to accelerated deterioration and loss of viability.

Seeds of corn (*Zea mays*), wheat (*Triticum aestivum*) and rye (*Secale cereale*) frequently receive injuries on primary roots (Fig. 4.4). Such injuries, however, are usually not considered as critical as similar injuries in sorghum. Unlike sorghum, seeds of small grains and corn produce lateral roots that permit retention of viability, even if the primary root is critically injured. It should be emphasised that even minor injuries tend to delay germination, reduce seedling vigour, encourage infection and hasten loss of viability.

The brittleness of embryos of certain varieties and crops is also commonly reflected by extensive loss of viability due to injuries. Seeds of certain varieties of snap beans are very susceptible to fracturing,

D. Maize. Seedlings from the initial lot of dry seed from field picking at 23 per cent moisture plus drying before shelling, as shown in B (right).

E. Crown vetch. Abnormal seedlings resulting from scarification injuries near hypocotyl attachment to cotyledons or to lower part of the radicles.

F. Wheat. Root development from embryos showing different amounts of mechanical injury at base of embryos. Seedling on right is sound. Primary root inactive in other seedlings as a result of staining patterns represented by the two embryos on the right of the upper row in G.

G. Wheat, and H Rye embryos. Tetrazolium stained. Illustrating topographic staining patterns associated with viable (+) and non-viable (−) seeds. (From Lakon, 1949b.)

especially when harvested or handled while dry and brittle.

Variations in seed coat characteristics of different varieties have considerable influence upon susceptibility to mechanical injuries. In this connection, Atkin (1958) noted that seed coats of snap bean varieties that are resistant to injuries generally adhere much more tightly than do those of susceptible varieties. He reported that white-seeded varieties are generally more subject to injury than are coloured varieties. In support of these observations, I have observed that white-seeded varieties are also highly susceptible to rapid water uptake and water damage. The white coats are usually thin, adhere loosely to cotyledons, and are more permeable to water than dark-coloured coats.

Seed moisture and mechanical injuries

The moisture content of individual seeds within a seed lot at the time of mechanical impact exerts an influence upon the nature and serious-ness of injuries (Fig. 4.4–B, C, D). The more moist seeds tend to bruise during impact, whereas the dryer seeds tend to fracture. Bruis-ing usually does not have such an immediate serious effect upon seed soundness and viability as fracturing. The high level of moisture associated with seeds that bruise, however, greatly hastens the rate of deterioration, especially if seeds are not dried promptly and properly.

In evaluating the injuries caused by air lift method of conveying, Bunch (1960) found, by using cold tests, that corn samples mechani-cally impacted at 14, 16 and 18 per cent moisture were injured less than samples at 8, 10, 12, or 20 per cent moisture. Soyabean samples similarly impacted at 12–16 per cent moisture germinated satisfac-torily, whereas those impacted at 8–10 per cent and at 18–20 per cent moisture germinated unsatisfactorily.

Seeds with the optimum levels of moisture are, no doubt, dry enough to prevent cell rupturing and release of destructive, hydrolytic liquids upon impaction, and yet not dry and brittle enough to promote fracturing. Uniformity of moisture content of individual seeds within a lot deserves special consideration. The average moisture content can be suitable even though seeds are present that are either too moist or too dry for acceptable resistance to impact damage.

The rate of drying of seeds shortly before an impact can possibly alter the seriousness of an impact. Iljin (1935, 1957), for example, found that cell membranes tend to stretch excessively and break during rapid drying. There is evidence from studies on peanut seeds to indicate that mechanical impacts can be particularly destructive to cell membranes under drying stresses. We know that an impact can bring about disorganisation of cellular contents that results in processes leading to premature death. We are less knowledgeable as to the basic causes for early death from minor pressure injuries at different moisture contents.

Mechanical injuries and infections

The prevalence of injuries and subsequent loss of viability by infections vary widely among crops, regions, processing techniques, etc. Time of harvest, moisture of individual seeds at harvest and time in storage as well as conditions of storage cause wide differences in extent and seriousness of injuries (Fig. 4.4–B, D).

The number and intensity of impacts to which seeds are subjected greatly influence the extent and seriousness of mechanical damage. Snap bean seeds are especially susceptible to repeated impacts. In hand-harvested samples of several varieties, Atkin (1957) reduced germination percentage from above 93 per cent to a range of 38–90 per cent by dropping seeds 15 times onto a steel plate from a height of 28 inches. Newer varieties were found to be more severely injured than older, more fibrous varieties. Atkin also noted that the germination of commercial seeds from production fields in an arid region was reduced 5–19 per cent by dropping the seeds only three times.

In studying the influence of combine cylinder speeds upon seed coat breakage of soya beans at time of combining, Moore (1957) noted differences in damage as reported in Table 4.1.

Harvests at 13.5 and 12.2 per cent moisture were made at approximately 10 a.m. and 1 p.m. respectively on the same day. These data indicate the importance of minimum cylinder speed and timely adjust-

TABLE 4.1 *Influence of combine cylinder speeds upon seed coat breakage of soya beans harvested at two moisture levels. (From Moore, 1957.)*

Revolutions per minute	Seed moisture at harvest	
	13.5 per cent	12.2 per cent
	seeds with broken coats, %	
700	4	5
900	5	24
1155	12	48

TABLE 4.2   *Seedling conditions of lupine seed samples showing different percentages of injured seed.* (*From Effman, 1963.*)

Seedling conditions	Injured seeds, %		
	1.2	16.5	29.6
		Germination, %	
Normal	95	83	67
Abnormal	2	11	23
Diseased	3	6	10

ments as harvesting conditions change. Seed infection in growth tests were related to the amount of injury received in harvest.

Effmann (1963) in evaluating abnormal seedlings in lupine (*Lupinus luteus*) in various samples, noted the relationship between the percentage of injured seeds and germination as shown in Table 4.2.

The extensiveness of externally recognised injuries on seeds harvested for certification has been studied by Moore (1956a and b). These studies pointed out that it is not uncommon to find one seed in three with obvious symptoms of damage severe enough to hasten the loss of viability. The failure to provide fungicide treatment prior to testing often resulted in unexpectedly low and irregular germination results.

It is commonly known that the presence and locations of injuries exert considerable influence upon susceptibility of seed to loss of viability by infections (Fig. 4.3–F). Systems for evaluating the seriousness of injuries are not readily available. Crosier (1958), however, reports the following system for grading the location of pericarp breaks on maize in relation to severity of infections:

Severe crown – 10; over plumule – 5; slight crown – 3; edge of germs – 3; over radicle – 2; other areas – 2; tip cap – 1.

My own studies using tetrazolium tests and comparing the growth of seedlings have emphasised the importance of protecting embryo necroses against infection by prompt drying, favourable storage, and by suitable fungicides before planting. I have also noticed that the need for fungicide can have considerable influence in determining whether an injured, viable seed decays or remains sufficiently sound to produce a countable plant.

Mechanical injuries and field emergence

Seed lots that consist of high percentages of embryos that are pre-dominantly sound will usually germinate well under a wide range of environmental conditions. Such lots have a stable germination ten-dency as discussed in detail by Lakon (1952), i.e. they will perform well under a wide range of conditions. Infection is usually un-important.

Seed lots containing high percentages of moderately or heavily damaged, viable embryos have an unstable germination tendency. Even slight adversities tend to promote infections, abnormal seed-lings, and loss of viability, and consequently such unstable seed lots are highly dependent upon fungicidal protection and a favourable set of germination conditions at the time of planting. Each seed lot differs in regard to the percentages of seeds at various levels of embryo soundness. For this reason, seed lots do not necessarily retain the same ranking of soundness when evaluated at different levels of adversity. Even the so-called very low quality seed lots often contain many high quality seeds.

The influence of injuries upon germination tendencies, as well as upon immediate viability, is becoming of increasing practical concern. The hybrid corn industry in particular has made extensive use of cold and tetrazolium tests for exposing injuries resulting from various harvesting, processing, treating and handling procedures.

Combination cold-test and tetrazolium-test studies have been made by Moore and Goodsell (1965) for gaining some idea of the nature of injuries that cause sensitive seeds to become non-viable in cold tests. A correlation coefficient of +0.96 between cold test results and a ger-mination-energy test based on tetrazolium treatment indicates the value of the tetrazolium treatment for visualising the internal nature of embryos of fungicide-treated seeds that were surviving the cold-test conditions. Other comparative tests revealed that different standards were needed for seeds that were not treated with a fungicide.

Upon using cold tests to emphasise weaknesses in mechanically injured seed corn, Gregg (1954) obtained results shown in Table 4.3. Although he did not specify, the seeds used in Gregg's study were probably not treated with fungicide.

In my own studies of the influence of seed coat fractures in soya beans upon field emergence, I observed emergences of 96, 72 and 52 per cent for seeds with non-broken, lightly-broken and moderately-broken seedcoats, respectively (Moore, 1957) (Fig. 4.3–C).

TABLE 4.3   *The influence of type of damage to maize upon cold test results. (From Gregg, 1954.)*

Type of damage	Germination, %
One puncture per kernel over the germ	26
Two punctures per kernel over the germ	13
Tip scraped on germ side	75
Pericarp cut at one side of germ	48
Crown pericarp and part of starch cut off	56
Undamaged kernels	97

Mechanical injuries and retention of viability in storage

The immediate influence of mechanical injuries upon viability is occasionally serious. The delayed effects, however, are usually more troublesome and of much greater economic importance, especially when bruising is the major kind of damage. During storage, the injured areas serve as centres of infection and result in accelerated ageing that shortens duration of viability. Injured areas, in addition to dying early, also promote rapid weakening and early death of surrounding normal tissues. The large and deep-seated injured areas, by being in contact with extensive amounts of non-injured tissues, are much more destructive during early stages of storage than are small injuries with only minor peripheral contact with sound tissues.

If an initial injury is non-critical in that it has no immediate effect on viability, but is located on or near an essential part of an embryo structure, a seed can readily become non-viable with only a minor amount of additional deterioration. Injuries near the point of attachment of cotyledons to the embryonic axis, or on most other vital parts of the embryonic axis (radicle, epicotyl and plumules), usually bring about a more rapid loss of viability during storage than will injuries of similar size located in less important areas of a seed.

An example of storage life of scarified alfalfa has been published by Graber (1922). He reported that samples of non-scarified seed germinated 70.6 per cent at time of storage, and 74.4 per cent after two to three years of storage in a cool, dry place. Samples of scarified seed, on the other hand, germinated 86.4 per cent shortly after scarification, and only 40.4 per cent after similar storage. Tetrazolium tests usually reveal extensive mechanical damage, especially on hypocotyls of mechanically scarified seeds.

It is commonly known that the quality of different seed lots varies widely under comparable storage conditions. Lakon (1949a and b), and Moore (1963) have attempted to clear up some mysteries by pointing out the manner in which necroses play a major role in accelerating seed deterioration.

Jones, McFarland and Midyette (1955), in studying causes for rapid deterioration of fungicide-treated wheat in storage, found that grains with intact seed coats germinated 95 per cent and those with coats broken over the germs, 53 per cent. They noted that 41 per cent of the seeds with breaks over the germs produced injured sprouts. In reference to their investigations, it is worthwhile mentioning that embryo injuries usually accompany broken seed coats.

Water injuries

A major source of trouble often mistaken for mechanical injury results from alternate moistening and drying of mature seeds. The most common examples I know of involve snap bean (*Phaseolus vulgaris*), lima bean (*Phaseolus limensis*), soyabean (*Glycinè max*), cowpea (*Vigna sinensis*) and lupine (*Lupinus*).

During early phases of tetrazolium studies that led to the initially recognised symptoms of water damage, the condition was identified by Moore as 'natural crushing' (Moore, 1960, 1963, 1965). Subsequent studies revealed the possibilities of membrane rupturing and other disturbances. The investigations of Iljin (1935, 1957), for example, suggest the involvement of plasmolysis-deplasmolysis injuries. His studies indicate the destruction of cell membranes either by rapid loss of water by moist, living tissues, or by rapid uptake of water by dry, living tissues.

Water damage involves numerous types of disturbance associated with water uptake and water loss. In large-seeded legumes in particular, extensive water damage occurs both internally and externally. Water damage is especially prevalent in some newer varieties of snap beans (*Phaseolus vulgaris*) and cowpeas (*Vigna sinensis*). Such damage is not so obvious in small-seeded crops. The earliest evidence in dry seeds of seed coat reaction to water uptake is reflected by sections of coats that failed to shrink sufficiently upon drying to make contact with the embryo. These expanded sections, especially in snap bean, are brittle and can be easily broken in harvesting and processing. Internal drying stresses also possibly increase susceptibility to mechanical injuries. Fractures are usually most prevalent and destructive on radicles and within cotyledon attachment regions. Often, only one cotyledon attachment breaks. The epicotyl or one or both leaf petioles are frequently fractured. Fractures likewise occur along edges of cotyledons and across the midsection of inner surfaces.

A common cause for initiation of disturbance on the outer surfaces of radicles and cotyledons is traceable to unequal expansion and folding of seed coats during the early stage of water uptake. Tissue disturbances are initiated when moist, inner surfaces of seedcoat folds touch limited streaks of dry embryo tissues. Embryo tissues at the boundary of an expanded wet and a tightly adhering dry section of a coat also receive considerable disturbance.

Unless fracturing occurs, the first symptoms of water damage are usually minor. The presumably non-fracture damage is often reflected in tetrazolium tests by narrow streaks or areas of surface tissues that initially stain darker than the normal tissue. With time for further deterioration, such disturbances gradually expand and deepen.

Water damage can usually be separated from mechanical damage by appearance and often by circumstantial evidence, especially when embryos are evaluated after being stained by tetrazolium. The moist folded coat that initiates one type of damage, tends to disturb the embryos in regular patterns. The earliest patterns usually consist of a series of rather uniformly spaced bands or streaks of disturbances (Fig. 4.3–A). In growth tests, the disturbances, if extensive or critically located, can either prevent germination or result in seedlings that reveal different degrees of abnormalities (Fig. 4.3–B). The true source of trouble is generally not obvious. It is often incorrectly diagnosed as mechanical injury.

A recent report by Kietreiber (1969) suggests drought damage as the cause for many seed and seedling abnormalities which are described here as water damage. She discusses such familiar abnormalities as shortened and twisted hypocotyls with longitudinal cracks and splits, primary leaves that are small and deformed, and cotyledons with fractures that cause loss of sections, or differential shrinkage. The damaged seeds gave poor emergence or produced weakly-developed seedlings in field plantings.

Regardless of the terms used to describe embryo disturbances involving variations in moisture, the cause should be recognised as being distinctly different from mechanical damage. These injuries are occasionally of greater economic importance than disturbances from mechanical forces. Either type of injury hastens the loss of viability. Preventive measures are distinctly different. Water damage, however, often preconditions seed to excessive amounts of damage during harvest. Knowledge of both types of injuries will often explain some of the mysteries surrounding unpredictable behaviour of seed lots.

Conclusions

Certain mechanical injuries cause immediate loss of viability. Other injuries indirectly influence vigour and viability. Such injuries serve as centres from which accelerated ageing processes advance, or as infection centres for saprophytic fungi. They also cause various degrees of abnormalities in seedling development.

Literature citations concerning mechanical injuries commonly report the use of adverse growth tests for exposing injuries that are not readily revealed in standard growth tests. The tetrazolium test has been found to be the most informative test for revealing the presence and nature of injuries.

Damage to mature seeds prior to harvest by rapid uptake or loss of water has often been mistaken for mechanical injury. Such injury not only reduces seed soundness and viability but also promotes increased mechanical damage and further loss of viability during harvest and processing. Injuries in general require that special care be given to favourable storage, seed treatment and planting conditions so as to protect the injured areas and thus seed viability. The presence and nature of injuries greatly influence the extent to which seed viability can persist in storage or under different planting conditions.

References to Chapter 4

ANDERSEN, A. M., 1957. Evaluation of normal and questionable seedlings of species of Melilotus, Lotus, Trifolium, and Medicago by greenhouse tests. *Proc. int. Seed Test. Ass.*, **22**, 237–58.

ATKIN, J. D., 1957. Bean seed injury and germination. *Fm. Research*, **23** (2), 10–11.

ATKIN, J. D., 1958. Relative susceptibility of snap bean varieties to mechanical injury of seed. *Proc. Am. Soc. hort. Sci.*, **72**, 370–73.

BAINER, R., and BORTHWICK, H. A., 1934. Thresher and other mechanical injury to seed beans of the lima type. *Calif. Bull.*, 580.

BULAT, H., 1969. Keimlingsanomalien und ihre Feststellung am ruhenden Samen im Topographischen Tetrazoliumverfahren. *Saatgut-wirt. SAFA*, **21**, 575–79.

BUNCH, H. D., 1960. Relationship between moisture content of seed and mechanical damage in seed conveying. *Seed Wld.*, **86**, 14, 16, 17.

COBB, R. D., and JONES, L. G., 1960. Germination of alfalfa as related to mechanical damage of seed. *Proc. Ass. Offic. Seed Analysts, N. Am.*, **50**, 104–8.

CROSIER, W. F., 1958. Relation of pericarp injuries of corn seed to cold seed germination. *Proc. Ass. Offic. Seed Analysts, N. Am.*, **48**, 139–44.

DELOUCHE, J. C., STILL, T. W., RASPET, M., and LIENHARD, M., 1962. The tetrazolium test for seed viability. *Miss. Tech. Bull.*, **51**.

EFFMANN, H., 1963. Beurteilung anomaler Keime bei *Lupinus luteus*. *Proc. int. Seed Test. Ass.*, **28**, 61–69.

GRABER, L. F., 1922. Scarification as it affects longevity of alfalfa seed. *Agron. J.*,

14, 298–302.

GREGG, R., 1954. Rough handling kills seeds. *Seedsmen's Dig.*, **5**, 12.

ILJIN, W. S., 1935. Die Veränderung des Turgors der Pflanzenzellen als Ursache ihres Todes. *Protoplasma*, **22**, 299–311.

ILJIN, W. S., 1957. Drought resistance in plants and physiological processes. *A. Rev. Pl. Physiol.*, **8**, 257–74.

JONES, J. S., MCFARLAND, A. G., and MIDYETTE, J. W., Jr., 1955. Seed coat injury as a contributing factor to mercury damaged wheat seed. *Proc. Ass. Offic. Seed Analysts, N. Am.*, **55**, 120–21.

JUSTICE, O. L., 1950. The testing for purity and germination of seed offered for importation into the United States. *Proc. int. Seed Test. Ass.*, **16**, 156–72.

KAMRA, S. K., 1967. Detection of mechanical damage and internal insects in seed by X-Ray radiography. *SVENSK bot. Tidskr.*, **61**, 43–48.

KIETREIBER, M., 1969. Abnormale Sprossentwicklung bei Bohnenkeimlingen. *Bodenkultur*, **20**, 38–45.

KISSER, J., und STASSER, R., 1930. Untersuchungen über die bei der Keimung geschalter Leguminosensamen auftretenden Wurzel – und Hypokrümmungen. *Beitr. Biol. Pfl.*, 161–84.

LAKON, G., 1940. Die Topographische Selenmethode, ein neues Verfahren zur Feststellung der Keimfähigkeit der Getreidefrüchte ohne Keimversuch. *Proc. int. Seed Test. Ass.*, **12**, 1–18.

LAKON, G., 1949a. The topographical tetrazolium method for determining the germination capacity of seeds. *Pl. Physiol.*, **24**, 389–94.

LAKON, G., 1949b. Biochemische Keimprüfung nach dem Lakonschen, 'Topographischen Tetrazolium – Verfahren' zur Feststellung der Keimfähigkeit bzw. Keimpotenz von Getreide und Mais. In *Methodenbuch Band V. Die Untersuchung von Saatgut*, S. 37–38 und Tafeln III–IV. Neumann-Neumdamm, Hamburg.

LAKON, G., 1952. Über Keimpotenz und labile Keimtendenz bei Pflanzensamen insbesondere bei Getreidefrüchten. *Saatgut-wirt.*, **4**, 210–13. (Reprint of a 1918 article.)

LA RUE, D., 1935. Regeneration in monocotyledonous seedlings. *Am. J. Bot.*, **22**, 486–92.

MOORE, R. P., 1956a. Mechanically damaged seed seriously reduce crop stands. *Sth. Seedman*, **19**, 44, 73.

MOORE, R. P., 1956b. Slam-bang harvesting is killing our seeds. *Seedsmen's Dig.*, **7**, 10–11.

MOORE, R. P., 1957. Rough harvesting methods kill soybean seeds. *Seedsmen's Dig.*, **17**, 14–16.

MOORE, R. P., 1960. Soybean germination? *Seedsmen's Dig.*, **11**, 12, 52, 54, 55.

MOORE, R. P., 1963. Previous history of seed lots and differential maintenance of seed viability and vigor in storage. *Proc. int. Seed Test. Ass.*, **28**, 691–99.

MOORE, R. P., 1965. Natural destruction of seed quality under field conditions as revealed by tetrazolium tests. *Proc. int. Seed Test. Ass.*, **30**, 995–1004.

MOORE, R. P., and GOODSELL, S. F., 1965. Tetrazolium test for predicting cold test performance of seed corn. *Agron. J.*, **57**, 489–91.

MOORE, R. P., 1969. History supporting tetrazolium seed testing. *Proc. int. Seed Test. Ass.*, **34**, 233–42.

MUNN, M. T., 1928. The behaviour during germination of cracked and broken seeds from badly threshed red clover seed. *Proc. Ass. Offic. Seed Analysts, N. Am.*, **20**, 68–69.

TATUM, L. A., and ZUBER, M. S., 1943. Germination of maize under adverse conditions. *J. Am. Soc. Agron.*, **35**, 48–59.

TOOLE, E. B., and TOOLE, V. K., 1951. Injury to seed beans during threshing and processing. *US Dept Agric. Circ. No. 874.*

CHAPTER 5

Effects of Environment Before Harvesting on Viability

R. B. Austin

It is well known that the germination and viability of seeds of culti-vated plants can vary greatly from year to year, dramatically affecting the value of the seed for sowing. To reduce the risks of crop failure that can result from sowing poor seed, most countries now have laws prohibiting the sale of seed lots unless they have a germination greater than a statutory minimum percentage, which varies according to the species. Germination tests, carried out according to the widely used International Rules (Anon., 1959) reveal only the percentage of the seeds that are viable under near-ideal conditions and there is usually a considerable discrepancy between the results of germination tests in the laboratory and the emergence of the seeds in the field, the discrepancy usually being greater the lower the per cent germination (for examples see Perry, 1967 and Austin, 1963).

Much of this variation in germination and viability in the field is the direct or indirect result of variation in the weather before and at harvest time; hot, dry periods at this time generally giving good seed. In England seed growers recognise 'seed years' in much the same way as wine growers recognise vintage years. Regions of the world having hot, dry weather at the time when seeds ripen have become recognised as being favourable for seed production and, in some of them, seed production is an important sector of the agricultural economy.

Thus it is important to know how environmental factors, acting on seeds before harvest or indirectly on them through the parent plant, can affect their viability, both in ideal conditions and in the field. Although not an environmental factor, the state of maturity of seed crops when harvested is known to be a major factor responsible for part of the variation in viability and size of seed, and the decision when to harvest, which is difficult to make for crops with complex inflorescences and in deteriorating weather, is therefore of great im-portance. Although the harvest dates for obtaining optimum yield and viability can be determined empirically, a knowledge of the development of the seed from fertilisation to maturity is valuable for interpreting the results of such experiments. This knowledge is also valuable because the effects of environmental factors on viability are likely to vary as the seed develops.

In this chapter, therefore, I shall first briefly examine the gross changes in dry weight and moisture content during the course of the development of seed, and the associated changes in fine structure and chemical composition. Next, the effects of environmental factors on seed structure and composition will be considered. Finally the effects of harvest date as it determines seed maturity and size, and the effects of environmental factors on seed viability will be considered against this background.

Gross changes in weight, moisture content and respiration during seed development and maturation

Loewenberg (1955) has described the growth in weight, cell number and the course of accumulation of nitrogen and phosphorus in *Phaseolus vulgaris*. From flowering to maturity – a period of 6–7 weeks – growth in dry weight followed an approximately logistic pattern (Fig. 5.1a), and the same was true for the total nitrogen and phosphorus content of the seed. Growth in fresh weight followed a similar pattern up to five weeks from flowering, but thereafter, because of loss of moisture, fresh weight declined (Fig. 5.1b). Cell division in the cotyledonary tissue was complete three weeks after anthesis (Fig. 5.1d). Oxygen uptake per bean was proportional to the fresh weight of the seed, reaching a maximum some four weeks after flowering, and declining to a very low rate at dry seed maturity (Fig. 5.1c). Howell, Collins and Sedgwick (1959) found that the respiration of maturing soya beans (*Glycine max*) was closely correlated with moisture content. The Q_{10} of oxygen uptake was 1.3–1.4, similar to that for diffusion of oxygen into water gels, suggesting that respiration was limited by diffusion of oxygen into the tissue.

Carr and Skene (1961) found a generally similar pattern of seed growth in *P. vulgaris* but their data indicate the occurrence of a pause of 3–5 days in the progress of logistic growth, and they pointed out that there is a similar pause in the growth of pea (*Pisum sativum*) seeds as observed by Bisson and Jones (1932). With peas the lag was associated with a decrease in the amount of sucrose per seed, and with *Phaseolus* it was associated with a change in the ratio of the relative growth rates of the embryo and testa. Since seeds are composed of several tissues each of which have their own characteristic patterns of growth, which may not be in phase with each other, it can only be

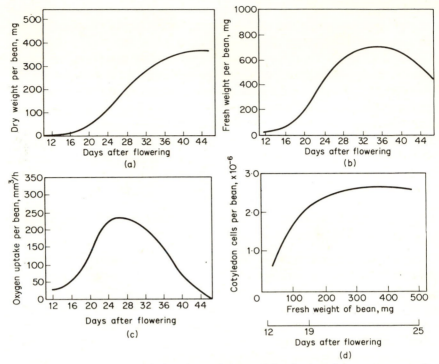

FIGURE 5.1   Changes with time in seed dry weight, fresh weight, oxygen uptake
and cell number during the growth of seeds of *Phaseolus vulgaris*. (After
Loewenburg, 1955.)
(a) mean dry weight per bean, mg.
(b) mean fresh weight per bean, mg.
(c) mean oxygen uptake per bean, mm³/h.
(d) mean number of cotyledonary cells per bean plotted against fresh weight per
      bean.

approximately true that the overall growth of the entire seed is logistic.
Nutman (1939) made a detailed quantitative study of the growth of
the embryo sac and embryo in winter rye (*Secale cereale*). Until 22
days after fertilisation the embryos grew exponentially, cell divisions
occurring throughout the embryo once per day. Exponential growth
ceased at a time coincident with the beginning of desiccation of the
fruit.

    The growth of seeds of many other species has been studied:
Brenchley and Hall (1909, 1912) described grain growth in wheat
(*Triticum aestivium*) and barley (*Hordeum vulgare*); Kersting, Steckler
and Pauli (1961) that of *Sorghum*; Grabe (1956) that of *Bromus*, and
Leininger and Urie (1964) that of safflower (*Carthamus tinctorius*).
All these studies show a pattern of seed growth generally similar to

that of *Phaseolus*, although desiccation is not a constant feature and in some species it may be lethal, for example in *Acer* (Jones, 1920) and *Citrus* (Barton, 1943).

Leininger and Urie (1964) and Kersting *et al.* (1961) found a steady decline in the percentage moisture of ripening seeds from an initial 80–90 per cent (fresh weight basis) to 10–20 per cent. Austin, Longden and Hutchinson (1969) found with seeds of carrot (*Daucus carota*) that, after a period during which the moisture content declined to 50–60 per cent, there was a further period before shedding during which there was a diurnal variation in the moisture content, typical values being 52 per cent in the morning and 25 per cent in the afternoon, the actual percentages being correlated with the prevailing water vapour pressure deficit close to the umbels. No further growth in embryo size was apparent during this phase of fluctuating moisture content, but it is possible that development or deterioration at a cellular level might have occurred. Moore (1963) found that in *Phaseolus* such diurnal changes caused differential swelling and contraction within the cotyledonary tissues of the seed, resulting in mechanical damage. Such mechanical damage may lead to death of parts of the cotyledonary tissue and to reduced vigour of the seedlings, an effect which is well known in peas (Perry, 1967; Matthews and Bradnock, 1967). With many species, the mature seeds reach a moisture content which is in equilibrium with that of the surrounding air. The equilibrium moisture content varies with species (Finn-Kelcey and Hulbert, 1957) but is little affected by temperature (for further information on equilibrium moisture contents see pp. 47–50 and Appendix 4).

Although the patterns of growth of individual seeds are similar within a species or crop variety, the time of onset of growth can vary greatly according to the habit of the plant, as also can the parameters of the curves of growth of individual seeds, and this is the origin of part of the variation in seed size on a plant. Plants with large complex inflorescences, such as carrot (Borthwick, 1932) and rape (*Brassica napus*) (Havstad, 1964), in which flowering occurs over a considerable period of time, produce seeds in which the variance of weight of mature dry seed is large (a Coefficient of Variation of 50–60 per cent is normal), whereas in plants with a more determinate growth habit, such as dwarf varieties of peas, the variance of seed weight is much smaller (Coefficient of Variation 10–25 per cent) (Longden, 1967).

CHANGES IN FINE STRUCTURE DURING RIPENING

Whereas most plant tissue is functional only at moisture contents of 80 per cent or more and cannot survive desiccation, seeds are notable for being able to resist desiccation to 20 per cent or less, and some can survive drying to a moisture content as low as 2 per cent, and also temperatures of 70–100°C for several days, for example, carrot (Austin, unpublished) or even weeks, *Lycopersicon* (Rees, 1970). Other species do not acquire the capacity to germinate unless they undergo a period of desiccation. It seems likely that to withstand such extremes, major biochemical and organisational changes within the cell are required. Klein and Pollock (1968) have investigated the changes in fine structure that occur in the seeds of *Phaseolus lunatus* during ripening. In this species, desiccation of the cotyledons and embryonic axis did not occur until sufficient morphological development had taken place to give full viability, as judged by the ability of the undried embryonic axes to grow. The bulk of the cotyledons in this species consists of parenchymatous ground tissue. While the tissue moisture content is above 60 per cent (the phase of 'ripening'), these cells are well vacuolated and contain chloroplasts with well-developed grana structure and starch grains, and polysomes associated with the endoplasmic reticulum. As the moisture content decreases from 60 per cent (the 'maturation' phase), protein bodies take the place of vacuoles, the endoplasmic reticulum becomes less prominent and the polysomes, which are no longer associated with it, gradually disappear. Mitochondrial structure remains more or less intact until the end of the ripening period; the mitochondria then rapidly lose their elongated shape and become rounded. The chloroplasts undergo the largest changes: they become globular or bell-shaped with frequent invaginations, the internal membrane structure is lost and the grana disappear. These changes in fine structure seem to be a consequence of the cessation of intensive protein synthesis by the end of the ripening phase, and it seems that the seeds acquire resistance to desiccation only when the structures associated with protein synthesis undergo degradation and become inactive. It follows that cessation of physiological activity is a prerequisite for the cells to become resistant to the effects of desiccation, rather than the loss of water being the cause of seed inactivity, and resistance to the effects of further desiccation. A similar pattern of events was found for the soya beans by Bils and Howell (1963).

Buttrose (1963) has described the changes in fine structure that occur during the development of the endosperm in wheat. In this tissue, free cell division takes place after fertilisation before the formation of cells, when the organelles in the tissue lack any recognisable structure. After wall formation is complete, about two days after fertilisation, the cells increase rapidly in volume, and recognisable plastids, mitochondria, Golgi bodies and endoplasmic reticulum structure appear. During the further development of the grain, the cells fill almost completely with starch and protein deposits.

BIOCHEMICAL CHANGES DURING SEED RIPENING

The changes in the composition of seeds during ripening have been studied by many investigators, frequently with the aim of devising chemical tests for determining the optimum harvest date. Early studies, such as that of Woodman and Engledow (1924) on wheat, established the changes in the major constituents with time and these findings have subsequently been confirmed and extended by Jennings and Morton (1963a and b). The maturation of peas has been investigated by McKree and Robertson and their colleagues (for reference to this series of five papers see Rowan and Turner, 1957), and Stoddart (1964a and b) has described the changes in certain constituents that occur during ripening of the seed of four grass species.

In young developing seeds, the concentrations of simple nitrogen and phosphorus compounds is high, associated with the active metabolism in such tissue. With maturation, the concentrations of such compounds, including amino acids and reactive phosphoryl groups, and in the starchy seed, of monosaccharides, decreases, and that of inositol phosphates ('phytin') increases. At maturity, the majority of the phosphorus in many seeds is in the form of phytin. Thus in corn (*Zea mays*) Earley and De Turk (1944) found that immature grains contained less than 10 per cent of their phosphorus as phytin, while in the mature grain 90 per cent of the total phosphorus in the grain was in the form of phytin. According to the type of seed, there is also a concomitant increase in the amount and concentration of starch, or protein or fat.

Skene and Carr (1961) found that changes in the gibberellin content of developing *Phaseolus vulgaris* seeds were closely coincident with their growth rates, but were unable to deduce whether or not there was a causal relationship between the growth rate and gibberellin concentration in the seeds.

Environmental effects on seed structure and composition

Environment markedly influences the composition of mature seeds, and site-to-site and year-to-year variations in composition are well documented, especially for cereals where the quality of wheat for baking and of barley for malting vary greatly (Kramer, Post and Wilten, 1952). The effects of changes in single environmental factors, however, have not been studied so extensively.

MINERAL NUTRITION

Mineral deficiencies predominantly affect the number of seeds produced but, unless the deficiency is severe, have relatively minor effects
on seed composition. As with the vegetative tissues of plants, application of a mineral fertiliser causes manifold changes in the elemental
composition of seed. These effects can be seen from the results of a
sand culture experiment with watercress (*Rorippa nasturtium-aquaticum*) given different levels of phosphorus in the nutrient solutions.

TABLE 5.1   Rorippa nasturtium-aquaticum. *Effects of three levels of supply of
phosphorus on plant and seed weights and mineral composition, at 20 weeks. Means
of three experiments. (From Austin, 1966a.)*

Solution type	P1	P2	P3
P, mE/l	0.2	1.0	4.0
Mean seed weight, mg.	0.235	0.214	0.213
Nitrogen in seeds, percentage D.M.	5.19	5.35	4.62
Phosphorus in seeds, percentage D.M.	0.47	0.84	0.95
Potassium in seeds, percentage D.M.	0.51	0.64	0.77

The severity of the phosphorus deficiency is indicated by the plant
yields, which show that reducing the phosphorus in the nutrient solution from 4 to 1 mg equivalents/l reduced plant weights by less than
10 per cent whereas a further reduction in the phosphorus supply to
0.2 mg equivalents/l reduced plant yield to 10 per cent of that from
the high phosphate culture. In general, only when the deficiency was
severe was seed size and composition affected, low phosphorus cultures giving seed with decreased phosphorus, but increased nitrogen
and potassium concentration.

The experiment in Hoosfield, Rothamsted, where plots given
various annual dressings of fertilisers since 1852, and cropped continuously with barley, illustrates well the effects of the treatments on
mineral composition and size of the grains (Table 5.2). Although the
composition of the grain did not vary as much as in the sand culture
experiment with watercress (Table 5.1), grain from the with-phos-
phorus plots contained a lower concentration of nitrogen than the
corresponding no-phosphorus plots, and the effects of the continued
applications of nitrogen and of nitrogen, potassium, sodium and
magnesium were to decrease the phosphorus content of the grain, as
compared with that from the phosphorus-only plots.

TABLE 5.2   *Barley. Size and mineral composition of grains from Hoosfield experiment, Rothamsted. (Austin, previously unpublished.)*

Fertiliser applied	Mean seed weight mg	N, %	P, %	K, %	Ca, %	Mg, %	Na, %
NIL	33.8	1.54	0.29	0.51	0.065	0.10	0.39
P	37.5	1.44	0.36	0.45	0.055	0.10	0.27
K, Na, Mg	40.1	1.44	0.36	0.52	0.050	0.12	0.18
P, K, Na, Mg	42.2	1.40	0.36	0.54	0.050	0.11	0.16
N	35.7	1.90	0.28	0.45	0.055	0.11	0.42
N, P	38.5	1.65	0.34	0.42	0.055	0.10	0.47
N, K, Na, Mg	45.4	1.62	0.30	0.46	0.045	0.11	0.20
N, P, K, Na, Mg	43.8	1.35	0.32	0.53	0.050	0.10	0.23

Rates of application of fertiliser, per ha
N, as ammonium sulphate 48 kg N
P, as superphosphate 88 kg P_2O_5
K, as potassium sulphate 135 kg K_2O
Na, as sodium sulphate 35 kg Na_2O
Mg, as magnesium sulphate 21 kg MgO

Austin and Longden (1966b) applied N, P and K fertilisers to experimental crops of carrots for seed and found that nitrogen (158 kg/ha) had the largest effect on the composition of the seed, increasing the nitrogen content by 0.62 to 4 per cent, but decreasing the phosphorus and potassium content by 0.072 to 0.593 per cent and by 0.09 to 1.06 per cent respectively. Dressings of 132 kg P_2O_5/ha and 250 kg K_2O/ha had only very slight effects on the N, P and K concentrations in the seed although all fertilisers gave substantial increases in seed yields.

In pot experiments with Australian soils giving large responses to applied phosphorus fertiliser, Lipsett (1964) found that the phosphorus concentration in wheat grain was increased from 0.20 per cent (control) to 0.45 per cent by applications equivalent to 150 kg P_2O_5/ ha. Varieties differed in their behaviour in these phosphorus-deficient soils, some giving more grains per ear with a lower phosphorus concentration than others.

In pot experiments with peas, Austin (1966b) found that phosphorus deficiency reduced the phosphorus concentration in the seed to 0.30 per cent (control: 0.59 per cent). The mean dry weight of the seeds from the phosphorus-deficient plants was smaller, 0.20 g/seed (control 0.24 g/seed) and they contained slightly greater concentrations of nitrogen and potassium. Generally similar results were obtained by Szukalskii (1961) with flax (*Linum usitatissimum*) and rape.

Finney, Meyer, Smith and Fryer (1957) found that foliar sprays of urea solutions affected the protein and water-soluble nitrogen concentrations in wheat grains. Sprays, supplying 154 kg N/ha, had the greatest effect when applied just before or after flowering and increased the protein content of the grain to 15 per cent (unsprayed controls: 10 per cent), but had little or no effect when applied earlier than 40 days before flowering, or when the grain was ripe. Although most of the applied nitrogen was converted into protein when sprayed early, the quality of the protein, as assessed by the volume of loaves baked with the grain flour by a standard procedure, varied with time of spraying. As the grain approached maturity, less of the urea that was absorbed was converted into protein.

Foliar application of urea solutions were used by Austin (1966a, 1967a) on a radish (*Raphanus raphanistrum*) seed crop in an attempt to break the negative correlation between the concentrations of nitrogen and phosphorus observed in experiments where the fertilisers were applied at sowing or transplanting. Eight main plots (a 2^3 combination of zero and high application rates of N, P and K fertilisers) were split, one half of each plot receiving three sprays of urea at flowering, the sprays supplying a total of 196 kg N/ha. Seed from plots to which no nitrogenous fertiliser had been applied had a nitrogen content of 4.24 per cent. This was increased to 5.31 per cent by the nitrogen fertiliser dressing, to 4.9 per cent by the urea spray alone, and to 5.69 per cent by both together. Irrespective of the time of application of the nitrogen, increases in nitrogen content of the seeds were accompanied by a concomitant reduction in their phosphorus content, which varied in this experiment from 0.98 to 0.67 per cent.

Extreme deficiency of the minor elements copper, zinc and molybdenum was shown by Hewitt, Bolle-Jones and Miles (1954) to give seed containing greatly reduced contents of these elements. Seed from pea plants grown in conditions of manganese deficiency has brown necrotic areas on the inside (adaxial) surfaces of the cotyledons, a symptom known as marsh spot. This condition has occurred widely in England and is of concern because it affects the suitability of the peas for processing (Reynolds, 1955). Manganese deficiency produces similar symptoms in other large seeded legumes.

RAINFALL AND SOIL MOISTURE

It has long been known that the protein nitrogen content and quality of grain is lower in years of high rainfall than in drier years, and from

TABLE 5.3   *Effect of irrigation on mineral concentration in grain. (From Greaves and Carter, 1923.)*

Element	Increase (+) or decrease (−) over controls, %		
	Wheat	Barley	Oats
Nitrogen	−21	−19	−40
Phosphorus	+55	+30	+35
Potassium	+35	+14	+31
Calcium	+155	+41	+22
Magnesium	+32	+9	+65

irrigated as compared with dry land. Greaves and Carter (1923) found that a high rate of irrigation decreased the nitrogen content of wheat, barley and oats in Utah, but increased the phosphorus, potassium, calcium and magnesium content (Table 5.3). Shutt (1935) found that irrigation decreased the protein content of Red Fife wheat grain by 2.8 per cent; the dry land controls containing 17.8 per cent protein.

Russell and Voelcker (1936) used the results of fifty years' experiments at the Woburn Experiment Station, and the corresponding rainfall data, to deduce the effects of rainfall at various stages of growth of the barley crop on the nitrogen content of the harvested grain (Fig. 5.2). The results show clearly that above-average rainfall during May and June and early July reduced the nitrogen content, and that rainfall above the average in August increased, but only slightly, the nitrogen content.

The ways in which rainfall affect composition have not been fully investigated. Thus, it is not clear whether the primary effect is on mineral absorption by the roots, or on their transfer from the plant to the seed or on the rate or extent of 'filling up' the grain with carbohydrates and the concomitant dilution of the basic cell constituents.

Swanson (1946) surveyed the effects of rainfall during harvest on the quality of wheat in Kansas. The main effects of wetting the grain after it had become dry enough to combine-harvest were: a decrease in bulk density due to a roughening of the bran coat and a swelling of the kernel as a whole and an increase in the percentage of the kernels with a mealy texture. The milling and baking qualities of the wheat were not adversely affected by wetting unless sprouting had occurred. Skazkin and Khvan (1962) applied water at various stages during the

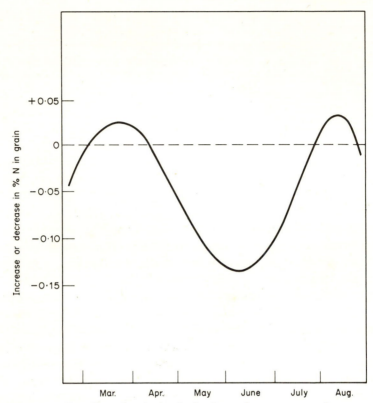

FIGURE 5.2   The effect of an inch of rain above the average on the percentage of nitrogen in barley grain (*Hordeum vulgare*), for the months February to August. (After Russell and Voelcker, 1936.)

ripening period of barley. The water was applied to the soil, or was applied as a spray to the stems and leaves, or to the spikes or to the whole plant. However applied, water at or after the stage of milky ripeness reduced mean grain weight. The reduction was greatest when the entire plant was sprayed.

Howell *et al.* (1959) attributed weight loss of soya beans in wet weather to high respiration rates persisting for longer than normal because of the slower drying of the seeds. Over the range 21–37°C the respiration losses at a moisture content of 55 per cent (fresh weight basis) were equivalent to 0.03–0.05 per cent of the seed dry weight per hour. Leaching of materials from the seeds while in the pods was an insignificant cause of weight loss.

Ogawa (1961) found that fruit set of *Veronica persica*, and *Polygonum* spp. was severely reduced by spraying the flowers with water, simulating rain. With carrot, onion (*Allium cepa*) and *Cancalis scabra* the stigmas were receptive for 3–4 days after flower opening and rain-fall treatments impaired pollination and consequent seed set only when they were continued for more than 3–4 days.

Salter and Goode (1967) have reviewed the evidence for moisture-sensitive stages during the growth of crops. They concluded that for many cereals and other annual crops, the period when the floral organs are developing is particularly sensitive to drought. Cereals experiencing drought during the early stages of flower initiation produce abnormal and sterile pollen grains and although the gynoecia are little affected, fewer ovules are fertilised and grain number per ear is reduced. Mean seed weight can be reduced by drought experienced after fertilisation.

TEMPERATURE

Experimental studies of temperature effects on seed structure and composition are few. Robertson, Highkin, Smydzuk and Went (1962) grew pea plants in a range of controlled temperatures. When photoperiod and light intensity were held constant, the growth of the seeds was much more rapid at high (23°C) than at low (10°C) temperature, their results suggesting that the final seed size attained was greatest at 17°C. Total sugar content declined rapidly at 14°C and higher, but remained high, with only a very slow rate of starch and protein synthesis, at 10°C. These studies, however, were not continued until dry seed maturity, and it is possible that the large differences in composition, attributable to the different temperatures in which the plants were grown, would not have been evident in the dry seed.

Howell and Carter (1958), in a controlled environment study, found that the oil content of soya bean seeds depended on the temperature during seed ripening: at 21°C the seeds contained 19.5 per cent oil, whereas at 30°C the oil content was 23.2 per cent. During the ripening period, which lasted five weeks, when the plants were grown at a day temperature of 21°C and a night temperature of 18°C, a one-week period of high temperature (30°C day, 18°C night) had most effect on oil content when given at the beginning of the ripening period, increasing the oil content by 2.4 per cent (controls 19.6 per cent). Zhdanova (1969) subjected either the flowers or the whole plants of flax and sunflower (*Helianthus annus*) to temperatures of 13–18°C or 25–35°C. In both species, the oil concentration in the seeds was greatest when the entire plants experienced the low temperatures. For flax, mean seed weight was greatest when the flowers or whole plants were kept at the low temperature. In sunflower, these temperatures did not produce appreciably different seed weights. Zhdanova found that the content and concentration of oil in the seeds increased rapidly as the seeds dried out, but the results are difficult to interpret because the oil concentration is given only as a function of seed moisture content. Zhdanova's results appear to conflict with those obtained for soya beans by Howell and Carter (1958), but it may well be that the effects of temperature on seed weights and oil content differ for the species covered by these investigations.

Nagato and Ebata (1960) found that high night temperature accelerated kernel development and maturation in rice (*Oryza sativa*), giving 'chalky' kernels. At low night temperatures the kernels were 'milky white'. High night temperatures given early in the period of kernel development increased the size of the aleurone cells and the thickness of the bran layer.

The effects of environment before harvest and of the stage of maturity at harvest on seed performance

There is little evidence to show at what stage during growth the various environmental factors can produce effects on seed viability, although it is likely that the mineral composition of seeds (which can affect their viability) is influenced by the mineral nutrition of their parent plants from an early age. Laude (1962), in a brief report, has claimed that heat stress experienced by the mother plants while at the seedling stage can affect the dormancy of the seed harvested from the mature plants, while Schwabe (1963) considers that winter rye may be receptive to vernalisation from the time of fertilisation, and does not exclude the possibility that the receptivity may extend back to the time when meiosis, leading to megaspore formation, occurs.

Pollination and pollen tube growth are under strong environmental influence. In addition, chance variations in the time of pollination and in turn variation in the age at which ripe ovules are fertilised, can affect the subsequent growth of the embryo. In corn, the stigmatic surfaces, the 'silks', are receptive for up to 19 days, although Peterson (1942) found that up to eight days from silking the mean seed set was 91 per cent and the Coefficient of Variation of seed set 10 per cent, whereas from 9–19 days the set was only 50 per cent and the Coefficient of Variation 42 per cent. This increase in variability suggests the processes of pollen tube growth and fertilisation were more senstive to environmental fluctuations in old than in young silks.

The decision when to harvest a seed crop is usually a compromise which takes into account past and expected weather, the availability of machinery and labour as well as the state of maturity of the crop.

In some crops, especially the grasses and *Beta*, heavy losses can occur by shedding of mature seed at or during harvest and these can be partly avoided by harvesting before all the seeds are mature, but the proportion of immature seed, of potentially low viability, can then rise to an uneconomic level.

Thus, seed maturity, usually reflected in mean seed weight or size has been the subject of numerous studies. As with structure and composition, the effects of environmental factors on seed viability and performance have been much less thoroughly studied.

In the following discussion 'percentage germination', unless otherwise defined, is the percentage of seeds or embryos, which, when sown under favourable temperature, moisture and light conditions, give complete normal seedlings. Test conditions are not described here, but the authors quoted have generally used conditions similar to those specified in the ISTA rules (Anon., 1959). A 'percentage viability' is the percentage of seeds which when sown in arbitrary conditions, often in the field or in soil or compost in a glasshouse, produce complete, normal seedlings. 'Field emergence' or 'compost emergence' will be used as convenient terms to refer to the percentage viability as determined in the field or in compost in a glasshouse.

EFFECTS OF SEED MATURITY AND SIZE

Practically all studies show that, until the attainment of full maturity, defined as time when there is no further increase in dry weight, full germination capacity and viability are not attained. The exceptions to this are when adverse weather intervenes and causes damage, and frost, excessive rainfall or drought can all impair seed viability in various ways.

Many studies have been carried out on corn. Walker (1933), in Manitoba, harvested a sweet corn variety Burbank Golden Bantam at various ages from 13 to 55 days after silking, in an attempt to determine whether full germination and viability were obtained before dry seed maturity, before which killing frosts frequently occur. The results (Table 5.4) show the changes in seed weight with age from silking, and the corresponding germination and field emergence of the seeds. Maximum germination capacity and field emergence were attained by the time that seed weight had reached 70 per cent of its maximum, after which there was no significant increase or decrease. Rush and Neal (1950) in a similar study found, however, that the germination of seeds in adverse, cold conditions did not reach its maximum value until some five weeks after the time when the germination under near-optimum conditions had reached almost 100

TABLE 5.4   Zea. *Mean seed weight, percentage germination and field emergence of seed harvested at various ages after appearance of silks. (From Walker, 1933.)*

Ages in days from silks	13	15	21	31	39	42	43	44	46	48	50	51	52	55
Mean seed weight, mg	34	29	59	135	145	146	190	135	181	166	189	201	219	201
Germination, %	22	16	72	99	93	99	98	97	97	96	98	96	90	98
Field emergence, %	—	23	86	96	95	97	99	92	89	89	91	94	83	88

per cent (Table 5.5). Culpepper and Moon (1941) also noted the large discrepancy between the germination of seeds and their emergence in the field, and that the size of seedlings from small, immature seed was much less than of seedlings from mature seed. Sprague (1936) found the germination of immature, dried seed was much less variable than that of comparable seed which had not been dried, and suggested that a beneficial, irreversible, change took place as a result of the drying. The study of Klein and Pollock (1968) on the changes in fine structure of *Phaseolus lunatus* during desiccation (see p. 118) suggests, however, that characteristic changes in fine structure take place before the seeds acquire resistance to subsequent desiccation, and it is possible that it is these changes rather than desiccation itself that are beneficial to subsequent germination. Although seed weights vary considerably over the corn ear, Bell (1954) found that seed of a sweet corn variety had 5 embryonic leaves regardless of position on

TABLE 5.5   Zea. *Moisture percentage, laboratory germination and cold test germination of seed harvested at 10-day intervals. Means of five different hybrids. (From Rush and Neal, 1951.)*

	Harvest dates in 1948					
	30th Aug.	9th Sept.	19th Sept.	29th Sept.	9th Oct.[a]	19th Oct.[b]
moisture, %	72.2	53.3	39.7	35.5	28.4	25.9
laboratory germination, %	88	96	97	98	98	98
cold test germination, % (untreated seed)	0.9	7.0	36.3	64.8	80	57

[a] Mean of 4 hybrids.
[b] A frost occurred two days before this harvest. Mean of 2 hybrids.

the ear, and that this stage of development had been reached well before the attainment of final dry weight of the seed.

Several studies have been made of maturity effects on grass seeds. McAlister (1943) harvested seeds of species of *Agropyron*, *Bromus*, *Elymus* and *Stipa* at four stages of maturity, viz. at pre-milk, milk, dough and at full maturity. The emergence in soil tests was determined from test sowings made at intervals up to 58 months from harvesting. For all genera except *Bromus*, the seed harvested earlier than the dough stage had low viability, especially after storage, as compared with that harvested when mature. In *Bromus* the very immature seed had initially a viability practically equal to that of the mature seed, and even after 58 months' storage, the viability of the seed harvested at the milk stage was equal to that of the mature seed. Kneebone and Cremer (1955), in a study with five grass species, graded seed lots by size. The largest grade within a lot had a mean seed weight approximately twice that of the smallest grade. Samples of each grade were sown in soil in a glasshouse. For all five species, within a lot, large seeds were superior in days to 50 per cent emergence, in the final percentage emergence, in seedling height and fresh weight. In a similar study of ten grass species, Kittock and Patterson (1962), found that within a species, the correlation between field emergence at three weeks and seed weight was 0.98. In *Agropyron desertorum*, Rogler (1954) found that seeds in different weight classes differed little in field emergence when sown at various depths up to 5 cm. The emergence when sown at depths greater than 6 cm, however, was strongly correlated with mean seed weight (correlation coefficients of 0.85–0.95 were obtained).

Black (1959) has reviewed seed size effects in herbage legumes, principally small seeded species like those of the genera *Trifolium*, *Melilotus* and *Medicago*. Practically all studies reviewed show that, as with the grasses, within a species and seed lot, large seed gave superior field emergence, especially when the seeds were sown deeply. An exception to this generalisation was noted by Moore (1943) who found that the emergence of the largest seeds of a sample of *Trifolium incarnatum* sorted into 5 size grades was less than that of the smaller seeds, at all depths of sowing. The cause of this was not ascertained by Moore, but it may have been due to genetic abnormality or to mechanical damage during harvesting which may have affected the large seeds in a sample more than the small seeds, as with *Phaseolus* (Faris and Smith, 1964). Hewston (1964) also found that the largest seeds in commercial lots of radish (*Raphanus raphanistrum*) seed had a lower percentage germination and field emergence than seeds of an intermediate size. This result too, may have been the consequence of mechanical damage, for the siliquae of the radish when being threshed require high drum velocities and close concave settings for efficient seed removal, and damage to the seed readily occurs, especially when the pods and seed are very dry. As with grasses, seed weight sets a lower limit to the depth from which a seedling can emerge in given conditions, and during early vegetative growth, seedling size is correlated with seed weight.

In a study of seed size and maturity effects on carrots, Austin and Longden (1967b) harvested seed at different stages of maturity. Seed from each harvest was graded into four size grades, and the percentage germination and field emergence determined. Typical results from some of the experiments are given in Table 5.6. They show that at each seed harvest date percentage germination and, to a more marked extent, field emergence increased with seed weight. Within a seed size grade there was very little variation in seed weight, but the germination and field emergence were poorer for seed harvested on the earlier, than on the later occasions, and this effect was more marked in the smaller size grades, which contained the seed from the higher order umbels which flowered later than the primary umbel and was therefore 'younger' (Borthwick, 1932).

Failure to produce viable seed will obviously set a limit to the distribution of a species. Thus it is possible that the onion does not occur wild in England because of failure to produce good seed each year. Austin (1963) showed that year to year differences in the weather during ripening led to variation in seed weight and percentage germination but proportionally much greater variation in field

TABLE 5.6   *Daucus carota. Effect of harvest date, and within a harvest, of seed size on seed weight, percentage germination and field emergence. (From Austin and Longden, 1967b.)*

Size grade, mm	Mean seed weight, mg				Germination, %				Field emergence, %			
	5th Sept.	14th Sept.	25th Sept.	5th Oct.	5th Sept.	14th Sept.	25th Sept.	5th Oct.	5th Sept.	14th Sept.	25th Sept.	5th Oct.
1.00–1.25	0.71	0.68	0.69	0.62	32	40	51	49	26	28	38	39
1.25–1.50	1.02	0.96	0.97	0.93	45	56	58	53	42	45	50	43
1.50–1.75	1.38	1.30	1.25	1.25	60	69	64	65	53	59	64	56
1.75–2.00	1.78	1.59	1.57	1.68	64	62	61	55	64	58	60	59
s.e. of a mean and d.f.	0.019 (33)				3.1 (33)				3.0 (33)			

emergence, confirming that ripening in the field to give commercially acceptable onion seed does not usually occur in England.

With larger seeded dicotyledonous species, seed maturity effects seem to be essentially similar to those of the smaller seeded species. Inoue and Suzuki (1962) harvested *Phaseolus vulgaris* seeds from 15 to 35 days after anthesis. Germination rose from nil for seeds harvested at 15 days and rose progressively to 100 per cent when the seeds were harvested fully mature. Considerable increase in seed weight took place when the plants were harvested at 20 days, well before dry seed maturity, and the seeds allowed to develop on the drying plant, and such seed has almost 100 per cent germination. A similar effect was observed by Kolev and Georgiev (1964) for *Allium porrum* inflorescences harvested before the seed had reached full maturity, and these authors found that the greater the length of stalk attached to the inflorescence, the greater the mean seed weight finally attained.

Studies with fleshy fruits show that the harvesting and extracting of seed before full fruit maturity is detrimental to viability. Cochran (1943) found that seed extracted from 30-day old fruits of pimiento pepper (*Capsicum frutescens*) had an emergence from compost of 5–6 per cent. Fruits picked at the same age, but stored for a further 30 days before the seed was extracted, gave seed with an emergence of 95 per cent. Seed extracted from fully ripe, 60-day old fruits, had an emergence of 92 per cent. Kerr (1963) found that providing tomato (*Lycopersicon esculentum*) fruits were at least at the mature green stage, maturity had little effect on germination until they were over ripe.

Seed size effects on plant growth and yield have been investigated for many species and no attempt will be made here to do more than state the general conclusions from such work. The relative growth rates of seedlings are nearly always found to be independent of seed or embryo weight. As a consequence, during the exponential phase of growth, seedling weights at a given time are directly proportional to seed weights provided that the times of emergence do not differ for the different size classes. In wheat, where the embryo is a small part of the whole seed, although there is a strong correlation between seed weight and embryo weight, the main determinant of seedling weight is seed weight rather than embryo weight (Bremner, Eckersall and Scott, 1963). The early hypothesis to explain heterosis in plants, which linked heterotic effects to initial differences in embryo size, has not been supported by subsequent investigations. If larger embryo size was the only, or a major, advantage possessed by hybrids exhibiting heterosis, the greatest difference between the hybrids and the mid-parental weights would be in the same ratio as that of their embryos. Typically however, such hybrids have higher relative growth rates than the mid-parent values and so any initial differences in embryo size tend, both absolutely and proportionally, to increase. Many studies, for example those of Luckwill (1939) and Hatcher (1940) on tomato and Righter (1945) on *Pinus* show that there is no general correlation between inherent vigour, which can be taken as the relative growth rate during the phase of exponential growth, and embryo or seed weight, within closely related taxonomic groups. Thus, selection for vigour within a progeny is ineffective when based on size of seed (see also pp. 215–220).

Many studies, notably those of Black (for a review, see Black, 1959) have shown that in a competitive situation or in a limiting environment (such as would exist if inherently large plants were grown in small containers) and when genetic differences are not confounded with seed size, plants from large seed, being larger than those from small seed, compete with each other, or become limited by the environment sooner than those from small seed. In such situations, therefore, the relative advantages of large seed size decline with time, eventually vanishing as the yields approach an asymptotic value.

Seed maturity is not invariably correlated with size. Thus Table 5.6, from Austin and Longden's (1967b) study on carrots, shows that for seed in the small size grade, germination and field emergence increased with harvest date, except at the last harvest. Within a size grade, maturity effects on plant weights 6 weeks after sowing were evident for all sizes of seed, and were of a similar order to the size effects (Table 5.7). As with seed size effects, the maturity effects tended to disappear with crop age.

TABLE 5.7   Daucus carota. *Mean plant fresh weight at six weeks from sowing, as affected by harvest date, and seed size within a harvest date. (From Austin and Longden, 1967b.)*

Seed size	Seed harvest date				Mean
	5th Sept.	14th Sept.	25th Sept.	5th Oct.	
1.00–1.25	0.26	0.25	0.34	0.32	0.29
1.25–1.50	0.30	0.38	0.46	0.38	0.38
1.50–1.75	0.48	0.42	0.54	0.43	0.47
1.75–2.00	0.40	0.49	0.61	0.52	0.50
Mean	0.36	0.39	0.49	0.41	

s.e. of an entry in the body of the table 0.04 (8 d.f.)
s.e. of a marginal mean 0.02 (8 d.f.)

EFFECTS OF MINERAL NUTRITION

Ozanne and Asher (1965) have pointed out that the amounts of potassium in seeds set a limit to the extent to which seedlings can grow in a minus-potassium culture media. In 21 species studied, most of the variation in potassium content (2×10^4 fold) was due to variations in seed size which itself varied by 80×10^4 fold, rather than to variation in potassium concentration which varied by only 3.4 fold. Species having a seed weight of 0.2 mg could penetrate minus-potassium sand to a depth of only 3–4 cm, while those weighing 150 mg could penetrate to at least 90 cm. Krigel (1967) found that normal seeds of *Trifolium subterraneum* suffered deficiency if calcium was withheld for more than 7 days, phosphorus for 10 days, nitrogen for 14 days or potassium for 21 days. Thus, it is clear that mineral-deficient seeds will be at a disadvantage as compared to normal seeds unless sown in a medium which is nutritionally adequate, especially during early growth.

Harris (1912) seems to have been one of the first to investigate the effects of nutrient deficiency on seed performance. Three pure lines of *Phaseolus vulgaris* were grown either in a field which 'bore a moderately heavy crop' or one in which 'the plants germinated well but were all exceedingly small'. Both fields had received similar cultivation, but Harris does not ascribe the differences in growth to lack of any specific nutrients. Seeds from plants which had been grown on the infertile field for one or two generations gave plants, which in a 'comparison field', had slightly, but consistently, fewer pods per plant than those from the fertile field. Subsequent studies have dealt mainly with the consequencies of deficiencies of single mineral elements, but as has been discussed (p. 120) a deficient supply of one element usually produces correlated changes in the concentration of other elements as well as in the one varied experimentally, frequently making it difficult to attribute any effects on viability and seedling growth to differences in seed reserves of that element alone.

Nitrogen

Williams (1965) and Hill-Cottingham and Williams (1967) have shown that nitrogen given in the summer of the previous season to young apple plants (*Pyrus malus*) markedly improved the 'quality' of the flowers, as compared with those which received no nitrogen at this time. Stigmas of the treated plants were receptive for longer than those of the untreated control plants. After pollination, pollen tube growth proceeded at similar rates in both types of stigma. Cross pollination of control plants was successful only up to the second day after anthesis, fertilisation taking place some 6–7 days afterwards, when the pollen tube reached the egg apparatus. In treated plants, cross pollination was successful for up to 6 days after anthesis. After fertilisation, the rate of egg sac and ovule growth was more rapid in the treated than in the control plants. A higher proportion of apparently fertilised eggs in control flowers failed to develop normal embryos within the fruit.

Sneddon (1963), in field experiments with sugar-beet (*Beta vulgaris*) found that for fruits in a given size grade, the numbers of embryos which germinated per fruit was significantly lower for those from plots which had received 260 kg N/ha than for those from control plots which had not received nitrogenous fertiliser. Similar results were obtained by Tolman (1943). Scott (1969) found that nitrogenous fertiliser applied to the seed crop depressed germination in sugar-beet only in years when the crop ripened late. Nitrogen delayed the ripening of the crop and thus the seed from the plots fertilised with nitrogen was less mature at harvest than that from the plots which received less, or no nitrogen. Thus the reduction in germination was a nitrogen-induced maturity effect. In *Beta* germination is influenced by inhibitors present in the perisperm in which the embryos are embedded, dark seed being associated with high levels of inhibitor. I have noted (unpublished) that fruits were darker from plants grown in high- than in low-nitrogen cultures, and it is possible that the effects observed by Sneddon (1963) and Tolman (1943) and Scott (1969) were the result of the effects of nitrogen on the concentration of germination inhibitors in the fruits.

Harrington (1960) produced seed of various species under severe nitrogen deficiency. Deficient plants gave very low yields of seed as compared with those from control plants, and much of the seed was abnormal. The percentage germination of normal seed from the deficient and control plants was, however, similar.

Fox and Albrecht (1957) collected samples of wheat grown in Nebraska and found that those with a high crude protein content (14.4 per cent) germinated and emerged more rapidly, and gave greener and more vigorous seedlings than low protein (11 per cent) seeds. In this test, where the seeds were sown 7.5 cm deep in sand, emergence was 91 per cent from the high protein, and 86 per cent from the low protein seed (LSD 2.4 per cent). In a less favourable seed year, an increase (from 10 to 15 per cent) in crude protein content, brought about by applying sprays of urea, decreased the emergence of the seed. In another experiment in the same year, similar increases in crude protein content brought about by soil dressings of ammonium nitrate had no effect on emergence.

Phosphorus

Several studies indicate the importance of seed phosphorus reserves for obtaining vigorous seedlings. In none of the species studied by Harrington (1960) was germination affected by the phosphorus nutrition of the parent plants. With watercress, however, Austin (1966a) found that freshly harvested seed from phosphorus-deficient plants had a slower rate of germination and a lower final percentage germination than normal seed. Seed more than two months old, or that after-ripened by moist storage at 20°C, showed no such differences. With peas, Austin (1966b) found marked differences in rates of emergence of dry seed both in glasshouses and field experiments, phosphorus-deficient seed emerging more rapidly than normal seed. When sown at a moisture content of more than 12–15 per cent no such differences were apparent.

Szukalski's (1961a and b), Birecka and Wlodkowski's (1961) and Austin's (1966a and b) experiments show that when sown in phosphorus-deficient cultures, low-phosphorus seed gives smaller plants than those from normal seed. Representative results from these experiments are given in Table 5.8. In high-phosphorus cultures the differences are usually small and not significant, presumably because the plants rely on seed phosphorus reserves for only a very short time. Nevertheless, sown in the field in a soil where responses to phosphatic fertilisers were not usually obtained, the phosphorus-deficient seed gave smaller plants and lower yields of peas than otherwise compar-

TABLE 5.8   Rorippa nasturtium-aquaticum. *Effects of three levels of phosphorus supply on the dry weights (g/pot) of plants from P1, P2 and P3 seed, at 7–9 weeks. Means of four experiments. See Table 1 for details of P1, P2 and P3 seed. (From Austin, 1966a.)*

Second generation treatments (G2)		P1	P2	P3
	P1	0.0953	2.44	3.35
First generation treatments (G1)	P2	0.139	2.76	3.59
	P3	0.181	2.56	3.41
s.e. a difference between two means for a given G2 treatment (96 d.f.)		0.0126	0.244	0.324

TABLE 5.9   Pisum sativum. *Effect of phosphorus nutrition of the parent plant on seed composition and yields of haulms and peas from field sowings of the seed. (From Austin, 1966b.)*

Seed type	1963 experiment		1964 experiments[b]	
	O and N	P and NP	O and N	P and NP
Elemental composition of seeds (% dry weight)				
Phosphorus	0.30	0.59	0.29	0.58
Nitrogen	3.72	3.75	3.98	3.81
Potassium	1.46	1.33	1.60	1.40
Mean air-dry weight of a seed, g	0.20	0.24	0.23	0.27
Yields				
Mean weight of haulm, g/plant	58.3 ± 2.4 (9)	73.0 ± 2.4 (9)	118.7[c] ± 2.5 (27)	139.2[c] ± 2.5 (27)
Mean weight of peas, g/plant	81.6[a] ± 4.2 (9)	96.2[a] ± 4.2 (9)	15.4 ± 0.5 (27)	19.3 ± 0.5 (27)
Tenderometer value	not determined	not determined	120.2 ± 1.12 (27)	126.2 ± 1.12 (27)

[a] Weight of peas plus pods.
[b] Data given are means from four experiments.
[c] Weight of haulm plus shelled pods.

able, control seed (Table 5.9). In these experiments extreme deficiency, obtained by supplying nutrient solutions containing only 5–10 per cent of the optimum phosphorus concentration, was required to give seed low enough in phosphorus to produce the growth differences, and it seems unlikely that any commercial crop of seed would experience phosphorus deficiency acute enough to cause such seed-phosphorus deficiency. Among 16 commercial samples of pea seed var. Dark Skinned Perfection, Austin (1966b) found that only one contained less than 0.4 per cent P, a concentration below which this seed could be regarded as deficient. Lipsett (1964), however, reported that 20 representative samples of wheat from phosphorus-deficient sites in Australia contained an average of only 0.25 per cent P, and fertiliser dressings which gave optimum yields increased the phosphorus concentration only to about 0.3–0.35 per cent; dressings of 150 kg P_2O_5/ha being required to achieve a concentration of 0.4 per cent (wheat from Europe normally contains about 0.5 per cent P). Thus, in areas where phosphorus deficiency is widespread, it would probably be advantageous to ensure that seed for sowing was produced on land of high phosphorus status to ensure that it contained the desirable high phosphorus content.

Potassium

Harrington (1960) found that severely potassium-deficiency plants of *Capsicum frutescens* gave a high proportion of abnormal seed with dark-coloured embryos and seed coats. Both normal and abnormal seed from such plants had a lower percentage germination than that from control plants and its viability declined more rapidly in storage than that from normal plants. Iwata and Eguchi (1958) found that seed from moderately potassium-deficient plants of *Brassica chinensis* was not inferior to that from normal plants.

Other elements

Deficiencies of the minor elements, calcium, boron and manganese produce characteristic damage to seeds, especially those of large seeded species, and when such damage is extensive the viability of the seed is impaired. Leggatt (1948) found that peas harvested from a boron-deficient area appeared quite normal but, when germinated in sand, they produced pale and stunted plumules lacking the normal curvature. A trace of borax added to the sand, however, resulted in the production of entirely normal plumules from deficient seeds. Damage in peanut (*Arachis hypogea*) kernels was attributed by Cox and Reid (1964) to calcium and boron damage. Lack of either element produced characteristic symptoms, and deficiencies were often found together. Deficiency of molybdenum in plants is difficult to obtain experimentally, at least in large seeded species such as *Phaseolus vulgaris* (Meagher, Johnson and Stout, 1952) because the seeds contain sufficient molybdenum to support the growth of several generations of plants. However, seed deficient in this element gave poorer plants than normal seed in molybdenum-deficient cultures. Similar results were obtained with copper, zinc and molybdenum deficiencies by Hewitt, Bolle-Jones and Miles (1954). Eaton (1942) found that sulphur-deficient seeds of *Helianthus annus*, containing only about 40 per cent of the amount of this element present in normal seeds, gave smaller plants both when sown in S-deficient and in normal sand cultures.

EFFECTS OF TEMPERATURE

The temperature during seed ripening can influence the subsequent performance of seeds in various ways. In cereals the vernalisation requirements can be met partly or entirely by low temperature during seed ripening. Gregory and Purvis (1936) kept ears of Petkus winter rye (*Secale cereale*) at 1–1.5°C for 24 days, the remainder of the plant being kept at higher temperature. Ears which were so treated during the middle period of ripening and then allowed to complete their ripening normally, proved to be vernalised. Kostjucenko and Zaburailo (1937) grew seed of several winter wheat varieties at two sites at latitudes 67° and 40°N in the Soviet Union. Seeds were harvested and a vernalised and an unvernalised series sown at the 67°N site. By September, plants from non-vernalised seeds from 40°N were only tillering, while those from 67°N were blooming. Vernalised seeds from 67°N were wax ripe at this stage while those from 40°N were only at the milk ripe stage. Riddell and Gries (1958) grew Warden, a late maturing cultivar of spring wheat responsive to vernalisation, in different years at Purdue, Indiana, and found that the responsiveness to vernalisation of plants grown from the seed varied. It appeared that this variation occurred because the vernalisation requirements for maximum rate of development were partially satisfied during seed-ripening, to an extent which varied from year to year. Differences in rates of development could be entirely eliminated by a 60-day vernalisation period at 0–2°C. Schwabe (1963) found that the rates of development, as measured by the number of days to anthesis, were progressively less for Petkus winter rye as the temperature during seed formation and ripening was reduced over the range from 16° to 9°C. Devernalisation did not occur when the seed was dried at temperatures of 20–35°C in the later stages of maturation, or during storage for up to three years. Winter rye, which is probably typical of most cereals having a vernalisation requirement, is thus receptive to vernalisation for some period up to an intermediate stage of seed-ripeness, possibly commencing at meiosis or fertilisation.

In many genera of the Rosaceae the embryos are anatomically mature at shedding, but require a period of after-ripening at low temperatures ('stratification') for normal seedling development. In the peach (*Prunus persica*) germination will occur without such stratification, but dwarf, distorted seedlings are produced (Pollock, 1962). In the cherry (*Prunus avium*) Braak (1962) showed that the extent to which the embryo grows in the seed before harvest is greater in years when flowering is early than when it is late. Von Abrams and Hand (1956), showed that for *Rosa* hybrids, the stratification requirements depended on the mean temperature during the 30 day period preceding harvest. Harvesting was done on 1st November in five successive years, and the stratification requirements of intact seeds were less in years with warm, than those with cool Septembers. Von Abrams and

Hand do not give any details about the extent of embryo development in the different years which, if it differed, might have accounted for the differences in stratification requirements of the seed.

In some species, temperature during ripening can affect the degree or duration of post-harvest seed dormancy, or the range of environments in which such dormancy is expressed. The best known example of this is *Lactuca sativa* cv. Grand Rapids. Thompson (1936) found that in immature seed of this variety, germination was only 7 per cent in the dark at 25°C. In diffuse daylight at 15–20°C germination was 81 per cent. The germination of mature seed in the dark was 44 per cent at 25°C, and 92 per cent at 10–15°C. In the light at 15–20°C the germination of mature seed was 98 per cent. It is now well known that light acts on this seed via the reversible photosensitive pigment, phytochrome (Borthwick, 1965). Whether in mature seed the reaction catalysed or 'switched on' by the far-red form of phytochrome has already taken place or the need for it disappeared, or whether with maturity some change has taken place in the phytochrome system itself does not appear to have been firmly established. Koller (1962) studied the effect of environment during maturation on subsequent germination of Grand Rapids. Maturation at high temperature gave seed which, when mature and after at least one month's storage, proved to be less dormant when put to germinate in the dark at 26°C than seed which had matured at a lower temperature.

A similar temperature effect but in an opposite direction occurs with *Anagallis arvensis*, which requires light for germination at 25°C. Grant Lipp and Ballard (1963) showed that seed ripened at 20/15°C (day and night temperature respectively) was almost completely dormant for 10 weeks from harvest. Seed ripened at 25/20°C showed much less dormancy, while that ripened at 30/25°C showed none.

The tendency of wheat grain to sprout in the ear is an aspect of dormancy that may be determined partly by temperature during ripening. Van Dobben (1947) and Kramer, Post and Wilten (1952) found that high temperature in the early stages of ripening reduced the tendency of the grains to sprout. Laude (1962) reported an effect of high temperature on seed dormancy in *Bromus*, but in this case exposure of the seedlings to a high temperature, 55°C for 3–5 hours, or 32°C for 1–3 weeks, gave plants the seed of which exhibited post-harvest dormancy lasting from three to six months; increased duration of heat stress resulted in a greater degree of dormancy. Stearns (1960) showed that when seed of *Plantago aristata* was matured at 22°C, the rates of growth of the seedlings from the seed were considerably greater than for comparable seed matured at 15°C; an effect which was not attributable to post-harvest dormancy or to any effect of the temperatures on the size of the ripe seed. The differences in growth persisted for at least 120 days.

During ripening of the seed of corn, temperatures below 0°C can cause freezing injury. Rossman (1949a and b) found that the extent of the damage produced depended on the type of seed, single-gene differences being associated with markedly different susceptibilities. For a given genotype, the susceptibility to damage depended on the state of maturity of the seed, of which moisture content was a good index. Serious injury was produced by freezing at −6°C for 8 hours at seed moisture contents about 40 per cent, the percentage emergence from sand and mean seedling weight both being reduced when compared to non-frozen control seed.

A more esoteric temperature effect has been described by Highkin (1958) for the pea cultivars Unica and 'L5'. When grown in a controlled environment at a constant day and night temperature, the rate of growth, expressed in mm per day, declined with successive generations. This effect was obtained at constant temperatures of 10°, 17° and 20°C and was nearly at its maximum after three generations. It was not immediately reversed by growing plants in an environment with a fluctuating temperature and it required 2–3 generations for complete reversal. This phenomenon is probably of an entirely different nature to the other effects of temperature described here, since it is inherited for several generations. It might be due to the progressive accumulation or reduction of a seed transmitted cytoplasmic particle or a virus, or to a change similar to that described for flax by Durrant (1962), but as yet these or other possibilities do not appear to have been investigated.

EFFECTS OF PHOTOPERIOD

Lona (1947) found that the photoperiod in which seeds of *Chenopodium amaranticolor* matured on their parent plant influenced the subsequent germination of the seed. When matured in long days the seed coats were much thicker than those of seed ripened in short days (6–8 hours). Seeds with thick coats could imbibe water readily when put to germinate but only a small proportion of the embryos were able to rupture the abnormally thick seed coat. Treatment of the coats with strong sulphuric acid weakened them sufficiently to permit germination. At temperatures of 25–28°C this long-day effect was much more marked than at temperatures of 15–16°C. *Chenopodium polyspermum* shows a somewhat similar dormancy imposed by the seed coat, the thickness of which is under photoperiodic control (Jacques, 1957).

Germination of seeds of the crucifer *Diplotaxis Harra* and of the legume *Ononis sicula* is also sensitive to the photoperiod in which they matured on the parent plant. Seeds of *D. Harra* require light for germination. Seed matured on the parent plant in long days shows a much greater light requirement than that matured in short days (Evenari, 1965), and this light sensitivity is apparently associated with the phytochrome system. Seeds of *O. sicula* are affected in an entirely different way. According to Evanari, Koller and Gutterman (1966), if they remain on the plant until fully mature, the seed coats are highly impermeable to water and this impermeability is retained for 80–90 days after moistening the seed at 20°C. When matured in long days, the seed is fully mature before natural dehiscence takes place, but in short days, the fruits dehisce before the seeds are fully mature, and in this condition they do not develop impermeable seed coats.

In all these cases the primary effect of photoperiod is on the maternal tissue associated with the embryo, rather than the embryo itself.

Conclusions

In addition to carrying genetic information, the seed is, in the widest sense, the inheritor of all the environmental influences which have acted on it before sowing. It is apparent that, unless it is very extreme, environmental variation during seed development has very little effect on the viability of the seeds of most species, provided that the ripening processes are not interrupted by premature harvesting. For wild species, this resistance to the effects of environmental variation during ripening aids survival. In many species, especially those which grow in harsh environments, such as deserts, dormancy mechanisms have been evolved to give a spread of germination in time, and in this way at least a proportion of the seeds will germinate when the environment is favourable for survival (Koller, 1955). These dormancy mechanisms are generally not affected by the maternal environment, and, in very diverse ways, act to permit germination only under favourable conditions. The relatively few exceptions to this probably occur where it is ecologically desirable for survival that dormancy should occur only when seed ripens at a particular time of the year.

In cultivated species which are normally propagated annually from seed, there will have been an unconscious selection against dormancy, except where it is required to prevent premature germination and deterioration on the plant during ripening. Accordingly, there will be little opportunity for the maternal environment to influence dormancy. Possible exceptions to this are the relatively recent cultivated leguminous herbage plants where hard seededness is influenced by the maternal environment and by the state of maturity of the seed at harvest. Within strains of a species, however, there exist large differences in the occurrence of hard seededness and so it should be possible to eliminate it in new varieties by appropriate breeding, and this may have already occurred by unconscious selection in the older crop plants. In a similar way, there will have been a selection against genotypes which display large maternal influences, as this will have led to undesirably variable performance. Even in the highly bred crop, wheat, however, careful trials have shown that such influences may not have been entirely eliminated (Kinbacher, 1962; Quinby, Reitz and Laude, 1962); consequently these trials suggest that significant improvements in yield could be expected if the causes of maternal influences were understood and an appropriate breeding programme were undertaken to secure the desired responses.

Seed size and maturity, which can be important sources of variation in viability affecting the establishment and yield of crops, can be controlled within limits by varying harvesting techniques. Although for some crops where the economic yield is that of the seed itself, optimum harvesting procedures have been established by experiment, in many others, particularly horticultural crops such as carrots, onions and lettuce, little attention has been paid to these problems of seed production. In this type of crop, large seededness, although generally a desirable character, has not consciously been bred for as it is not a 'final' character. Seed size is known to be highly heritable and so there is opportunity for improvement in these crops by breeding for large seed.

Mineral deficiencies in seeds are likely to be rare and the consequences for yield even more rare. Conditions that would lead to the production of deficient seed are likely to be recognised easily and action taken to avoid them, and thereby, any possible consequences. There is very little evidence as to whether the amounts of nitrogenous and phosphatic compounds and of mineral elements present in seeds are optimal for the subsequent growth of the seedlings. The results of Finney *et al.* (1957) and Fox and Albrecht (1957) suggest however, that for nitrogen supra-normal amounts are not beneficial. My own unpublished experiments suggest that although seeds can be soaked in mineral salt solutions and will take up considerable amounts of nitrogen, phosphorus, potassium and calcium, and can then be dried without incuring any measurable damage, there is no evidence that such treatment is beneficial. Russian literature contains many papers suggesting that soaking seeds in solutions of salts of trace elements confers significant benefits in terms of subsequent performance, but these claims do not seem to have been confirmed by work done elsewhere. Recent reports (Ries, Schweizer and Chmiel, 1968) show that sub-herbicidal doses of simazine (4-chloro-4,6-bis(ethylamino)s-triazine) can increase the protein content and concentration both of species regarded as tolerant and intolerant to this herbicide. Fox and Albrecht's (1957) finding that emergence and seedling size of wheat grains are related to their crude protein content suggests that seed from simazine treated plants might possess advantages over normal seed containing a lower protein concentration, provided that it does not contain harmful concentrations of simazine.

For many species the effects of premature harvesting on viability can be related satisfactorily to existing knowledge of their embryogeny and seed development. In only a few of the studies of the effects on seed viability of environmental factors before harvesting can it be shown whether the sensitivity of the seed to environmental influences changes with age, and at what stage the seed or its precursor tissues first become susceptible to environmental influences. Further research would benefit from relating the stage and duration of exposure to environmental treatments to the embryogeny and developmental processes taking place in the seed at the time of treatment: there is a considerable body of relevant knowledge of plant embryogeny (see Maheshwari, 1950 and Johansen, 1950 for sumaries) which could be used as a guide in further work of this kind.

In this review I have not been concerned with one of the most important aspects of the maternal environment which affect the viability of seeds, namely diseases of the seed crop, many of which are seed-borne. Almost every crop has one or more recognised seed-borne disease which can be serious in their effect on seed viability. The incidence of these diseases, caused by fungi, bacteria and viruses is largely determined by the environment in which the mother plant is grown. Although important new systemic fungicides and bactericides can lessen the damage caused by some of the diseases, they probably remain an important cause of uncontrolled variation in seed viability, as affected by the maternal environment.

References to Chapter 5

ANON., 1959. International Rules for Seed Testing. *Proc. int. Seed Test. Ass.*, **24**, 475–584.

AUSTIN, R. B., 1963. Yield of onions as affected by place and method of seed production. *J. hort. Sci.*, **38**, 277–85.

AUSTIN, R. B., 1966a. The growth of watercress (*Rorippa nasturtium-aquaticum* L. (Hayek)) from seed as affected by the phosphorus nutrition of the parent plant. *Pl. Soil*, **24**, 113–20.

AUSTIN, R. B., 1966b. The influence of the phosphorus and nitrogen nutrition of pea plants on the growth of their progeny. *Pl. Soil*, **24**, 359–68.

AUSTIN, R. B., and LONGDEN, P. C., 1966a. Seed production. *Rep. natn. Veg. Res. Stn, Wellesbourne, 1965*, 44.

AUSTIN, R. B., and LONGDEN, P. C., 1966b. The effects of manurial treatments on the yield and quality of carrot seed. *J. hort. Sci.*, **41**, 361–70.

AUSTIN, R. B., and LONGDEN, P. C., 1967a. Seed production. *Rep. natn. Veg. Res. Stn, Wellesbourne, 1966*, 50.

AUSTIN, R. B., and LONGDEN, P. C., 1967b. Some effects of seed size and maturity on the yield of carrot crops. *J. hort. Sci.*, **42**, 339–53.

AUSTIN, R. B., LONGDEN, P. C., and HUTCHINSON, J., 1969. Some effects of 'hardening' carrot seed. *Ann. Bot.*, **33**, 883–95.

BARTON, L. V., 1943. The storage of citrus seeds. *Cont. Boyce Thompson Inst.*, **13**, 47–55.

BELL, M. E., 1954. The development of the embryo of *Zea* in relation to position on the ear. *Iowa State Coll. J. Sci.*, **29**, 133–40.

BILS, R. F., and HOWELL, R. W., 1963. Biochemical and cytological changes in developing soybean cotyledons. *Crop Sci.*, **3**, 304–8.

BIRECKA, A., and WLODKOWSKI, M., 1961. Influence of phosphorus content in seeds on the nitrogen accumulation and the growth of peas and yellow lupines. *Rocz. Nauk. Roln.*, Ser. A, **84**, 346–67.

BISSON, C. S., and JONES, H. A., 1932. Changes accompanying fruit development in the garden pea. *Pl. Physiol., Lancaster*, **7**, 91–105.

BLACK, J. N., 1959. Seed size in herbage legumes. *Herb. Abstr.*, **29**, 235–41.

BORTHWICK, H. A., 1932. Carrot seed germination. *Proc. Am. Soc. hort. Sci.*, **28**, 310–14.

BORTHWICK, H. A., 1965. Light effects, with particular reference to seed germination. *Proc. int. Seed Test. Ass.*, **30**, 15–27.

BRAAK, J. P., 1962. Influence of growth conditions on fruit and embryo development in apple and cherry. *Meded. Inst. Vered. TuinbGewass.*, **182**, 57–65.

BREMNER, P. M., ECKERSALL, R. N., and SCOTT, R. K., 1963. The relative importance of embryo size and endosperm size in causing the effects associated with seed size in wheat. *J. agric. Sci.*, **61**, 139–45.

BRENCHLEY, W. E., and HALL, A. D., 1909. The development of the grain of wheat. *J. agric. Sci.*, **3**, 197.

BRENCHLEY, W. E., and HALL, A. D., 1912. The development of the grain of barley. *Ann. Bot.*, **26**, 903–28.

BUTTROSE, M. S., 1963. Ultra structure of the developing wheat endosperm. *Aust. J. biol. Sci.*, **16**, 305–17.

CARR, D. J., and SKENE, K. G. M., 1961. Diauxic growth curves of seeds with special reference to French Beans (*Phaseolus vulgaris*). *Aust. J. biol. Sci.*, **14**, 1–12.

COCHRAN, H. L., 1943. Effect of stage of fruit maturity at time of harvest and method of drying on the germination of pimiento seed. *Proc. Am. Soc. hort. Sci.*, **43**, 229–34.

COX, F. R., and REID, P. H., 1964. Calcium-boron nutrition as related to concealed damage in peanuts. *Agron. J.*, **56**, 173–76.

CULPEPPER, C W., and MOON, H. H., 1941. Effect of maturity at time of harvest on germination of sweet corn. *J. agric. Res.*, **63**, 335–43.

DURRANT, A., 1962. The environmental induction of heritable change in *Linum*. *Heredity*, **17**, 27–61.

EARLEY, E. B., and DETURK, E. E., 1944. Time and rate of synthesis of phytin in corn grain during the reproductive period. *J. Am. Soc. Agron.*, **36**, 803–14.

EATON, S. V., 1942. Sulphur content of seeds and seed weight in relation to effects of sulphur deficiency on growth of sunflower plants. *Pl. Physiol., Lancaster*, **17**, 422–34.

EVENARI, M., 1965. Physiology of seed dormancy, after ripening and germination. *Proc. int. Seed Test. Ass.*, **30**, 49–71.

EVENARI, M., KOLLER, D., and GUTTERMAN, Y., 1966. The effects of environment of the mother plant on germination by control of seed-coat permeability to water in Ononis sicula. *Aust. J. biol. Sci.*, **19**, 1007–16.

FARIS, D. G., and SMITH, F. L., 1964. Effect of maturity at time of cutting on quality of dark red kidney beans. *Crop Sci.*, **4**, 66–69.

FINN-KELCEY, P., and HULBERT, D. G., 1957. The relationship between relative humidity and the moisture content of agricultural products – *Preliminary report. Electrical Research Association Tech. Rep. W/T33*, 23.

FINNEY, K. F., MEYER, J. W., SMITH, F. W., and FRYER, H. C., 1957. Effect of foliar spraying of wheat with urea solution on yield, protein content and protein quality. *Agron. J.*, **49**, 341–47.

FOX, R. L., and ALBRECHT, W. A., 1957. Soil fertility and the quality of seeds. *Res. Bull. Mo. agric. Exp. Stn, No. 619*, 23.

FUNK, C. R., ANDERSON, J. C., JOHNSON, M. W., and ATKINSON, R. W., 1962. Effect of seed source and age on field and laboratory performance of field corn. *Crop Sci.*, **2**, 318–20.

GRABE, D. F., 1956. Maturity in smooth bromegrass. *Agron J.*, **48**, 253–56.

GRANT LIPP, A. E., and BALLARD, L. A. T., 1963. Germination patterns shown by the light sensitive seeds of *Anagallis arvensis*. *Aust. J. biol. Sci.*, **16**, 572–84.

GREAVES, J. E., and CARTER, E. G., 1923. The influence of irrigation water on the composition of grains and the relationship to nutrition. *J. biol. Chem.*, **58**, 531–41.

GREGORY, F. G., and PURVIS, O. N., 1936. Vernalisation of winter rye during ripening. *Nature, Lond.*, **138**, 973.

HARRINGTON, J. F., 1960. Germination of seeds from carrot, lettuce and pepper plants grown under severe nutrient deficiencies. *Hilgardia*, **30**, 219–35.

HARRIS, J. A., 1912. A first study of the influence of starvation of the ascendants upon the characteristics of the descendants. *Am Nat.*, **46**, 313–43.

HATCHER, E. S., 1940. Studies on the inheritance of physiological characters V. Hybrid vigour in tomato Pt. III. A critical examination of the relation of embryo development to the manifestation of hybrid vigour. *Ann. Bot.*, **4**, 735–64.

HAVSTAD, J., 1964. Investigations on the generative phase in swede turnips (*Brassica napus*) with special regard to determine correct harvesting stage and the best method of drying the seed crop. *Meld. Norg. Landbrhogsk.*, **43**, No. 15.

HEWITT, E. J., BOLLE-JONES, E. W., and MILES, P., 1954. The production of copper, zinc and molybdenum deficiencies in crop plants with special reference to some effects of water supply and seed reserves. *Pl. Soil.*, **5**, 204–22.

HEWSTON, L. J., 1964. *Seed size studies on some vegetable species.* M.Sc. Thesis, University of Birmingham.

HIGHKIN, H. R., 1958. Temperature induced variability in peas. *Am. J. Bot.*, **45**, 626–32.

HILL-COTTINGHAM, D. G., and WILLIAMS, R. R., 1967. Effect of time of application of fertiliser on the growth, flower development and fruit set of maiden apple trees, var. Lord Lambourne, and on the distribution of total N within the trees. *J. hort. Sci.*, **42**, 319–38.

HOWELL, R. W., and CARTER, J. L., 1958. Physiological factors affecting composition of soybeans II. Responses of oil and other constituents of soybeans to temperature under controlled conditions. *Agron. J.*, **50**, 664–67.

HOWELL, R. W., COLLINS, F. I., and SEDGWICK, V. E., 1959. Respiration of soybean seeds as related to weathering losses during ripening. *Agron. J.*, **51**, 677–79.

INOUE, Y., and SUZUKI, Y., 1962. Studies on the effects of maturity and after ripening on seed germination in snap bean *Phaseolus vulgaris* L., *J. Japan Soc. hort. Sci.*, **31**, 146–50.

IWATA, M., and EGUCHI, Y., 1958. Effects of phosphorus and potassium supplied for various stages of growth on the yield and quality of seeds of Chinese cabbage. *J. hort. Assoc. Japan*, **27**, 171–78.

JACQUES, R., 1957. Quelques données sur le photoperiodisme de *Chenopodium polyspermum* L., influence sur la germination des graines. *Colloque Intern. de l'U.I.S. B. Parma, M.*, 125–30.

JENNINGS, A. C., and MORTON, R. K., 1963a. Changes in carbohydrate, protein and non-protein nitrogenous compounds of the developing wheat grain. *Aust. J. biol. Sci.*, **16**, 318–31.

JENNINGS, A. C., and MORTON, R. K., 1963b. Changes in nucleic acids and other phosphorus containing compounds in developing wheat grain. *Aust. J. biol. Sci.*, **16**, 332–41.

JOHANSEN, D. A., 1950. *Plant Embryology.* Chronica Botanica: Waltham, Mass., p. 305, 1950.

JONES, H. A., 1920. Physiological study of maple seeds. *Bot. Gaz.*, **69**, 127–52.

KERR, E. A., 1963. Germination of tomato seed as affected by fermentation time, variety, fruit maturity, plant maturity and harvest date. *Rep. hort. Exp. Stn Prod. Lab. Vineland, 1962*, 79–85.

KERSTING, J. F., STICKLER, F. C., and PAULI, A. W., 1961. Grain sorghum caryopsis development I. Changes in dry weight, moisture percentage and viability. *Agron J.*, **53**, 36–38. II. Changes in chemical composition. *Agron J.*, **53**, 74–77.

KINBACHER, E. J., 1962. Effect of seed source on the cold resistance of pre-emerged Dubois Winter Oat seedlings. *Crop Sci.*, **2**, 91–93.

KITTOCK, D. L., and PATTERSON, J. K., 1962. Seed size effects on performance of dryland grasses. *Agron J.*, **54**, 277–78.

KLEIN, S., and POLLOCK, B. M., 1968. Cell fine structure of developing lima bean seeds related to seed desiccation. *Am. J. Bot.*, **55**, 658–72.

KNEEBONE, W. R., and CREMER, C. L., 1955. The relationship of seed size to seedling vigour in some native grass species. *Agron. J.*, **47**, 472–77.

KOLEV, N., and GEORGIEV, D., 1964. The dynamics of movement and accumulation of nutritive matter in the inflorescence and seed of leeks after cutting the seed stalks. *Gradinarska i Lozarska Nauka.*, **1**, 70–75.

KOLLER, D., 1955. The regulation of germination in seeds (review). *Bull. Res. Counc. Israel, D.*, **5**, 85–108.

KOLLER, D., 1962. Pre-conditioning of germination in lettuce at the time of fruit ripening. *Am. J. Bot.*, **49**, 841–44.

KOSTJUCENKO, I. A., and ZABURAILO, T. J., 1937. Vernalisation of seed during ripening and its significance. *Herb. Rev.*, **5**, 146.

KRAMER, C., POST, J. J., and WILTEN, W., 1952. *Brouwgerst en Klimaat*. Utrecht, Kemink & Zoon.

KRIGEL, I., 1967. The early requirement for plant nutrients by subterranean clover seedlings (*Trifolium subterraneum*). *Aust. J. agric. Res.*, **18**, 879–86.

LAUDE, H., 1962. Fresh seed dormancy in annual grasses. *Calif. Agric.*, **16** (4), 3.

LEGGATT, C. W., 1948. Germination of boron deficient peas. *Sci. Agric.*, **28**, 131–39.

LEININGER, L. N., and URIE, A. L., 1964. Development of safflower seed from flowering to maturity. *Crop. Sci.*, **4**, 83–87.

LIPSETT, J., 1964. The phosphorus content of grain of different wheat varieties in relation to phosphorus deficiency. *Aust. J. agric. Res.*, **15**, 1–8.

LOEWENBURG, J. B., 1955. The development of bean seeds (*Phaseolus vulgaris* L.). *Pl. Physiol., Lancaster*, **30**, 244–49.

LONA, F., 1947. L'influenza delle condizioni ambientali, durante l'embriogenesi, sulle caratteristiche del seme e della pianta che ne deriva. *Lav. Ist. bot. Univ. Milano I*, 313–52.

LONGDEN, P. C., 1967. *The extent, source and significance of variation in seed weight in vegetable crops*. M.Sc. Thesis, University of Nottingham.

LUCKWILL, L. C., 1939. Observations on heterosis in Lycopersicum. *J. Genet.*, **37**, 421–40.

MAHESHWARI, P., 1950. *An introduction to the embryology of angiosperms*. McGraw-Hill, New York.

MATTHEWS, S., and BRADNOCK, W. T., 1967. The detection of seed samples of wrinkle-seeded peas (*Pisum sativum* L.) of low planting value. *Proc. int. Seed Test. Ass.*, **32**, 553–63.

MCALISTER, D. F., 1943. The effect of maturity on the viability and longevity of the seeds of Western range and pasture grasses. *J. Am. Soc. Agron.*, **35**, 442–53.

MEAGHER, W. R., JOHNSON, C. M., and STOUT, P. R., 1952. Molybdenum requirements of leguminous plants supplied with fixed nitrogen. *Pl. Physiol., Lancaster*, **27**, 223–30.

MOORE, R. P., 1943. Seedling emergence of small seeded legumes and grasses. *J. Am. Soc. Agron.*, **35**, 370–81.

MOORE, R. P., 1963. Previous history of seed lots and differential maintenance of viability and vigour in storage. *Proc. int. Seed Test. Ass.*, **28**, 691–99.

NAGATO, K., and EBATA, M., 1960. Effects of temperature in ripening periods upon development and qualities of lowland rice kernels. *Proc. Crop Sci. Soc., Japan*, **28**, 275–78.

NUTMAN, P. S., 1939. Studies in vernalisation of cereals. VI The anatomical and cytological evidence for the formation of growth promoting substances in the developing grain of rye. *Ann. Bot.*, **3**, 731–57.

OGAWA, T., 1961. Studies on seed production of onion. I. Effects of rainfall and humidity on fruit setting. *J. Japan Soc. hort. Sci.*, **30**, 222–32.

OZANNE, P. G., and ASHER, C. J., 1965. The effect of seedling potassium on emergence and root development of seedlings in potassium deficient sand. *Aust. J. agric. Res.*, **16**, 773–84.

PERRY, D. A., 1967. Seed vigour and field establishment of peas. *Proc. int. Seed Test. Ass.*, **32**, 3–12.

PETERSON, D. F., 1942. Duration of receptiveness of corn silks. *J. Am. Soc. Agron.*, **34**, 369–71.

POLLOCK, B. M., 1962. Temperature control of physiological dwarfing in peach seedlings. *Pl. Physiol., Lancaster*, **37**, 190–97.

QUINBY, J. R., REITZ, L. P., and LAUDE, H. H., 1962. Effect of source of seed on productivity of hard red winter wheat. *Crop Sci.*, **2**, 201–3.

REES, A. R., 1970. Effect of heat treatment for virus attenuation on tomato seed viability. *J. hort. Sci.*, **45**, 33–40.

REYNOLDS, J. D., 1955. Marsh spot of peas: a review of present knowledge. *J. Sci. Fd Agric.*, **6**, 725–34.

RIDDELL, J. A., and GRIES, G. A., 1958. Development of spring wheat. III. Temperature of maturation and age of seeds as factors influencing their response to vernalisation. *Agron. J.*, **50**, 743–46.

RIES, S. K., SCHWEIZER, C. J., and CHMIEL, H., 1968. The increase in protein content and yield of simazine-treated crops in Michigan and Costa Rica. *Bio Sci.*, **18**, 205–8.

RIGHTER, F. J., 1945. Pinus: The relationship of seed size and seedling size to inherent vigour. *J. For.*, **43**, 131–37.

ROBERTSON, R. N., HIGHKIN, H. R., SMYDZUK, J., and WENT, F. W., 1962. The effect of environmental conditions on the development of pea seeds. *Aust. J. biol. Sci.*, **15**, 1–15.

ROGLER, G. A., 1954. Seed size and seedling vigour in crested wheat grass. *Agron. J.*, **46**, 216–20.

ROSSMAN, E. C., 1949. Freezing injury of maize seed. *Pl. Physiol., Lancaster*, **24**, 629–56.

ROSSMAN, E. C., 1949. Freezing injury of inbred and hybrid maize seed. *Agron J.*, **41**, 574–83.

ROWAN, K. S., and TURNER, D. H., 1957. Physiology of pea fruits. V. Phosphate compounds in the developing seed. *Aust. J. biol. Sci.*, **10**, 414–25.

RUSH, G. E., and NEAL, N P., 1951. The effect of maturity and other factors on stands of corn at low temperatures. *Agron. J.*, **43**, 112–16.

RUSSELL, E. J., and VOELCKER, J. A., 1936. *Fifty years of field experiments at the Woburn Experiment Station*. London.

SALTER, P. J., and GOODE, J. E., 1967. *Crop responses to water at different stages of growth*. Commonwealth Agricultural Bureaux, Farnham Royal.

SCOTT, R. K., 1969. The effect of sowing and harvesting dates, plant population and fertilizers on seed yield and quality of direct drilled sugar-beet seed crops. *J. agric. Sci. Camb.*, **73**, 373–85.

SCHWABE, W. W., 1963. Studies in vernalisation of cereals. XV. After effects of temperature and drying during seed ripening and the origins of vernalisation requirements in Pektus winter Rye. *Ann. Bot.*, **27**, 671–83.

SHUTT, F. T., 1935. The nitrogen content of wheat as affected by seasonal condition. *Trans. R. Soc., Canada, Sect.*, 3, **29**, 37–39.

SKAZKIN, F. D., and KHVAN, A. V., 1962. Effects of rain during seed maturation on the quality and yield of grain. *Dokl. – Botan. Sci. Sect.*, (Eng. transl.), **140**, 169–71.

SKENE, K. G. M., and CARR, D. J., 1961. A quantitative study of the gibberellin content of seeds of *Phaseolus vulgaris* at different stages in their development. *Aust. J. biol. Sci.*, **14**, 13–25.

SNEDDON, J. L., 1963. Sugar beet seed production experiments. *J. natn. Inst. agric. Bot.*, **9**, 333–45.

SNYDER, F. W., 1959. Effect of nitrogen on yield and subsequent germinability of sugar beet seed. *J. am. Soc. Sug. Beet Technol.*, **10**, 438–43.

SPRAGUE, G. F., 1936. The relation of moisture content and time of harvest to germination of immature corn. *J. am. Soc. Agron.*, **28**, 472–78.

STEARNS, F., 1960. Effects of seed environment during maturation on seedling growth. *Ecology*, **41**, 221–22.

STODDART, J. L., 1964a. Seed ripening in grasses. I. Changes in carbohydrate content. *J. agric. Sci.*, **62**, 67–72.

STODDART, J. L., 1964b. Seed ripening in grasses. II. Changes in free amino acid content. *J. agric. Sci.*, **62**, 321–25.

SWANSON, C. O., 1946. Effects of rains on wheat during harvest. *Tech. Bull. Kans. agric. Exp. Stn.*, No. 60, 92 pp.

SZUKALSKI, H., 1961a. The influence of a high posphorus content of the seeds on the development and yield of plants. I. Investigations on Rape. *Rocz. Nauk. Roln.*, Ser. *A*, **84**, 463–92.

SZUKALSKI, H., 1961b. The influence of a high phosphorus content of the seeds on the development and yield of plants. II. Investigations on flax. *Rocz. Nauk. Roln.*, Ser. *A*, **84**, 789–810.

THOMPSON, R. C., 1936. Some factors associated with dormancy of lettuce seed. *Proc. Am. Soc. hort. Sci.*, **33**, 610–16.

TOLMAN, B., 1943. Sugar beet seed production in Southern Utah, with special reference to factors affecting yield and reproductive development. *Tech. Bull. U.S.D.A.*, No. 845.

VAN DOBBEN, W., 1947. De invloed von Klimaat op de gevolisherd von tarwe von schot. *J. versl. cent. Inst. landbouwk. Onderz.*, 40–43.

VON ABRAMS, G. J., and HAND, M. E., 1956. Seed dormancy in *Rosa* as a function of climate. *Am. J. Bot.*, **43**, 7–12.

WALKER, J., 1933. The suitability of immature sweet corn for seed. *Sci. Agric.*, **13**, 642–45.

WILLIAMS, R. R., 1965. The effect of summer nitrogen applications on the quality of apple blossom. *J. hort. Sci.*, **40**, 31–41.

WOODMAN, H. E., and ENGLEDOW, F. L., 1924. A chemical study of the development of the wheat grain. *J. agric. Sci.*, **14**, 562–86.

ZHDANOVA, L. P., 1969. Effect of temperature on the synthesis of fat in maturing seeds of oil bearing plants. *Fiziologiya Rast.*, **16**, 488–97.

CHAPTER 6

Effects of Environment after Sowing on Viability

Bruce M. Pollock

There have been frequent reviews on the subject of germination. Those reviews by Edwards (1932), Koller, Mayer, Poljakoff-Mayber and Klein (1962), Lang (1965), and Toole, Hendricks, Borthwick and Toole (1956) are especially useful as background reading to this chapter, which is written in an attempt to emphasise those aspects of the environment after sowing where affects on viability appear at present to be of the greatest importance.

To place the seed within the conceptual framework outlined in Chapter 12, we start at the moment of fusion of the egg and sperm nuclei, the time when maximum potential for the new plant is established. This potential is a blueprint in the form of information encoded in the DNA from the parent gametes. Not only the form and maximum productivity of the plant, but also the environmental boundaries within which plant productivity can be expressed are determined by this information.

During seed development the blueprint first establishes the form of the embryo and then specifies the metabolic machinery which provides the embryo with a supply of stored nutrients. During germination, the metabolic machinery and the nutrient reserve provide the tools and building materials necessary for construction of a photosynthetically autotrophic plant according to the DNA blueprint.

While the embryo is still dependent on the mother plant, the formation of its metabolic machinery and its nutrient reserves may be limited by environmental conditions resulting in an embryo less well-constructed than the potentiality its DNA blueprint specifies. The embryo is still largely functional, but its potential for development is less than the maximum possible.

After separation from the mother plant, other environmental factors, such as threshing and cleaning equipment (Chapter 4), saprophytic fungi (Chapter 3), and adverse storage conditions (Chapter 2), can act to modify still further the developmental potential of the seed. These environmental factors continue to act until the moment the seed enters the soil, e.g. planting machinery can cause mechanical damage.

In addition to its endogenous reserves, the seed needs to obtain only water and oxygen under suitable environmental conditions to germinate and establish the new plant. As the seed enters the soil, it may possibly have less than the full complement of genetic information (Chapter 9) but probably, even more important, it may have less than a full complement of undamaged machinery and construction materials to build to that genetic blueprint; thus its response to environmental conditions may be greatly modified from its original genetic potential. The degree of modification may differ widely among the individual seeds in a seed lot.

Primary environmental factors which influence germination are water, oxygen, temperature, light, soil structure and microorganisms. The soil structure and microflora are constantly in a state of flux as a result of the changes in temperature and in the oxygen and water supply. Temperature, water and oxygen supply change with time; these environmental conditions and changes can vary greatly from one locality in the soil to another. These changing and variable environmental conditions act on seeds whose modification from their original genetic potential differs according to the history of the individual seed. It is not surprising to find that the results of these complex interactions are sometimes unpredictable and the causes of seed response to environment difficult to identify.

Temperature

RADICLE EMERGENCE IN RELATION TO SEEDLING EMERGENCE FROM THE SOIL

The effect of temperature provides a good example of the experimental details which must be considered in relating environment to germination. The work of Edwards (1934) on soyabean (*Glycine max*) is especially useful in illustrating some of the basic temperature-germination relationships. He measured germination as radicle protrusion through the seed coat, and counted the number of seeds germinating at 2 hour intervals. In Fig. 6.1 his data are plotted as cumulative per cent germination against time. For each germination temperature, the data fall on a typical sigmoid curve. The curves are all similar but the point of origin changes with germination temperature.

Such curves are common in germination literature and frequently are interpreted incorrectly in terms of rate of germination. In Edwards's experiment, rate was measured by the slope of the sigmoid

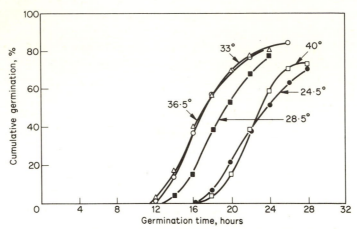

FIGURE 6.1   Germination of soyabean seeds at different temperatures (°C)
plotted as cumulative germination per cent against germination time. (Data from
Edwards, 1934.)

curves and was approximately the same for all temperatures used.
The major effect of temperature was not on germination rate but on
the time that germination began.★

In this experiment, each seed was counted as germinating at the
time that the radicle protruded through the seed coat. In obtaining
these data, Edwards terminated the germination counts when the rate
of germination became low. However, he recognised that true rate
curves (dg/dt) plotted from these data, are positively skewed by the
slow germinating seeds. This is true for most other germination data
(Nichols and Heydecker, 1968). Edwards considered that the slow
germinating seeds were low in vigour, and he discussed the possibility
of making genetic selections from a seed population based on rate of
germination; but there would be difficulties (see p. 224).

Germination is frequently measured after seedlings emerge from
the soil, or after they have grown sufficiently to permit observation
of seedling structure. However, it is difficult to speak in a meaningful
way of the effect of temperature on emergence. Emergence is the
summation of the effects of time and rate of germination plus growth

★The term 'rate' is used here in a special sense and is a measure of the uniformity
of the time taken for the individual seeds in the test to germinate – the steeper the
slope of the germination curve (dg/dt), the more uniform is the time taken for the
individual seeds to germinate. Other workers would use the term rate in a different
sense to describe the velocity of germination defined as the reciprocal of the time
taken for germination to take place. For example, equation (8.1) (Chapter 8) defines
the Coefficient of Velocity of Germination which is the reciprocal of the average time
for germination to occur. (Editor)

of the seedling from radicle protrusion until the time of observation; temperature may affect each of these phases of growth independently. These aspects of germination were studied by Edwards, Pearl and Gould (1934) who measured the growth of cantaloup (*Cucumis melo*) seedlings growing in the dark. Their data (Fig. 6.2) show that the

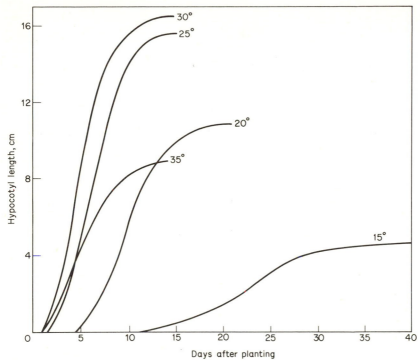

FIGURE 6.2   Height growth of *Cucumis melo* seedlings as a function of time at different temperatures (°C). (Data from Edwards *et al.*, 1934.)

hypocotyl growth curves are sigmoid. Replotting the data as growth rates (Fig. 6.3) clearly shows that both the maximum rate of growth and the time at which maximum growth occurs are controlled by temperature.

These data are replotted in Fig. 6.4 to show the effect of temperature on total height and maximum growth rate. As the temperature increases, each of these curves increases to a maximum and then decreases. This maximum occurs at the 'optimum' temperature. Extrapolation to lower or higher temperatures should provide a minimum and a maximum beyond which growth cannot occur. Minimum, optimum and maximum temperatures are referred to as the 'cardinal' temperatures.

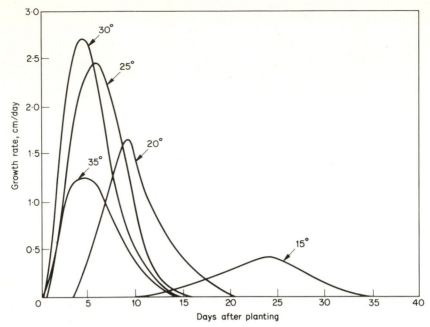

FIGURE 6.3   Data from Fig. 6.2 replotted as rate of growth against time.

However, cardinal temperatures cannot be considered to be absolute values. One of the reasons can be seen by replotting the data from Fig. 6.1 to show the temperature optimum (Fig. 6.5). It is clear that, while the temperature optimum remains in the range 33–36°C as germination proceeds, the minimum and maximum temperatures will depend on the time allowed for the germination test, since the

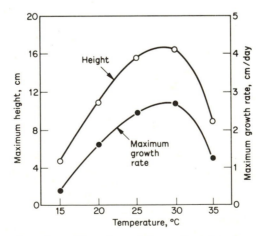

FIGURE 6.4   Data from Figs. 6.2 and 6.3 replotted to show temperature optimum.

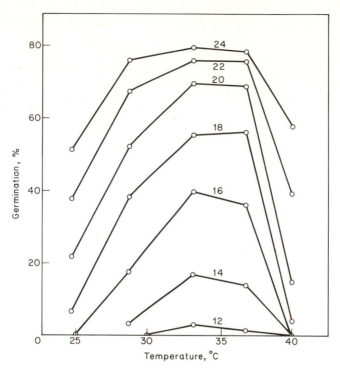

FIGURE 6.5   Data from Fig. 6.1 replotted as per cent germination against temperature for each of the times (in hours) at which data were recorded. The optimum temperature is 33–36°C; the curves fall more steeply at higher temperatures than at lower.

seeds at the sub-optimal and supra-optimal temperatures will not have had a chance to germinate if the period of the test is relatively short. Consequently, longer test periods will give the effect of extending the extreme ranges at which germination can take place.

In this case, the temperature optimum remained the same as germination proceeded. In other plants, optimum temperature may shift to either a higher or lower value as germination time increases. Furthermore, the temperature optimum is normally not the same for the different organs of the seedling. At the end of the experiment, shown in Figs. 6.2–6.4, the hypocotyls and roots were weighed; the data (Fig. 6.6) show a temperature optimum of 30°C for the hypocotyl. However, root weight was maximal at 20°C and decreased with temperature. Presumably, the root had a temperature optimum of 20°C or below.

TEMPERATURE-TIME RELATIONSHIPS

The concept of cardinal temperatures is based on experiments performed at constant temperatures. However, seeds germinating in the

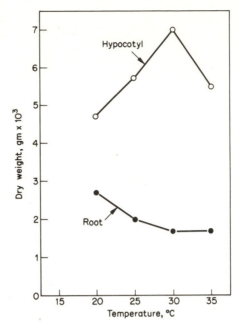

FIGURE 6.6   Dry weights of hypocotyls and roots of *Cucumis melo* seedlings at
the end of the sigmoid hypocotyl growth curve. (Data from Edwards *et al.*, 1934.)

field are exposed to diurnal temperature fluctuations. In the laboratory, such diurnal changes are known to be essential for certain species
and are routinely used for regulatory seed testing (International Seed
Testing Association, 1966). For example, Harrington (1923) found a
number of species which required a daily temperature alternation
(Table 6.1). For Johnson grass, he found a great difference in temperature response between different seed lots, with poorly germinating lots being much more sensitive to temperature. The response was
not restricted to radicle emergence because at high temperatures some
seeds produced normal seedlings, while others produced only slight
radicle or coleoptile elongation and normal growth ensued when the
temperature was reduced. The influence of alternating temperatures
on breaking seed dormancy is considered further in Chapter 11.

The rules for seed testing fail to specify the beginning temperature
in an alternating series of temperatures. Elkins, Hoveland and Donnelly (1966) found for *Vicia* species that the first temperature of the
alternation (the temperature to which seeds were exposed during
imbibition) was critical to both the rate of germination and the final
germination percentage. This is also an example of a delayed type of
temperature response, in contrast to the previously discussed direct

TABLE 6.1 *Percentage of seeds germinating*[a] *at either constant temperature or a daily temperature alternation of 16 hr at 20°C and 8 hr at 30°C. (From Harrington, 1923.)*

	Germination temperature		
	20°	30°	20–30°
Celery (*Apium graveolens*)	65	0	70
Kentucky bluegrass (*Poa pratensis*)	15	12	63
Bermuda grass (*Cynodon dactylon*)	0	0	30[b]
Johnson grass (*Sorghum halepense*)	8	32	58

[a]16 days for Johnson grass, 21 days for Bermuda grass and Kentucky bluegrass, 28 days for celery.
[b]79 per cent for 20–35 alternation, with 3 per cent at 35° constant.

temperature responses in which plant development is controlled during the temperature exposure. In another example, Knapp (1957) found that the germination temperature influenced subsequent plant development in *Senecio vulgaris, Agrostemma githago* and *Galinsoga parviflora* even though the plants were grown at identical temperatures following germination. Highkin and Lang (1966) found a similar effect for peas. Pollock (1962) found that the development of dwarfing symptoms in non-after-ripened peach (*Prunus persica*) was controlled by temperature during the first few days of germination. Within this period, germination at 22°C produced almost entirely normal seedlings; germination at 25°C resulted in severe dwarfing. The dwarfing effect was expressed several months after the critical temperature exposure.

Many seeds or seedlings are subject to chilling injury during germination. Injury symptoms are expressed as (a) failure of plant establishment, (b) slow growth, or (c) development of morphological abnormalities following a chilling period. Pollock and Toole (1966) found that the imbibition of germinating lima beans was especially temperature-sensitive. Exposure to a relatively low (15°C) temperature at the beginning of imbibition resulted in a large decrease in the number of seeds surviving and in reduced growth of the surviving seedlings. Similar results have been obtained for other species (see later discussion on water relations). These results emphasise the importance of the timing of environmental exposure during germination and plant development. Time, as an environmental factor, has been almost entirely overlooked in earlier research.

TEMPERATURE SENSITIVITY DIFFERS WITH SPECIES,
ORIGIN AND HISTORY OF THE SEEDS

In the past, most workers studying temperature relationships were
restricted by the number of temperature chambers available; hence,
the number of points which could be obtained for a temperature curve
were limited. To avoid this problem, several workers have used a
thermal gradient bar. One end of the metal bar is heated while the
other end is cooled; the heat differential produces a linear temperature
gradient between the two extremes. Moist germination paper is placed
on the bar and evaporation is minimised by a tight cover. Using this
type of equipment, Elliott and French (1959) showed the effect of
crossed gradients of light and temperature on germination of lettuce
(*Lactuca sativa*) seeds. Biggs and Langan (1962), working with peach,
and Larsen (1965) with alfalfa (*Medicago sativa*), found that the temperature optimum for early germinating seeds in a sample was relatively high, while the optimum for later germinating seeds was lower.
Wagner (1967) with *Plantago* spp., and Thompson (1970) with *Pri-
mula* spp., found different temperature optima for species within a
genus; these optima correlated with habitat and season of germination
of the species. Thompson (1970) found that after-ripening seeds of
Silene conoidea in dry storage broadened their tolerance to germination temperature, especially to higher temperatures.

Dormancy, or rest, may be defined as the condition in which a seed
will not germinate, even though all environmental conditions normally
necessary for germination are otherwise satisfactory. It is, therefore,
treated as a special modification of germination, as in Chapter 11 of
the present volume. However, dormancy is not necessarily an absolute
block to germination, but may represent only a narrowing of environmental tolerance (Vegis, 1964). As such, it becomes a major factor in
relating germination and plant establishment to environmental conditions after sowing. Probably the best known case of 'relative' dormancy is found in certain varieties of lettuce which germinate well at
20°C or below, but which are dormant at higher temperatures (thermal dormancy) and require light for germination (Toole *et al.*, 1956).
Sensitivity to temperature is determined by environmental conditions
during seed development. Thompson (1936) found that seeds which
were immature when harvested tended to be sensitive to thermal
dormancy, while Koller (1962) found that seeds which matured under
high temperature conditions were more tolerant of high temperatures.

In addition to relatively complex interactions between seeds and
the germination temperature, soil temperature may vary from loca-

tion to location even over relatively short distances. For example, Shadbolt, McCoy and Little (1961) found that the soil temperature at 2.5 cm depth in lettuce beds (southern California, USA) was 3–4°C higher on the side with a southerly or easterly exposure to the sun. Some control over soil temperature is available for agricultural purposes. Where cold soil is a problem, soil temperature can sometimes be increased significantly by treatment with asphalt or plastic mulches (Takatori, Lippert and Whiting, 1964). Where high soil temperature is a problem, irrigation may reduce soil temperature as much as 10°C (Robinson and Worker, 1966).

Water and oxygen

The requirements for water and oxygen are much more difficult to study than are temperature relationships. This is, at least in part, the result of the complex interrelationships between the two. A supply of water is required to hydrate the dry seed and to maintain hydration as growth begins. Oxygen is required for respiratory processes. However, because of the low solubility of oxygen in water the two compounds compete for physical space in the environment of the seed. Thus, an excess of water limits oxygen availability.

WATER

When a dry seed is placed in a moist environment, it absorbs water in three stages: (1) an initial period of rapid uptake; (2) a lag period in which little water is absorbed; and (3) a second uptake stage which is associated with embryo growth. Most considerations of water uptake in seeds have been restricted to the initial stage (see review by Brown, 1965). This uptake is considered to be the result of physical absorption of water by colloidal materials in seeds. This uptake occurs in both living and dead seeds.

Recognition of the physical nature of the initial water uptake has long been accompanied by the tacit assumption that little of biological interest occurs during this period. As a result, many studies of seed germination start with soaked or partially imbibed seeds, and little information is provided on the conditions under which this initial water was absorbed. However, Pollock and Toole (1966) found that for lima beans (*Phaseolus lunatus*), the very first stage of imbibition was temperature-sensitive. Initial imbibition at a moderately low temperature (5–15°C) caused injury which was expressed in subsequent growth. Later, Pollock (1969) found that the critical factor was seed moisture at the beginning of imbibition of liquid water.

Imbibition temperature sensitivity occurred only if the initial seed moisture was below 12–14 per cent. This same result has been obtained by Pollock, Roos and Manalo (1969) for garden beans (*Phaseolus vulgaris*), Christiansen (1969) for cotton (*Gossypium* spp.), Phillips and Youngman (1970) for sorghum (*Sorghum bicolor*), and Obendorf and Hobbs (1970) for soya beans.

There are a number of other seed phenomena which show a critical relationship to water content at low moisture levels. In evaluating seed moisture at these levels, however, it is essential to recognise that the moisture percentage of different seeds cannot be directly compared. Differences in relative quantities of water-absorbing compounds, e.g., carbohydrates, proteins, nucleic acids, which have different affinities for water can change the total water content of the seed without changing the percentage of water absorbed to any one class of compound (hence the different hygroscopic equilibrium relationships shown in Appendix 4).

One of the most striking cases of the effect of slight changes in water content was reported by Conger, Nilan and Konzak (1968) for barley seeds (*Hordeum vulgare*) which had been injured by gamma rays. They found that, as seed moisture was increased from 10.7 to 11 per cent, the radiation-induced injury of seeds hydrated with oxygenated water decreased from 65 to 25 per cent. They attributed this result to free water which permitted mobility and consequent reactivity of radiation-induced free radicals; water in excess of that bound to cell constituents occurred only above the 10.7 per cent level. Minett, Belcher and O'Brien (1968) found a critical level of 11.6–11.8 per cent water in wheat (*Triticum aestivum*) for the breakdown of the insecticide, malathion. Linko and Milner (1959) recognised that several enzymes of amino-acid metabolism were activated in wheat at moisture levels of 15–18 per cent. In the food processing industry, there is a clear recognition (Acker, 1963) of the importance of enzyme reactions in seed products at low moisture levels. Thus, we can conclude that chemical reactions and environmental sensitivity are important in seeds from the very beginning of water uptake.

Most water is absorbed by seeds from the liquid form, and research has shown that an excess of water can be harmful, probably because this excess interferes with oxygen supply (see following section). In the soil, however, the supply of water is frequently sub-optimal and may restrict germination. Collis-George and Sands (1962) considered that the total potential of water included two components of importance in germination: (1) matric (suction or capillary) potential, and (2) osmotic (solute) potential. They found that germination could be retarded as the matric potential decreased from that of free water. However, Sedley (1963) and Collis-George and Hector (1966) found that if an area of the seed coat was wet before the matric potential was lowered, the effect of matric potential was greatly reduced, i.e., the contact between the seed surface and the soil water was of greater importance. Working with the large seeds of pea (*Pisum sativum*), Manohar and Heydecker (1964) showed that both the contact area and the anatomical localisation of the contact area were important. Later, Collis-George and Williams (1968) state that the matric potential limits germination not by its affect on free energy of water but by its influence on the isotropic effective stress in the solid framework of the soil.

Many workers have studied the effect of osmotic stress on germination, assuming that they could use osmotic stress to simulate drought conditions. However, Collis-George and Sands (1962) concluded that osmotic and matric potentials were not equivalent, and Manohar and Heydecker (1964) showed that permeability of seeds to osmotic agents is sufficiently complex that the interpretation of results of osmotic treatments becomes difficult to relate to soil-moisture stress.

Most methods of supplying water in laboratory germination experiments involve the use of free water. These methods are, therefore, difficult to interpret relative to the water supply available to the seed in the soil. Furthermore, under such laboratory conditions it is very difficult to avoid an excess of water, and certain physiological conditions of seeds (probably related to dormancy) involve a sensitivity to water. When germination containers dry out and are rewatered during an experiment, the seeds are subjected to a violent alteration of water stress followed by water excess. Negbi, Rushkin and Koller (1966) developed methods to study the interactions between water supply and other parameters of germination. One method involved a 'ladder' of filter paper, with seeds planted at different heights above a free-water surface. The other method utilised a constant amount of water but employed different numbers of sheets of filter paper per Petri dish. Applying these techniques to the germination of seeds of *Hirschfeldia incana*, they found that the timing of maximum sensitivity to light was independent of seed hydration, but that the degree of water sensitivity was determined by hydration or temperature during the initial hours of imbibition.

The problem of water supply to large seeds is even more complex than to small seeds because of the increased difficulty of obtaining uniform seed-substrate contact. To minimise this problem, Pollock and Manalo (1969) developed a method which involves the use of sand of different particle sizes. Seeds were placed in the sand, the sand and seeds were quickly immersed in water, and the excess water was removed from the container by suction applied from below. The amount of water remaining (and, hence, the oxygen supply) was a function of the size of the sand particles used and the amount of suction applied. Using this method, Pollock, Roos and Manalo (1969) found that seed lots of garden beans vary in their response to stress conditions. These stress conditions included the aeration and water supply in the substrate and the temperature during imbibition.

OXYGEN

The competitive relationship between water and air is the result of the low solubility of oxygen in water and the great difference in diffusion coefficient of oxygen in water and in air. The respiratory processes of the seed can be controlled by the rate at which oxygen reaches the mitochondria of the respiring cells. The combined effects of solubility and diffusibility reduce the rate of diffusion from 0.205 ml/cm^2 × sec. in air to 6.7×10^{-7} ml/cm^2 × sec. in water (Goddard and Bonner, 1960). The availability of oxygen to the respiratory enzymes is thus greatly reduced by any water between the air spaces of the soil and the seed, in addition to the limitations imposed by the seed coat and the bulk of the seed itself. For example, Ohmura and Howell (1960) found that the oxygen uptake of germinating soyabean cotyledons was reduced from 900 to 600 μl/hr/g by immersion in water, and for maize (*Zea mays*) scutella from 1,900 to 680.

Three different methods have been used to study the effects of oxygen and other gases on germination. One of the earliest was to soak seeds in water through which air, oxygen or other gases were bubbled. A number of interesting observations were made, including the fact that high concentrations of oxygen may be toxic. However, these data are difficult to interpret relative to the germination of seeds in soil because control seeds on filter paper, in atmospheres containing the same gases, failed to show the soaking response (Barton, 1950).

Another technique has been to germinate seeds on moist filter paper in a container in which the composition of the gas phase is controlled. A modification of this technique has been used with seeds planted in soil through which gas mixtures were passed. Using this technique, Grable and Danielson (1965) found that, independent of the aeration treatment used, germination growth of maize increased as soil-moisture suction decreased until the soil was saturated. At that point, growth stopped, probably because of reduced oxygen-diffusion rates. At lower soil-moisture levels, reduction in oxygen concentration from 20 to 7.5 per cent reduced root length by 20–30 per cent. Soya bean germination was adversely affected by soil moisture below that of full saturation, primarily because of invasion by micro-organisms.

A more recent technique has been to study the oxygen diffusion rate (ODR) to a platinum microelectrode which serves as an oxygen 'sink'. The ODR is then correlated with plant response in the same soil (see review by Stolzy and Letey, 1964). Using this technique, Wengel (1966) correlated percentage emergence of maize with ODR over the range 8 to 40×10^{-8} g cm^{-2} min.$^{-1}$. Perhaps one limitation to this technique for germination studies is that the platinum micro-electrodes more accurately simulate the size and geometry of a root than a bulky seed.

Sensitivity to oxygen availability changes with stage of germination. Ikuma and Thimann (1964) found that oxygen is not required during imbibition of lettuce seeds ('Preinductive phase') but is required for radicle emergence. Unger and Danielson (1965) found that radicle emergence of maize occurred over a wide range of oxygen concentrations, with restrictions only at zero oxygen where 85 per cent of the seeds germinated, and 130 cm Hg O_2 pressure where 81 per cent germinated. However, further root growth was sharply reduced by oxygen pressures below those of air.

The seed coat iself constitutes a barrier to oxygen diffusion in many seeds. It may, in fact, constitute a boundary between the seed and the external environment which almost universally tends to maintain the internal seed environment at oxygen and carbon dioxide levels different from those in the surrounding environment. Earlier work on plant metabolism was based on the assumption that enzyme systems in germinating seeds were fixed and that the seed response to altered environmental conditions could occur only from changes in equilibria among existing enzymes. It is becoming clear that changes in the environment of the seed, particularly the availability of oxygen, can serve to induce the production of new enzymes such as alcohol dehydrogenase (Kolloffel, 1968) and lactic dehydrogenase (Sherwin and Simon, 1969), and that rupture of the seedcoat by the radicle can alter the inducing conditions. Expansion of these observations offers the potential for explaining the complex responses of the germinating seed to its environment.

Light

Exposure to light stimulates the germination of many seeds, especially seeds of wild species and of agricultural plants such as lettuce, tobacco (*Nicotiana tabacum*) and tomato (*Lycopersicon esculentum*) which have not been subject to selection for seed characteristics. In most cases, germination is stimulated by red light (660 nm) and inhibited by far-red (730 nm) in a reaction mediated by phytochrome (Toole *et al.*, 1956). The degree of light sensitivity varies greatly between seed lots and is strongly related to the germination temperature. Frequently, this light sensitivity appears only in seeds germinated at higher temperatures. This condition can be considered to be a case of 'relative' dormancy because seeds in this condition will germinate only under a restricted range of temperature conditions.

Some of the aspects of light sensitivity can be illustrated by the work of Kasperbauer (1968a and b) with seeds of tobacco. He found that certain seed lots required light for germination at 20°, 25° and 30°C. Other lots germinated in the dark at low temperatures but required light at the higher temperatures; the ability to germinate in the dark tended to increase with seed age. Seeds of a light-requiring lot were planted in moist soil at different depths. Those on the soil surface germinated well, but the few seeds which germinated at a depth of 3 mm appeared to be associated with localised disturbances in the soil surface, and no seeds germinated from depths greater than 3 mm. However, after 30 days the soil was cultivated and large numbers of seeds germinated immediately. Light-indifferent seeds planted in the same way emerged from depths as great as 25 mm. The inheritance of the light requirement was studied by crossing selections which were either light-requiring or light-indifferent; both genetic and maternal influences were found to be operative.

The ecological significance of light sensitivity for native plant species is obvious. Seeds brought to the surface of disturbed soil are stimulated to germinate in a location where they have an optimal chance for establishment before competitive pressures from other plants become too great. Physiologists have studied mechanisms of light sensitivity in great detail. Unfortunately, the significance of many of these complex interrelationships in plant establishment is not clear at the present time. However, several recent observations may be of significance in explaining the ubiquitous nature of light sensitivity. The first of these is the observation that seeds which do not normally require light, such as tomato (Mancinelli, Yaniv and Smith, 1967) and cucumber (*Cucumis sativus*) (Mancinelli and Tolkowsky, 1968) can be made light-requiring by exposure to far-red illumination. Once germination is inhibited by far-red, the seeds can again be stimulated to germinate by exposure to red light, just as with seeds which normally require light for germination. Another observation is that some seeds which normally require low-temperature after-ripening to break dormancy can be stimulated to germinate by light. For example, Black and Wareing (1955, 1959) found that un-chilled seeds of *Betula pubescens* are sensitive to light, but the light requirement is removed by chilling or by breaking the seed coat, factors presumably increasing the oxygen supply to the embryo. Taylorson and Hendricks (1969) found that exposure of *Amaranthus retroflexus* seeds to red or far-red light during chilling influenced subsequent germination. This was apparently brought about through effects on the phytochrome system. Wesson and Wareing (1969), studying the induction of light-sensitivity in native species, suggested that a gaseous inhibitor from the seeds accumulated in the soil atmosphere and was responsible for the inhibition of germination. Further ecological aspects of light-sensitivity are discussed in Chapter 11 (p. 333) and physiological aspects in Chapter 12 (p. 372).

Soil

The importance of soil as an environmental factor in germination and plant establishment has long been recognised in an empirical way. In recent years, because of the importance of precision planting and stand establishment in agriculture, a number of workers have attempted to relate quantitatively plant establishment to soil structure. For example, Heydecker (1961, 1962) developed a test in which the emergence of vegetable seedlings was used to evaluate soils and their interactions with such other environmental factors as water. He found that field capacity could be obtained either by supplying water by capillarity from below or by soaking and then draining a soil sample. The soaking method, however, was very detrimental to seedling emergence in certain soils because it tended to move unstable soil particles to fill pores in the soil which were otherwise accessible to air. Wetting and then drying the surface of certain soils formed crusts which were very detrimental to seedling emergence. Different species responded differently to different soil types, and the response of any one species varied between seed lots.

Methods have been developed to measure quantitatively the energy required for emergence of a mechanical 'seedling' (Morton and Buchele, 1960). They found that the energy required for emergence varied both with the initial compaction pressure and with initial soil moisture and amount of surface drying. Parker and Taylor (1965) found that emergence of sorghum seedlings at a specified soil strength was affected by soil moisture tension and by planting depth. Temperature affected the rate of emergence but not the relationship between soil strength and final emergence. At a given soil strength, emergence of the dicotyledonous seedlings of guar (*Cyamopsis tetragonoloba*) and of cotton was much more drastically reduced than was the emergence of the monocot, sorghum. Taylor, Parker and Roberson (1966) found that oxygen concentration in the range from 7 to 42 per cent did not alter the relationship between soil-strength and emergence in sorghum. For most of the seedlings tested in the Gramineae, emergence decreased slightly as soil strength increased in the range 6–9 bars and was prevented in the range of 12–18 bars. Onion (*Allium cepa*), another monocot, was prevented from emerging by a soil strength of 2 bars.

However, increased soil compaction is not always detrimental to seedling growth. Working under dryland conditions of limited moisture supply, Dasberg, Hillel and Arnon (1966) found that preplanting compaction reduced seedling emergence of sorghum. By contrast, postplanting compaction significantly increased seedling survival and yield, presumably by minimising loss of available moisture.

Microorganisms

Perhaps the most complex of the environmental factors is the soil microflora. The importance of the soil microflora is well illustrated by the germination of maize. Regulatory germination test results (International Seed Testing Association, 1966) provide very little information from which field emergence can be predicted. The problem is the high mortality which results from the attack of soil microorganisms, especially in cold, wet soils (Harper, Landragin and Ludwig, 1955). In an attempt to evaluate probable seedling emergence under these conditions, a 'cold' test is commonly used (Tatum and Zuber, 1943). In this test, seeds are mixed with moist soil from a maize field which presumably will have a high pathogenic potential. This mixture is held for several days at a temperature of 5–10°C and then transferred to normal germination temperature. Although the test is very effective in predicting the relative emergence of different seed lots, it is very difficult to standardise because of the dynamic nature of the innoculum in the soil. The cold test is considered in further detail in Chapter 8 (p. 235).

The role of microorganisms in germination phenomena is frequently overlooked. For example in barley a phenomenon frequently associated with dormancy is 'water-sensitivity', in which the seeds fail to germinate in the presence of a slight excess of water in the germination environment. Gaber and Roberts (1969) found that water-sensitivity could be overcome by including antibiotics in the germination medium. A single antibiotic was not effective. A combination of an antibacterial with an antifungal agent was essential, thus suggesting that the microorganism population, rather than a specific microorganism, is involved.

Although the soil microflora must be considered as a dynamic, ever-changing part of the seed's environment, the seed cannot be considered as a static object waiting to be attacked. Seeds lose large amounts of such organic nutrients as carbohydrates, amino acids and coenzymes to the medium in which they germinate. These nutrients alter the activity of the soil microflora and may determine the population density of selected organisms in the vicinity of the seed. This may influence pathogenicity of such organisms as *Pythium* which causes pre-emergence damping-off (see review by Schroth and Hildebrand, 1964). The amount of leaching may be determined both by the quality of the seed and the environmental conditions to which it is exposed during germination. For example, Pollock and Toole (1966) and Pollock (1969) found that the amount of material lost by lima bean axes was determined by seed quality, imbibition temperature, and seed moisture at the beginning of imbibition. However, most research on this subject has emphasised either the dynamics of microbial populations or the effect of seed quality on leaching of nutrients. The complex interrelationships between the seed, the soil microflora, and the soil environment have not been studied in detail.

Summary

The literature contains a multitude of observations on the effects of temperature, water, oxygen, light, soil structure and microorganisms on germination and establishment of many plant species. These data show clearly that the actions of these environmental factors interact in complex ways. What we lack is the knowledge necessary to understand and to control these interactions. That knowledge will undoubtedly come from research on the biochemical mechanisms which sense environmental changes and control plant response. Already the broad outlines of such knowledge are becoming available through an understanding of the mechanisms of gene action, allosteric effects on protein structure and function, the synthesis and degradation of specific enzymes, and the compartmentalisation of functions within cells.

Technological advances in agriculture are centred on the increasing use of machinery to replace expensive hand labour. Efficient use of machinery requires a uniformity of plant development which does not exist in the native flora. Technologists recognise that uniformity of plant development is established at the time of seed germination, or before (see Chapter 10). For this reason, a great deal of applied research is presently directed towards the development of planting 'systems' and chemical treatments to modify germination response. Thus, the technological basis exists for immediate application of new basic knowledge of the mechanisms which integrate and control the response of the germinating seed to its environment.

References to Chapter 6

ACKER, L., 1963. Enzyme activity at low water contents. In *Recent Adv. Food Sci.*, eds J. M. Leitch and D. N. Rhodes, **3**, 239–47. Butterworths, London.

BARTON, L. V., 1950. Relation of different gases to the soaking injury of seeds. *Contrib. Boyce Thompson Inst.*, **16**, 55–71.

BIGGS, R. H., and LANGAN, M. C., 1962. Effect of temperature on germination of Okinawa peach seeds. *Fla. State Hort. Soc. Proc.*, **75**, 379–81.

BLACK, M., and WAREING, P. F., 1955. Growth studies in woody species. VII. Photoperiodic control of germination in *Betula pubescens* Ehrh. *Physiol. Plant.*, **8**, 300–16.

BLACK, M., and WAREING, P. F., 1959. The role of germination inhibitors and oxygen in the dormancy of the light-sensitive seed of *Betula spp.* *J. Exp. Bot.*, **10**, 134–45.

BROWN, R., 1965. Physiology of seed germination. In *Encyclopedia of Plant Physiology*, ed. W. Ruhland, **15** (2), 894–908. Springer-Verlag, Berlin.

CHRISTIANSEN, M. N., 1969. Seed moisture content and chilling injury to imbibing cottonseed. *1969 Beltwide Cotton Production Research Conferences*, 50–51.

COLLIS-GEORGE, N., and HECTOR, J. B., 1966. Germination of seeds as influenced by matric potential and by area of contact between seed and soil water. *Australian J. Soil Res.*, **4**, 145–64.

COLLIS-GEORGE, N., and SANDS, J. E., 1962. Comparison of the effects of the physical and chemical components of soil water energy on seed germination. *Australian J. Agr. Res.*, **13**, 575–84.

COLLIS-GEORGE, N., and WILLIAMS, J., 1968. Comparison of the effects of soil matric potential and isotropic effective stress on the germination of *Lactuca sativa*. *Australian J. Soil Res.*, **6**, 179–92.

CONGER, B. V., NILAN, R. A., and KONZAK, C. F., 1968. Post-irradiation oxygen sensitivity of barley seeds varying slightly in water content. *Radiat. Bot.*, **8**, 31–36.

DASBERG, S., HILLEL, D., and ARNON, I., 1966. Response of grain sorghum to seedbed compaction. *Agron. J.*, **58**, 199–201.

EDWARDS, T. I., 1932. Temperature relations of seed germination. *Quart. Rev. Biol.*, 7, 428–43.

EDWARDS, T. I., 1934. Relations of germinating soybeans to temperature and length of incubation time. *Pl. Physiol., Lancaster*, 9, 1–30.

EDWARDS, T. I., PEARL, R., and GOULD, S. A., 1934. Influence of temperature and nutrition on the growth and duration of life of *Cucumis melo* seedlings. *Bot. Gaz.*, 96, 118–35.

ELKINS, D. M., HOVELAND, C. S., and DONNELLY, E. D., 1966. Germination of *Vicia* species and interspecific lines as affected by temperature cycles. *Crop Sci.*, 6, 45–48.

ELLIOT, R. F., and FRENCH, C. S., 1959. Germination of light sensitive seed in crossed gradients of temperature and light. *Pl. Physiol., Lancaster*, 34, 454–56.

GABER, S. D., and ROBERTS, E. H., 1969. Water-sensitivity in barley seeds. II. Association with microorganism activity. *J. Inst. Brew.*, 75, 303–14.

GODDARD, D. R., and BONNER, W. D., 1960. Cellular respiration. In *Plant Physiology*, ed., F. C. Steward, 1A, 209. Academic Press, New York.

GRABLE, A. R., and DANIELSON, R. E., 1965. Effect of carbon dioxide, oxygen, and soil moisture suction on germination of corn and soybeans. *Soil Sci. Soc. Amer. Proc.*, 29, 12–18.

HARPER, J. L., LANDRAGIN, P. A., and LUDWIG, J. W., 1955. The influence of environment on seed and seedling mortality. I. The influence of time of planting on the germination of maize. *New Phytol.*, 54, 107–18.

HARRINGTON, G. T., 1923. The use of alternating temperatures in the germination of seeds. *J. Agr. Res.*, 23, 295–332.

HEYDECKER, W., 1961. The emergence of vegetable seedlings as a standard test of soil quality. *Adv. Hort. Sci.*, 1, 381–92.

HEYDECKER, W., 1962. From seed to seedling: factors affecting the establishment of vegetable crops. *Ann. appl. Biol.*, 50, 622–27.

HIGHKIN, H. R., and LANG, A., 1966. Residual effect of germination temperature on the growth of peas. *Planta*, 68, 94–98.

IKUMA, H., and THIMANN, K. V., 1964. Analysis of germination processes of lettuce seed by means of temperature and anaerobiosis. *Plant Physiol.*, 39, 756–67.

International Seed Testing Association, 1966. International rules for seed testing. *Proc. int. Seed Test. Ass.*, 31, 1–152.

KASPERBAUER, M. J., 1968a. Germination of tobacco seed. I. Inconsistency of light sensitivity. *Tobacco*, 166, 23–26.

KASPERBAUER, M. J., 1968b. Dark-germination of reciprocal hybrid seed from light-requiring and -indifferent *Nicotiana tabacum*. *Physiol. Plant.*, 21, 1308–11.

KNAPP, R., 1957. Über den Einfluss der Temperature während der Keimung auf die spätere Entwicklung einiger annueller Pflanzenarten. *Z. Naturforsch.*, 12b, 564–68.

KOLLOFFEL, C., 1968. Activity of alcohol dehydrogenase in the cotyledons of peas germinated under different environmental conditions. *Acta Bot. Neer.*, 17, 70–77.

KOLLER, D., 1962. Pre-conditioning of germination of lettuce at time of fruit ripening. *Amer. J. Bot.*, 49, 841–44.

KOLLER, D., MAYER, A. M., POLJAKOFF-MAYBER, A., and KLEIN, S., 1962. Seed germination. *Ann. Rev. Pl. Physiol.*, 13, 437–64.

LANG, A., 1965. Effects of some internal and external conditions on seed germination. In *Encyclopedia of Plant Physiology*, ed., W. Ruhland, **15** (2), 850–993. Springer-Verlag, Berlin.

LARSEN, A. L., 1965. The use of thermogradient plate for studying temperature effects on seed production. *Proc. int. Seed Test. Ass.*, **30**, 861–68.

LINKO, P., and MILNER, M., 1959. Enzyme activation in wheat grains in relation to water content. Glutamic acid-alanine transaminase, and glutamic acid decarboxylase. *Pl. Physiol., Lancaster*, **34**, 392–96.

MANCINELLI, A. L., YANIV, Z., and SMITH, P., 1967. Phytochrome and seed germination. I. Temperature dependence and relative PFR levels in the germination of dark-germinating tomato seeds. *Pl. Physiol., Lancaster*, **42**, 333–37.

MANCINELLI, A. L., and TOLKOWSKY, A., 1968. Phytochrome and seed germination. V. Changes of phytochrome content during the germination of cucumber seeds. *Pl. Physiol., Lancaster*, **43**, 489–94.

MANOHAR, M. S., and HEYDECKER, W., 1964. Effects of water potential on germination of pea seeds. *Nature, Lond.*, **202**, 22–24.

MINETT, W., BELCHER, R. S., and O'BRIEN, E. J., 1968. A critical moisture level for malathione breakdown in stored wheat. *J. Stored Prod. Res.*, **4**, 179–81.

MORTON, C. T., and BUCHELE, W. F., 1960. Emergence energy of plant seedlings. *Agr. Engng*, **41**, 428–31, and 453–55.

NEGBI, M., RUSHKIN, E., and KOLLER, D., 1966. Dynamic aspects of water-relations in germination of *Hirschfeldia incana* seeds. *Plant Cell Physiol.*, **7**, 363–76.

NICHOLS, M. A., and HEYDECKER, W., 1968. Two approaches to the study of germination data. *Proc. int. Seed Test. Ass.*, **33**, 531–40.

OBENDORF, R. L., and HOBBS, P. R., 1970. Effect of seed moisture on temperature sensitivity during imbibition of soybean. *Crop Sci.*, **10**, 563–66.

OHMURA, T., and HOWELL, R. W., 1960. Inhibitory effect of water on oxygen consumption by plant materials. *Pl. Physiol., Lancaster*, **35**, 184–88.

PARKER, J. J., Jr., and TAYLOR, H. M., 1965. Soil strength and seedling emergence relations. I. Soil type, moisture tension, temperature, and planting depth effects. *Agron. J.*, **57**, 289–91.

PHILLIPS, J. C., and YOUNGMAN, V. E., 1970. Effect of initial seed moisture control on emergence and yield of grain sorghum. Submitted to *Crop Sci.*

POLLOCK, B. M., 1962. Temperature control of physiological dwarfing in peach seedlings. *Pl. Physiol., Lancaster*, **37**, 190–97.

POLLOCK, B. M., 1969. Imbibition temperature sensitivity of lima bean seeds controlled by initial seed moisture. *Pl. Physiol., Lancaster*, **44**, 907–11.

POLLOCK, B. M., and TOOLE, V. K., 1966. Imbibition period as the critical temperature sensitive stage in germination of lima bean seeds. *Pl. Physiol., Lancaster*, **41**, 221–29.

POLLOCK, B. M., and MANALO, J. R., 1969. Controlling substrate moisture-oxygen levels during the imbibition stage of germination. *J. Amer. Soc. Hort. Sci.*, **94**, 574–76.

POLLOCK, B. M., ROOS, E. E., and MANALO, J. R., 1969. Vigor of garden bean seeds and seedlings influenced by initial seed moisture, substrate oxygen, and imbibition temperature. *J. Amer. Soc. Hort. Sci.*, **94**, 577–84.

ROBINSON, F. E., and WORKER, G. F., Jr., 1966. Factors affecting the emergence of sugar beets in an irrigated desert environment. *Agron. J.*, **58**, 433–35.

SCHROTH, M. N., and HILDEBRAND, D. C., 1964. Influence of plant exudates on root-infecting fungi. *Ann. Rev. Phytopath.*, **2**, 101–32.

SEDGLEY, R. H., 1963. The importance of liquid-seed contact during the germination of *Medicago tribuloides* Desr. *Australian J. Agr. Res.*, **14**, 646–53.

SHADBOLT, C. A., MCCOY, O. D., and LITTLE, T. M., 1961. Soil temperatures as influenced by bed direction. *Proc. Amer. Soc. Hort. Sci.*, **78**, 488–95.

SHERWIN, T., and SIMON, E. W., 1969. The appearance of lactic acid in *Phaseolus* seeds germinating under wet conditions. *J. Exp. Bot.*, **20**, 776–85.

STOLZY, L. H., and LEYTEY, J., 1964. Correlation of plant response to soil oxygen diffusion rates. *Hilgardia*, **35**, 567–76.

TAKATORI, F. H., LIPPERT, L. F., and WHITING, F. L., 1964. The effect of petroleum mulch and polyethylene films on soil temperature and plant growth. *Proc. Amer. Soc. Hort. Sci.*, **85**, 532–40.

TATUM, L. A., and ZUBER, M. S., 1943. Germination of maize under adverse conditions. *J. Amer. Soc. Agron.*, **35**, 48–59.

TAYLOR, H. M., PARKER, J. J., Jr., and ROBERSON, G. M., 1966. Soil strength and seedling emergence relations. II. A generalized relation for Gramineae. *Agron. J.*, **58**, 393–95.

TAYLORSON, R. B., and HENDRICKS, S. B., 1969. Action of phytochrome during prechilling of *Amaranthus retroflexus* L. seeds. *Pl. Physiol., Lancaster*, **44**, 821–25.

THOMPSON, P. A., 1970. Characterization of the germination response to temperature of species and ecotypes. *Nature, Lond.*, **225**, 827–31.

THOMPSON, R. C., 1936. Some factors associated with dormancy of lettuce seed. *Proc. Amer. Soc. Hort. Sci.*, **33**, 610–16.

TOOLE, E. H., HENDRICKS, S. B., BORTHWICK, H. A., and TOOLE, V. K., 1956. Physiology of seed germination. *Ann. Rev. Pl. Physiol.*, **7**, 299–324.

UNGER, P. W., and DANIELSON, R. E., 1965. Influence of oxygen and carbon dioxide on germination and seedling development of corn. *Agron. J.*, **57**, 56–58.

VEGIS, A., 1964. Dormancy in higher plants. *Ann. Rev. Pl. Physiol.*, **15**, 185–224.

WAGNER, R. H., 1967. Application of a thermal gradient bar to the study of germination patterns in successional herbs. *Amer. Midland Natur.*, **77**, 86–92.

WENGEL, R. W., 1966. Emergence of corn in relation to soil oxygen diffusion rates. *Agron. J.*, **58**, 69–72.

WESSON, G., and WAREING, P. F., 1969. The induction of light sensitivity in weed seeds by burial. *J. Exp. Bot.*, **20**, 414–25.

CHAPTER 7

The Measurement of Viability

D. B. MacKay

Viability is usually measured in order to assess the suitability of a seed lot for a particular purpose, most commonly to produce a crop but also for industrial purposes, especially the malting of barley. Techniques for its measurement are also essential in research. Different criteria apply according to the purpose of the examination. In order to judge the suitability of seed for sowing it is necessary to determine its ability to produce plants in the field, so that information is required about the development of the seedling following rupture of the seed coat. But for the assessment of malting quality tests are needed to indicate the number of live grains, the incidence of dormancy and the most suitable steeping procedures (Pollock, 1962); they take no account of seedling development after sprouting (Institute of Brewing, 1967) since this stage is not involved in the malting process. This chapter will deal with the measurement of viability for crop production purposes, and the special circumstances of industrial processing will not be considered further; nor will there be more than very occasional reference to tree seeds.

The results of germination tests are used in two ways: to determine the suitability of a seed lot for sowing and to compare the value of different lots, so providing a basis for trade in seeds. For the former it is essential that they provide a realistic assessment of field planting value; for the latter they must be completely standardised and reproducible within narrow limits when tests are repeated. The most commonly used test demonstrates the ability of a population of seeds to produce plants in the field by exposing it to conditions in the laboratory which are optimal for germination; the capacity of the seedlings produced to develop into normal plants is then judged on the basis of a careful examination of their root and shoot systems. Wellington (1965) has traced the origin and development of this type of test from the principles outlined by Nobbe in 1876, and has shown how the commercial concept of testing, with its emphasis on reproducibility, has been gradually reconciled with the agricultural concept, which stresses provision of the best estimate of planting value.

This has been achieved largely by the adoption of standard methods of seedling evaluation, including agreed definitions of normal and abnormal seedlings based on the development of essential structures, which are incorporated in International Rules for Seed Testing (International Seed Testing Association, 1966).

A number of indirect methods for measuring viability have also been devised, most of which depend on examination of the metabolic activity of the seed. Their development has been reviewed by Holmes (1951) and Barton (1965). The most widely used of these tests involves treatment of the seed with tetrazolium salts; these are colourless in solution but are reduced in living tissue by enzymes of the dehydrogenase group to the formazan, which is red-coloured, stable and non-diffusible (Bulat, 1961). An assessment of viability is based on examination of the distribution of stained and unstained areas of the embryo (Lakon, 1949).

Sampling and the preparation of seed

Whatever method of measurement is applied it is only practicable to test a very small proportion of the population about which information is required. Crop seed is rarely pure, so that sorting is necessary to obtain the material for test. Thus, if results are to be comparable and reproducible, accurate methods of sampling must be employed and the procedure for selecting the seed must be rigorously standardised.

Procedures for sampling crop seed are prescribed in International Rules for Seed Testing (ISTA, 1966) and embodied in the legislation of many countries. Mullin (1968) has outlined the principles and some of the practical problems which arise; the most commonly used equipment is described in Agriculture Handbook No. 30 (United States Department of Agriculture, 1952). Seed lots are never perfectly homogeneous; the variation inevitably present in a crop is not completely eliminated during harvesting, cleaning and processing. Further segregation of material may occur in bagging and handling, and viability may be differentially affected through mechanical injury, the uneven distribution of moisture during storage or the uneven application of pesticide treatments. In order to obtain a reasonably representative sample for test, primary samples must be drawn from a sufficient number of containers, at different depths and horizontal positions. When seed is sampled from a stream, as in a cleaning machine, the entire cross-section of the stream must be sampled and primary samples must be drawn at intervals throughout the process;

automatic samplers can be built into seed cleaning systems, and are satisfactory provided they fulfil these conditions.

The sample collected from the bulk is too large for testing and special laboratory methods are applied to obtain accurate working samples. A number of mechanical dividers may be used, including soil dividers and centrifugal dividers which are operated on the principle of mixing the seed and then halving it continuously until the appropriately sized working sample is obtained. For laboratory germination tests the final sampling and planting operations can be combined by the use of a vacuum planter. This consists of a flat plate with a series of equally spaced holes drilled in it; it is fitted over a hollow head connected to a vacuum source, with a valve which allows the vacuum to be switched on and off at will. Seed is scattered on the plate so as to cover the surface; the vacuum is then applied so that seeds which happen to rest over a hole are held, while the remainder can be shaken off. By reversing the plate over the seed bed and releasing the vacuum the required number of seeds are positioned on the substrate.

Many seed samples contain impurities such as weed seeds and seeds of other crop species. The viability of this material may be of importance in assessing the extent to which it might contaminate a crop, but it is usually impracticable to test it separately because of the small numbers present. Seed impurities, as well as other material such as chaff, soil, etc., are therefore removed before the pure seed is sub-sampled for the germination test.

While it is often easy to distinguish impurities there are occasions when decisions are difficult, for example in the case of broken seed, or immature, undeveloped or diseased material. The results of a germination test can be greatly influenced by decisions on whether or not such structures should be included. General definitions are prescribed in International Rules for Seed Testing (ISTA, 1966). These are based on the principle that any material which might conceivably germinate should be tested. Thus undersized, shrivelled, immature and sprouted seeds are included, but grass florets without a caryopsis are not. In order to obtain uniformity of interpretation an element of arbitrariness is inevitable: only broken seeds and caryopses larger than one half the original size are included, no attempt being made to determine the presence or absence of an embryo; clusters or pieces of cluster of *Beta* are tested, regardless of whether they contain fruits, provided they do not pass through a sieve with 1.5 mm × 20 mm slits. Detailed descriptions of structures which are classed as pure seed, and are therefore included in germination tests, have been published for all crop seed genera (MacKay, 1969).

The laboratory germination test

The result of a laboratory germination test indicates the percentage of pure seeds which have produced seedlings, capable of continued development into mature plants, when germinated under standardised conditions of substrate, moisture supply and temperature to ensure that the result is reproducible (Wellington, 1966). Its reliability, both in terms of measurement of agricultural planting value and reproducibility for commercial purposes, depends upon the extent to which the seed environment can be controlled at the optimal levels for rapid and complete emergence, and development of the germinated seedlings to a stage at which their condition can be correctly evaluated.

The principal environmental conditions which must be available for seed germination are an adequate water supply and suitable temperature and composition of gases in the atmosphere. Requirements vary according to species and are determined both by the conditions which prevailed during seed formation and even more by hereditary factors (Mayer and Poljakoff-Mayber, 1963). Populations of crop seeds are likely to be more physiologically and genetically uniform than seeds collected from a series of wild plants; a seed lot will commonly be harvested from a single crop, while the breeding of crop varieties and modern seed production methods favour stability and uniformity. Nevertheless the seeds within a sample may vary in their requirements for germination because of, for example, differences in maturity between seeds from different plants or from different positions on the same plant; they may also have been unequally affected by processing or storage. The optimum conditions for different stages of germination and seedling growth are not identical; while it might be possible to devise a programme for varying the environment to suit particular stages, it could not be applied to a population of seeds since their germination is not synchronised. Thus it is only possible to provide conditions covering the range which is most favourable to the majority of seeds in a sample, probably representing those levels which have the least limiting effect on each of the intermediate stages (Wellington, 1965). In practice it is necessary to determine conditions which are suitable for a particular species, since in routine seed testing it would be out of the question to determine experimentally the conditions to be applied to each sample. The aim has been to find a combination of conditions which will give the most regular, rapid and complete germination for the majority of samples of the same species (Brett, 1939).

SUBSTRATE

Natural soils vary greatly in their structure and in their chemical and biological content. Although it might be argued that tests made in soil would be undertaken in conditions more closely resembling those to which the seed would be exposed when planted in the field, to which the results would then be more relevant, the loss of reproducibility and the inability to obtain comparable figures for different seed lots outweigh any advantage (Saunders, 1923). Artificial media may be much more readily standardised.

Paper is used extensively in seed testing, and is particularly suitable for small seeds and for seeds which may require light for germination. It should have an open, porous formation and be free from defects and impurities which might affect its performance, or toxic substances which might injure the roots of germinating seedlings. It should be free from fungi or bacteria which might interfere with seedling growth; sterilisation may be necessary, but it should not be treated with chemicals which might suppress disease organisms on the seed as well as in the substrate. Its texture should be such that the roots of germinating seedlings grow on, and not into, the paper, so that they are clearly visible for evaluation (Colbry, 1965). Paper should be strong when wet and thick enough to supply adequate moisture. There must be a high degree of uniformity between different sheets and across the surface of individual sheets. Detailed specifications are included in International Rules for Seed Testing (ISTA, 1966).

Light requiring seeds are planted on top of moist paper, and placed either on a Copenhagen tank (Jacobsen apparatus), which provides a supply of water throughout the test period by means of a wick dipping into a reservoir (see later), or in an incubator. Since germination is dependent on water uptake from the substrate exceeding water loss to the atmosphere (Benton, 1965), it is essential to maintain a high relative humidity over the exposed surface of the seed. This is achieved by placing a bell jar or glass cover over tests on Copenhagen tanks, or raising the relative humidity of incubators by evaporation from water trays or by humidifying machinery. Seeds without a specific light requirement may also be planted on top of paper, or they may be placed between folded paper: this method brings a greater proportion of the seed's surface area into contact with the moisture-supplying medium. To prevent drying the folded papers can be inserted into envelopes or between sheets made from polyethylene film. A third method, most frequently used for larger seeds, involves the use of rolled paper. After planting the moist paper containing the seeds is rolled and placed in an upright or slightly inclined position; this has the advantage of permitting the roots to grow straight downwards in response to geotropism, so avoiding the tangling of roots which can occur when papers are placed horizontally unless the seed is very widely spaced (Munn, 1950).

Paper substrates provide conditions favourable to the development and spread of fungi present on the seeds, which may interfere with seedling growth; seedling evaluation may be difficult when it is not known whether the source of infection is the seed whose viability is being judged or another seed. The problem is usually more severe in folded paper than in top of paper tests; it can be reduced by spacing the seeds widely, by removing infected and clearly dead seeds at intermediate counts, and by transferring ungerminated seeds to clean papers at intervals during longer tests.

Fungi develop less freely on sand; seeds can be buried or pressed into the surface, so that a more effective mechanical barrier is formed between them than is possible on paper. This substrate is not suitable for very small seeds, the manipulation and recovery of which would prove difficult, but it is widely used for larger seeds. Seeds are planted either pressed into the surface of the sand or covered by it to a depth of 1–2 cm, so that a substantial part of their surface area can be brought into close contact with the water in the substrate, thereby increasing the rate of water uptake and germination (Manohar and Heydecker, 1964; Sedgley, 1963).

Sand may be readily sterilised by heat to destroy bacteria, fungi and foreign seeds. Moisture content is standardised by determining the water holding capacity and subsequently adding sufficient water each time the substrate is prepared to bring the content up to the required level: under International Rules for Seed Testing (ISTA, 1966) this is prescribed as 50 per cent of moisture holding capacity for most kinds of seed, but 60 per cent for maize (*Zea mays*) and large-seeded legumes. Moisture holding capacity is determined primarily by the particle size distribution and the corresponding pore size distribution; although there has been little detailed study of the ideal particle distribution (Kåhre and Wiklert, 1965), it is necessary to sieve out particles which are too large or too fine, and a range of 0.8 mm to 0.05 mm is recommended in International Rules. It is essential that toxic chemicals should be absent; these might either be present initially or could accumulate as a result of re-using sand in which seed treated with a pesticide has been previously tested.

Seedlings growing on artificial substrates can usually be readily evaluated, but sometimes they are affected by chemicals or pathogenic or saprophytic fungi, or exhibit physiological defects the extent of which is not readily apparent. The use of artificial media may exaggerate these conditions, and the substitution of soil or compost as substrate, under standard laboratory conditions, may provide a more accurate reflection of the seed's potential field performance (Tonkin, 1969). Thus seed treatment with fungicides and insecticides often gives rise to phytoxicity in seedlings grown on sand or paper (Brett and Dillon Weston, 1941; Lafferty, 1953; Thomson, 1954), but its incidence may be reduced when the seed is sown in the field or in a laboratory medium containing soil (Wellington, 1957). The tissues of seedlings of broad beans (*Vicia faba* var. *faba*) sometimes develop conspicuous blackened lesions in laboratory tests (Crosier, 1951), but the condition was alleviated when John Innes potting compost (Lawrence and Newell, 1939) or a proprietary soilless compost ('Levington' compost) was substituted for a sand substrate (Official Seed Testing Station for England and Wales, 1964, 1970). When doubts arise about the evaluation of seedlings grown on artificial media, tests can be repeated in compost; the seedlings can then be grown to a more advanced stage when the condition of the structures can be readily assessed.

The principal function of the substrate is to supply moisture. International Rules for Seed Testing prescribe the extent to which sand should be moistened, but there are no precise standards for the amount of water to be added to paper media, only a requirement that they should not be so wet that a film of water forms around the seeds. Collis-George and Sands (1961) have interpreted this as being at an indefinite suction approaching zero (free water), and point out that suction would not necessarily be constant throughout the test; they therefore recommend the use of tension plates and the prescription for each species of suctions optimal for germination.

The supply of oxygen in laboratory germination tests is largely determined by the moisture conditions: seed may not germinate when the water supply is inadequate (Chowings, 1970), but excessive water physically impedes the uptake of oxygen (Chetram and Heydecker, 1967; MacKay and Tonkin, 1965; Orphanos and Heydecker, 1967) which may also be competed for by fungi and bacteria in the covering structures (Gaber and Roberts, 1969; Roberts, 1969).

TEMPERATURE CONTROL AND GERMINATION EQUIPMENT

Mayer and Poljakoff-Mayber (1963) have defined the optimal temperature for germination as that at which the highest percentage of germination is attained in the shortest time, and the minimal and maximal temperatures as the highest and lowest temperatures at which germination will occur. In laboratory seed testing the temperature at the level of the seed must be controlled near to the optimum. The values are genetically determined, varying for different species, but are also influenced physiologically (Lang, 1965). The range over which complete germination is attained may differ with the age of the seed, usually becoming wider as the seed ages; a regular alternation of temperatures is sometimes more effective than a constant temperature. Both these phenomena, however, are probably due to dormancy, whose importance in germination testing will be discussed later. The temperature regime selected in the laboratory is also influenced by the relative effects on seed germination and fungal development; Fritz (1966) has shown how the use of particular temperatures within the range optimal for the germination of disinfected cereal seed can result in a lowering of germination capacity in untreated seed through the activity of parasitic fungi.

Appropriate temperatures for each of the principal crop and tree species are listed in International Rules for Seed Testing. Precise temperature control demands special equipment, incubators and water baths being the most widely used. Incubators are used for many purposes besides seed testing and innumerable models are available. However, Oomen and Koppe (1969) have prepared a specification for an automatic cabinet incubator to provide standardised germination conditions, eliminating the need for daily attention and watering. This includes air temperature control adjustable between 10° and 35°C, the level for each regime to be uniform within ±1 per cent throughout the controlled area over a number of days. To allow for temperature alternation, the changes from the high to low and low to high phases should be achieved within 30 minutes, a steady temperature being reached in one hour. Air humidity should be as high as possible, but never lower than 90 per cent, and air movement should be low to prevent seeds from drying. A single change of air per hour is considered sufficient. There should be uniform illumination of between 750 and 1,250 lux at the level of the seed. Condensation should not occur. Four prototypes designed to meet these conditions were systematically tested and their adequacy discussed.

The Copenhagen tank (Jacobsen apparatus) (Fig. 7.1) was developed specifically for seed germination (Jacobsen, 1910). It consists of a bath containing water whose temperature may be controlled thermostatically. Seed is planted on top of paper and placed on a metal or glass strip suspended above the bath; the upper end of a paper or cotton wick lies beneath the seed bed and the lower end dips into the

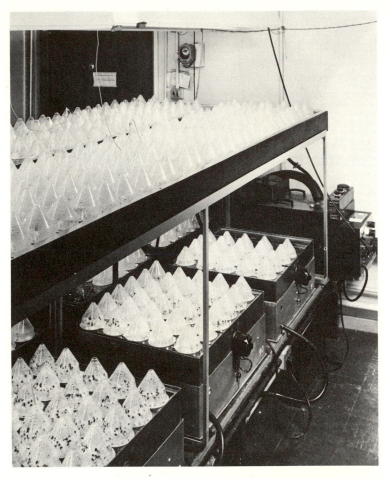

FIGURE 7.1   Copenhagen tanks for seed germination (Jacobsen apparatus). The upper tank is fully automatic, operated by the control box at the rear. (Photograph from the National Institute of Agricultural Botany.)

water. The moisture content of the substrate may be adjusted by altering the water level, and so increasing or decreasing the distance between seed bed and water (Kamra, 1968, 1969a, 1969b). High humidity is maintained around the seed either by covering the entire tank with a transparent lid or by placing bell jars over the individual replicates. The apparatus may be exposed to natural daylight, but artificial light is usually desirable, especially for light-sensitive seeds such as meadow grass (*Poa* spp.) and lettuce (*Lactuca sativa*); fluorescent tubes with a high spectral emission in the germination-promoting red region are recommended (ISTA, 1966).

One of the disadvantages of the standard Copenhagen tank is the absence of direct temperature control at the seed bed; this is influenced by the temperature not only of the water but also of the room, which must be controlled. Improved models which overcome this difficulty and permit rapid temperature alternation at seed level have been developed: Thomson (1962) incorporated panels of reconstituted wood pulp containing internal electric heating elements on which the paper substrate rested; Verhey (1955) and Overaa (1962) used hollow plates, through which water was circulated for heating and cooling, instead of the usual glass or metal strips. Marschall (1969) reduced the quantity of water whose temperate must be adjusted by placing small water trays, each large enough for the four replicates of one test, on standard central heating radiator plates, in order to accelerate heat transfer.

Improved seedling development occurred when the seeds were planted with the primary axis of the embryo vertically aligned, and apparatus has been developed in which they were attached to an inclined plane holding a paper substrate (Cobb and Jones, 1966; Jones and Cobb, 1963). A system for the germination of ryegrass seeds between paper on a vertical glass plate in order to determine the presence of fluorescent substances in their roots has also been described (Official Seed Testing Station for England and Wales, 1952).

When large numbers of tests are made whole rooms may be used as germinators, with temperature, humidity and light controlled. The precise control of large areas presents greater problems than arise with the use of cabinet incubators, and various methods have been applied (Clayton and MacKay, 1962; Verhey, 1955).

A system for the almost complete automation of germination testing has been described by Brandt (1964).

DORMANCY

The germination process can be blocked by various chemical and physical means anywhere along the chain of physiological events leading from imbibition to growth of the embryo (Evenari, 1961). This may be due to the absence of one or more conditions always essential for germination, such as moisture in seed kept under dry storage conditions (enforced dormancy); to incomplete after-ripening, the processes occurring within the seed subsequent to the harvest-ripe stage (innate dormancy); or to the imposition or re-imposition of a block as a result of exposure in the imbibed state to unfavourable conditions, such as excessive temperature (induced dormancy) (Harper, 1959). These three categories of dormancy are discussed in greater depth in Chapter 11 (p. 321) and the physiological nature of dormancy is discussed in Chapter 12 (p. 370). In laboratory germination tests enforced dormancy is eliminated by the provision of appropriate moisture, temperature and aeration, but live seeds may still not germinate because of innate dormancy unless the relevant blocks are removed through the use of external agents. Dormancy may also sometimes be induced during laboratory tests if conditions, especially temperature, are not adequately controlled.

A condition which delays or interferes with the full expression of a sample's germination capacity is troublesome in a seed testing laboratory. Since the degree of dormancy changes during after-ripening (Evenari, 1961) the precision with which test results can be reproduced at different times will be reduced unless special measures are taken to overcome the condition. But if the results of germination tests are to yield information on the field planting value of a seed lot removal of dormancy in the laboratory sample can only be justified if this condition is irrelevant to the field establishment of the remainder of the bulk at the normal time of sowing (Wellington, 1965). It has been generally assumed that after-ripening is complete by the time seed is sown in the spring following harvest, or that it will be exposed to conditions in the field, such as low temperature or exposure to light with shallow sowing, which overcome dormancy (Thomson, 1963b; Wellington, 1965); and so special treatments to overcome dormancy are applied in the laboratory whenever necessary (ISTA, 1966). The assumption is probably justified for the majority of traditional temperate crops, whose selection and breeding has tended to eliminate extensive dormancy (Evenari, 1961), but irregular establishment may sometimes occur when, for example, early autumn sowing of cereals closely follows harvest or when seed of pasture species more recently brought into cultivation from the wild is sown (Smith, 1968; Wellington, 1965; Whittet, 1952).

An exception to the general rule that dormancy should be removed in laboratory tests is made in the case of seeds with impermeable seed coats, the so-called 'hard seeds' which occur most frequently in the small seeded legumes. The field planting value of hard seed has been the subject of a number of investigations. After extensive tests Witte (1931, 1934, 1938) recommended that from 50 to 100 per cent, according to species, of the hard seed remaining at the end of a germination test should be added to the percentage germination figure. But Overaa (1960) found great differences between the laboratory germination of scarified hard seeds from different samples of red clover, and suggested that tests after scarification were necessary for a proper judgement of the real value of impermeable seeds. Harrington (1916) concluded that a large proportion of impermeable seeds would germinate in the soil during the first few months after planting, some of them early enough to be of importance to the crop; the remainder would constitute a reserve which, under favourable conditions, might improve thin areas of the stand, although this would depend on the extent to which such areas had been colonised by more rapidly growing weeds. Field emergence has usually been studied in plots sown only with hard seed, so that the effect of competition with seedlings produced from permeable seeds has been excluded. Zaleski (1957) has shown that, in lucerne (*Medicago sativa*), only those seeds which germinate rapidly contribute to the crop, later germinating seedlings, which would include those from hard seeds, being eliminated. In seed testing no attempt is now made to allocate a planting value to impermeable seed; seeds which have not absorbed water at the end of the test period are reported separately as a percentage of hard seeds (ISTA, 1966).

The methods for overcoming dormancy in the laboratory vary according to the nature of the germination block as well as to its intensity. Thus different procedures may be necessary for different species, in different seasons and in seed from different sources. The seed is either specially pre-treated, for example with low temperature, before the germination test, or special conditions, such as light, are applied during the test.

The application of low temperature to seed in the imbibed state in the laboratory (often referred to as 'stratification', after the horticultural practice) is a procedure analagous to the situation in nature where seeds are shed in the autumn and then subjected to low temperatures under moist conditions in the soil, during which afterripening occurs (Mayer and Poljakoff-Mayber, 1963). The method is effective for many species of agricultural and vegetable seeds (including the cereals), whose seed is first planted on the germination medium and then exposed to a temperature between 5° and 10°C for a period of up to seven days, after which it is transferred to the standard germination conditions for the full test period. It is usually necessary to instal refrigerators or cold rooms for this purpose, although some incubators are capable of operating at a sufficiently low temperature. Care is needed to ensure that the correct temperature is maintained for the full period required, since if it is too high dormancy may not be effectively broken, whereas if it is too low freezing injury may be inflicted (see pp. 36–37, and 140).

In most crop seeds dormancy is lost during after-ripening in dry storage. However, exposure to a short period of relatively high temperature may sometimes accelerate the process (Hite, 1923; Roberts, 1962, 1965; Stapledon and Adams, 1919; Wellington, 1956), although it is not always clear whether the effect is one of desiccation or of temperature (Hewett, 1958, 1959; Stokes, 1965). Seed is usually heated at a temperature not exceeding 40°C, with free air circulation, for up to seven days before being planted, although temperatures up to 47°C can be used satisfactorily for rice (*Oryza sativa*) (Roberts, 1965).

Etiolated seedlings are often difficult to evaluate and light may be applied in laboratory germination tests in order to influence seedling growth. However, this is quite distinct from the application of light to remove a block to germination in light-sensitive seed, for which the conditions are more critical. It is usually necessary to supplement daylight from an artificial source; fluorescent tubes with relatively high emmission in the red region are preferable since, for some seeds, incandescent-filament light and even diffused sunlight may be inhibitory because of their nearly equal red and far-red energies (Borthwick, 1965). An even intensity over the entire testing surface of 750–1,250 lux is prescribed in International Rules for Seed Testing (ISTA, 1966). Photo-dormancy may be induced in imbibed seeds kept in the dark at high temperature (Borthwick, Hendricks, Parker, Toole and Toole, 1952; Borthwick, Hendricks, Toole and Toole, 1954), so that it is necessary to ensure that temperature does not rise excessively during periods of darkness.

Dormancy often narrows the range of temperature over which germination takes place (Lang, 1965; Stokes, 1965). More complete germination may be obtained by exposing the seed to a temperature rather lower than the optimal for non-dormant seed: for example, under International Rules, cereal and clover seed may be tested at 15°C instead of at 20°C. Because germination may be slower at lower temperature the test period is extended. Lang (1965) has suggested that the promotion of germination by introducing a daily fluctuation of temperature does not reflect the removal of specific blocks, but rather an increase in the general physiological activity level of the seed. Cohen (1958) has discussed the possible mechanisms. Whatever its mode of action, the use of alternating temperatures often leads to more complete germination in laboratory tests, especially when a rapid fall from the high to the low phase is introduced. It is widely used, in conjunction with light during the high phase, in the testing of grasses and, with or without light, for many other species (ISTA,

1966). Tests may either be transferred between equipment maintained at the two prescribed temperatures, or the temperature may be varied in individual germinators. The high phase (which ranges from 25° to 35°C according to species) is maintained for eight hours and the low phase (between 10° and 20°C) for sixteen hours of each twenty-four.

A number of chemicals may influence the germination of dormant seeds. Thus potassium nitrate can replace the requirement for, or reinforce the effect of, other dormancy-breaking agents such as light and particular temperature regimes in large numbers of species (Evenari, 1965; Mayer and Poljakoff-Mayber, 1963; Stokes, 1965), and is widely used in seed testing. The substrate is moistened with a 0.2 per cent solution, prepared by dissolving 2 g in 1,000 ml of water, at the beginning of the test, but subsequently only water is used for further moistening (ISTA, 1966). The interactions between light, fluctuating temperatures and nitrate are considered in further detail in Chapter 11. At present potassium nitrate is the only chemical agent approved under International Rules, since until the introduction of gibberellic acid other chemicals were not generally suitable for breaking dormancy in seed testing (Wellington, 1965); but gibberellic acid has been shown to be capable of replacing requirements for light, low temperature and alternating temperatures in a number of species (Evenari, 1965; Nakamura, Watanabe and Ichihara, 1960; Tager and Clarke, 1961). On the basis of experiments with wheat (*Triticum* aestivum) and barley (*Hordeum vulgare*), Bekendam and Bruinsma (1965, 1966) have recommended pre-soaking the seed or treating the substrate with solutions of gibberellic acid at concentrations of from 0.02 to 0.2 per cent according to the intensity of dormancy. Kåhre, Kolk and Fritz (1965) obtained satisfactory germination of freshly harvested cereal seed when the substrate (sand) was moistened with a solution of gibberellic acid at a concentration of 200 ppm.

The dormancy of a range of small-seeded legumes may be broken by treatment of the imbibed seed with concentrations of carbon dioxide as low as 0.3–0.5 per cent by volume (Ballard, 1967; Grant Lipp and Ballard, 1959). If the seed is planted between folded paper, enclosed in a sealed polyethylene film envelope and exposed to standard germination conditions in a cabinet incubator, dormant seed with permeable testas is eliminated, apparently because of the retention of carbon dioxide from the respiration of non-dormant seeds in the sample (Thomson, 1965).

The dispersal unit in *Beta* spp. consists of the hardened receptacle and sepals of the flower cluster, in which the fruits are embedded.

This structure contains inhibitors (Battle and Whittington, 1969; Stout and Tolman, 1941; Tolman and Stout, 1940) which are water-soluble and interfere with germination in the laboratory, although probably not in the soil (MacKay, 1961). Before planting clusters are washed in water at 25°C for $\frac{1}{2}$ to 2 hours; regular changes of the water are desirable during this period to ensure the removal of the inhibitors. If a number of tests are to be made the process can be carried out in a machine consisting of two tanks: in the upper tank water is heated under thermostatic control by an immersion heater, and then flows to the lower tank in which the clusters are placed in small tubes with apertures to allow access to the water; when the water has reached the correct level it is siphoned off and the tank refills. After washing the clusters are dried before planting, since the germination of very wet seed may be reduced through interference with oxygen supply (Bekendam, 1968; Heydecker, Orphanos and Chetram, 1969; MacKay, 1961). Cuddy (1959) reported that pre-washing might be detrimental to germination in some samples which had been processed to produce single-seeded fragments, in which the inhibitor content is reduced through loss of cluster material (MacKay, 1961).

SEEDLING EVALUATION

Seed is exposed to the conditions of the germination test for a prescribed period, the duration of which depends primarily on the species and may vary from as little as 6 days for radish (*Raphanus sativus*) to as much as 35 days for Johnson grass (*Sorghum halepense*), or even longer in the case of certain tree seeds (ISTA, 1966), being fixed at a level where complete germination can be expected for the majority of samples under the prescribed environmental conditions; but the test can be extended for up to seven days if some seed is just beginning to germinate at the end of the period. Seedlings which have developed sufficiently to be correctly assessed are removed at intervals during the test to avoid over-crowding and restrict the spread of fungi. These interim counts were formerly regarded as having a bearing on the relative field performance of different seed lots (Brett, 1939), rapidly germinating lots with high interim counts being of greater value than those germinating more slowly. But the figures can be misleading since they are affected by dormancy and by the need to delay decisions on seedlings of questionable value until further development has taken place, so that the concept has now been generally abandoned (Verhey, 1960). Tests can be concluded before completion of the prescribed period if full assessment of all seeds and seedlings has been possible.

The definition of germination for seed testing purposes differs from the more strictly botanical definition quoted by Evenari (1961) as the processes starting with the imbibition of the dispersal unit and ending with the protrusion of the embryonic root which take place inside the dispersal unit and prepare the embryo for normal growth. Wellington (1965) has explained how it has been evolved in terms of development of the seedling to a stage where it can be inferred that it would have been capable of emerging through the soil and sustaining autotrophic growth, thus involving a degree of seedling growth after completion of the physiological process of germination: in International Rules for Seed Testing (ISTA, 1966) germination in a laboratory test is now defined as the emergence and development from the seed embryo of those essential structures which, for the kind of seed being tested, indicate the ability to develop into a normal plant under favourable conditions in soil. This involves not only the provision of optimal conditions to permit rapid and complete emergence, but also the detailed examination of the structures produced by each seedling and the formation of a judgement on the condition of each of these in terms of its potential influence on field performance. Wellington (1968) has outlined the development and structure of each of the essential structures for the main groups of crop plants, a knowledge of which is essential for the correct and uniform application of the principles of seedling evaluation.

To be classed as normal the seedling requires a well-developed root system. For plants whose commercial product is a swollen tap root, with or without stem tissue, such as turnip (*Brassica rapa*) or carrot (*Daucus carota*), an intact primary root is clearly indispensable to the development of a normal plant. But in a number of genera entirely satisfactory plants can be produced from seedlings with the primary root injured or absent, provided secondary roots have developed sufficiently to support the seedling in soil; these genera are listed in International Rules as *Pisum, Vicia, Phaseolus, Lupinus, Vigna, Glycine, Arachis, Gossypium, Zea* and all *Cucurbitaceae*. Many of the *Gramineae* species produce seminal roots, in which there is no clearly visible distinction between the primary and secondary roots; such seedlings must have at least two well-developed roots if they are to be classified as normal.

The hypocotyl, that part of the primary axis of the seedling lying immediately below the point of attachment of the cotyledons, represents the region of transition from root to stem. In species with hypogeal germination, such as *Vicia faba*, it remains below ground and is rather short and root-like; but in species with epigeal germina-

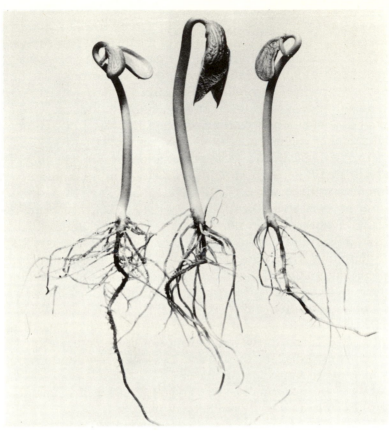

FIGURE 7.2   Dwarf bean seedlings at the conclusion of a germination test. Centre seedling shows normal development; the others show injury to the plumular bud precluding development into normal plants. (Photograph from the National Institute of Agricultural Botany.)

tion, such as *Phaseolus vulgaris*, it extends to carry the cotyledons above ground, and its upper part is stem-like in appearance. In a normal seedling from a laboratory test the hypocotyl should be well-developed and without damage to the conducting tissues, although superficial damage may be present provided it is limited in area.

The epicotyl extends between the point of attachment of the cotyledons and the stem apex. In species with hypogeal germination its lower part is below ground. In a normal seedling it should be intact, apart from limited superficial damage, and terminate in a normal plumular bud. This consists of an apical meristem and leaf primordia enclosed by the developing leaves. In many species detailed examination of the condition of the plumular bud is often impossible during the period of a germination test, and its ability to develop normally is assumed if, at the conclusion of the test, there is no evidence that the surrounding tissues are damaged or decayed (Wellington, 1970). But in some species, especially large-seeded legumes, the primary leaves may show injury which can affect plant performance (Verhey, 1961), so that the final assessment must not be undertaken until they are clearly visible (Fig. 7.2).

The cotyledons are the first leaves of the plant. In dicotyledons they may emerge and perform a photosynthetic function for a time as foliage leaves or be adapted for storage of food reserves. The single cotyledon of the monocotyledons may protect the emerging shoot during its passage through the soil, at the same time remaining in contact with the endosperm from which the food reserves are absorbed; or it may be substantially modified to form the scutellum of the *Gramineae*, which also functions in the utilisation of food reserves by the developing embryo. The presence of functioning cotyledons is thus necessary for normal seedling development, and must be determined in the germination test. In the case of dicotyledons one cotyledon is considered sufficient to support normal development.

In the *Gramineae* the shoot system arises from a plumule containing leaf primordia and the stem apex, surrounded by a coleoptile formed from the sheath of the first leaf. The coleoptile is negatively geotropic and also protects the developing leaves until they emerge from its tip above the level of the soil. For normal development the plumule must be intact, except for superficial local damage to the coleoptile which does not penetrate to the enclosed leaves; there should be a well-developed green leaf within or emerging through the coleoptile (Fig. 7.3).

In a laboratory test the majority of the normal seedlings are usually removed at interim counts, but the assessment of many of the doubtful and abnormal seedlings must be left until its conclusion to ensure that slower growing but otherwise normal seedlings are not incorrectly classified; but it is desirable to count and remove diseased seedlings as soon as their condition can be assessed, to prevent infection spreading to others. Seedlings with minor defects may still produce normal plants, and it is essential to distinguish between these and seedlings whose condition would preclude satisfactory further development. Essential structures may be physically damaged, deformed on account of weak or unbalanced development, or decayed due to attack by microorganisms, and these defects can arise from a number of causes at various stages during seed formation, ripening, harvesting, processing or storage; some examples are given below.

FIGURE 7.3   Analyst evaluating maize seedlings at the conclusion of a germination test. Both root and shoot systems are checked for normal development. (Photograph from the National Institute of Agricultural Botany.)

Lack of manganese in the parent plant when grown on deficient soils can lead to a condition in pea seedlings, known as 'marsh spot', in which the plumular bud is destroyed (Reynolds, 1955). Although shoots may develop in the axils of the cotyledons the resultant seedlings do not produce normal plants. Less severely affected seedlings exhibit only a necrotic area at the centre of the cotyledons, but these are classed as normal in laboratory tests.

Seed may be infected by pathogenic fungi in the field during its formation and ripening, and seedlings grown from it may be attacked during germination tests. Thus, for example, wheat seedlings infected with *Septoria nodorum* show brown spots, knobs or stripes on the coleoptile (Kietreiber, 1962, 1966), and are classed as abnormal if the discolouration has penetrated to the enclosed leaves (ISTA, 1968). Seedlings infected with *Alternaria, Ascochyta* or *Phoma* spp. may show decay or discolouration of the cotyledons; if this affects the area adjacent to the shoot apex, or covers more than half of the total cotyledon area, they are considered abnormal. The treatment of seed-borne disease in germination tests is one of the more controversial aspects of seed testing (Neergaard, 1965); however, the germination test is not designed to establish the presence of seed-borne infection, and special disease tests are available for more accurate determination.

Cereal seed produced in high latitudes is subject to low temperature before harvest and may exhibit freezing injury. Typical symptoms described by Andersen (1950) for wheat, oats and barley include malformed coleoptiles with few or no foliar leaves, spindly seedlings with the tissue of the coleoptile and foliar leaves split longitudinally or desiccated, shrivelled and withered, and seedlings with little or no root development; none of these would develop into normal plants.

Mechanical injury can occur during harvesting and threshing (see Chapter 4). Badly broken seeds are removed in cleaning, but the embryo may be bruised or fractured, when the condition does not become apparent until the seed germinates. Mechanical injury is found in most species, but is especially frequent in the *Leguminosae*. Seriously damaged seedlings are classed as abnormal since, although they may develop adventitious roots or shoots may grow from the axils of the cotyledons, they are unlikely to produce normal plants in a crop. Abnormal seedlings in this category include those with damage to the cotyledons reducing their total area to less than half that of two normal cotyledons (Ching and Pierpoint, 1957) or with the epicotyl or hypocotyl fractured or the plumular bud injured (Andersen, 1954; Verhey, 1961). Essential structures may have open splits or constrictions likely to affect the conducting tissues. Root damage is often found in cereal and grass seed in which only the plumule develops. MacKay and Flood (1968, 1969) have associated injury with embryo structure and the degree of exposure of the radicle in cereal seed and of the radicle, hypocotyl and distal end of the cotyledons in seed of red clover (*Trifolium pratense*). Cereal seed may also show injury to the coleoptile, which splits so that the leaves emerge from near the base instead of from the tip; although the leaves may be normal in a laboratory test their ability to reach the surface when released below ground level is restricted in the absence of the geotropic response supplied by the coleoptile.

It is often necessary to dry seed artificially after harvest in order to reduce its moisture content to a safe level for storage, but this may affect germination if the temperature is too high. Wellington and Bradnock (1964) have described the deformed seedlings produced in barley. In some only the coleorhiza emerged and there was no extension of the seminal roots; in others elongation of the seminal roots was restricted. Some seedlings showed normal development of the plumule, although in others the coleoptile elongated but the leaf was delayed, or the plumule did not elongate and failed to emerge from the covering layers. It is important to ensure that tests are not assessed too early, since some of these categories resemble early stages of normal germination.

Abnormal seedlings can also arise as the result of processing as when, for example, essential structures are fractured during the abrasive treatment used to produce single seeded units from sugar beet clusters (Tolman and Stout, 1944) or to overcome the hard-seeded condition in clover (Witte, 1928). Seedlings produced by seed treated with chemical fungicides or insecticides may be deformed, with roots and shoots stunted and swollen. The condition in cereals has been described by Brett and Dillon Weston (1941), Lafferty (1953) and Thomson (1954). The effect of artificial media in exaggerating these symptoms and the need to re-test affected samples in a medium containing soil has been discussed in the section dealing with substrates.

Germination capacity declines as seed ages during storage, but complete death is usually preceded by the production of abnormal seedlings whose development is weak or unbalanced because the loss of vital functions does not occur simultaneously in the different tissues. Typical symptoms which have been described include stunting of the plumule or failure of the first leaf to develop within the coleoptile in *Gramineae* (Griffiths and Pegler, 1964; Kearns and Toole, 1939; MacKay and Flood, 1969), breakdown of hypocotyl tissue giving a glassy or watery appearance and restricted root and shoot development in *Leguminosae* and *Cruciferae* (MacKay and Flood, 1969, 1970), and failure of the characteristic bend or 'knee' to develop in the cotyledon of onion (*Allium cepa*) (Clark, 1948). Under high moisture conditions seedlings in which the essential structures are decayed may be frequent (MacKay and Flood, 1968, 1969).

At the conclusion of the test dead seeds must be separated from those which are dormant. The most reliable criterion is the condition of the embryo, which is firm to the touch in dormant seed but soft and watery in dead seed, usually but not always accompanied by mould development. 'Hard' seeds are easily recognised because they have not imbibed.

RELEVANCE TO FIELD ESTABLISHMENT

The concept of germination capacity embodied in International Rules (ISTA, 1966) has been evolved in response to an appreciation of the function of seed testing in providing information on sowing quality. The principles of seedling evaluation have been steadily refined in an attempt to improve the relationship between the results of a laboratory test and the production of plants in the field.

Seed sown in the field is subject to hazards to which it is not exposed in the laboratory, so that precise reproduction of laboratory results cannot be expected; and because of the great differences in soil conditions depression of the laboratory figure is unlikely to be the same at different sites or seasons. On the basis of a review of data from many experiments in the USA and Northern Europe, Essenburg and Schoorel (1962) concluded that there were generally very high correlations between laboratory germination and field emergence but, whereas in some species the emergence percentage usually shows a fairly constant relationship to the germination capacity, others are much more sensitive to differences in soil conditions. Gadd (1932) reported a very high correlation between laboratory and field germination for both autumn and spring wheat, regardless of the variable climatic conditions from year to year. When seed rates were adjusted on the basis of the germination test to give equal numbers of viable seeds Abdalla and Roberts (1969) obtained almost identical densities with barley from storage treatments which produced germination percentages over the range 100 to 15 per cent; but Valle and Mela (1965) found better emergence from samples of wheat, barley and oats with higher germination (88–93 per cent) than with lower germination (63–70 per cent). MacKay and Tonkin (1965) reported that laboratory tests placed samples of sugar beet in the same order as their percentage field emergence, although tests in the greenhouse discriminated rather more clearly between samples of different field planting value.

In modern farming practice the establishment of plant populations at pre-determined densities, without the intervention of transplanting or hand thinning, is increasingly sought in order to obtain, at the lowest cost, maximum yields of produce of the precise quality demanded by the food processing industry. To achieve this, precision drills are used to place the seed accurately in the soil, and the seed itself may be mechanically processed or pelleted within an envelope of inert material in order to produce units nearer to the spherical shape which is ideal for this purpose. Seed rates are calculated on the basis of formulae incorporating the laboratory germination figure (Austin, 1963; Bleasdale, 1965). Since the majority of samples show high correlations in individual series of comparisons between laboratory germination and field emergence it should be possible to apply a correction to allow for the expected depression; but since the extent of the depression in soil varies, the 'field factor' to be included in the seed rate formula must be based on knowledge of local climatic and soil conditions.

There are some samples which do not follow the general pattern because, although they perform similarly under optimal conditions, they vary in their reaction to environmental factors such as soil moisture or pathogen content (Heydecker, 1960, 1962), and this situation is particularly apparent in certain species. When peas are sown under the adverse conditions of soil moisture and temperature encountered in early spring in the United Kingdom the laboratory test gives little indication of emergence (Matthews and Bradnock, 1967, 1968), though the relationship improves with later sowing (Perry, 1970). A similar situation arises with maize, in which differences in pre-emergence mortality are associated with differences in the ability to germinate rapidly and to grow at low temperatures (Harper, Landragin and Ludwig, 1955a). Mortality in both peas and maize may be reduced, but not eliminated, by treating the seed with a fungicidal dressing (Harper, Landragin and Ludwig, 1955b; Matthews and Bradnock, 1967), and it is necessary to supplement the laboratory germination test with a special test to indicate susceptibility to pre-emergence failure, such as the electro-conductivity test for peas (Matthews and Bradnock, 1967), or the cold test for maize (Harper and Landragin, 1955; Wernham, 1951).

The procedure by which total germination is reported, even when accompanied by an interim figure, has been criticised because it does not provide reliable information on the speed and evenness of germination which is of particular importance in precision drilled crops or in connection with the use of pre-emergence herbicides. Timson (1965) proposed the recording of germinated seedlings each day and the expression of the test result as the sum of each day's totals, but Heydecker (1966) showed that in some circumstances this would obscure important differences between samples. Nichols and Heydecker (1968) considered that the use of quartiles (the times to 25, 50 and 75 per cent of the ultimate number germinating), supplemented by the final germination percentage, would be valuable because it would yield information on the mean time to emergence and also the scatter around this time; but they found the data were not adequate for predicting establishment in the soil.

The laboratory germination test provides an appraisal of seed quality based on a knowledge of what proportion is incapable of growing and therefore worthless under all circumstances (MacKay, 1966). It seems probable that the additional information required for particular circumstances will have to come from specially devised supplementary tests.

REPRODUCIBILITY OF TEST RESULTS

The reproducibility of germination test results within fairly narrow limits is necessary because the figures are used not only to judge the suitability of a seed lot for a particular agricultural situation, but also for price determination and in the enforcement of seed quality legislation. However, unlike for example physical purity, viability is a dynamic property liable to change with time (Wellington, 1965); the speed and extent of this variation is determined by a series of genetic, physiological and environmental factors both before (Chapters 4 and 5) and during storage (Chapter 2) (see also Barton, 1961; MacKay and Flood, 1968, 1969, 1970; MacKay and Tonkin, 1967; Owen, 1956). Since storage conditions can usually be more readily controlled at a level favourable for the retention of viability in laboratory samples than in seed bulks, a change in percentage germination after a given period is likely to be less when both tests have been made on the same sample than when a second sample has been newly drawn from the bulk.

The extent to which test results vary is also influenced by the uniformity of the bulk. Sampling techniques are designed to produce representative samples from reasonably uniform lots, but they cannot take account of excessive variation (Chowings, 1968). Many of the factors causing loss of viability do not affect all parts of a bulk equally: initial differences in the condition of the standing crop may be reflected in uneven deterioration of the seed; variation in the severity of mechanical, heat or chemical injury and the uneven distribution of moisture taken up during storage may result in the level of viability being higher in some parts of the bulk than in others. Unless the seed is thoroughly mixed successive samples may then show wide differences in percentage germination. Miles (1962) proposed a procedure for measuring the heterogeneity of seed lots, and this has subsequently been incorporated in International Rules for Seed Testing (ISTA, 1966). It involves the separate analysis of a series of samples

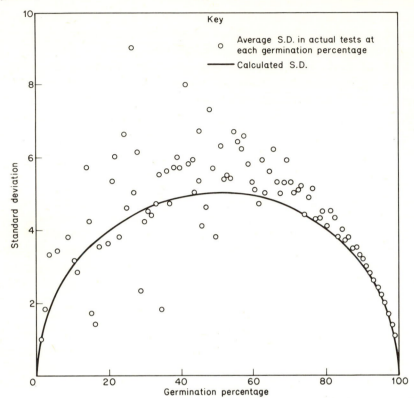

FIGURE 7.4 Average standard deviations of over 20,000 germination tests at the Official Seed Testing Station for Scotland compared with standard deviations calculated on the basis of random selection of seeds. (From Thomson, 1963.)

and the calculation of a 'heterogeneity value' (H) from the results based on the equation:

$$H = \frac{\text{actual variance}}{\text{theoretical variance}} - 1$$

The higher the value of H the greater is the variability of the bulk. Chowings (1968) has applied the test to 83 seed lots, showing that in crops where cleaning processes are highly effective, such as cereals, cleaning can remove evidence of heterogeneity, so that low H values for purity can be achieved without mixing, but the values for germination can remain high.

Variability may also be introduced as a result of imperfect sampling, or of differences in the conditions to which seed is exposed during the test or in seedling evaluation. On the basis of an analysis of standard deviations for over 20,000 tests Thomson (1963a) concluded that, in a single laboratory, at germination levels above 80 per cent, variability was due almost entirely to the random selection of seeds; but at lower levels variation exceeded values calculated on this basis because the seed was more sensitive to slight differences in test conditions and also produced more seedlings whose evaluation as normal or abnormal presented difficulties (Fig. 7.4). Variation between laboratories exceeds that accounted for by random sampling (Miles, 1961), since there is a greater likelihood of results being affected through differences in techniques, even when uniform rules are applied. Variability between laboratories has been measured and forms the basis of tables by means of which the compatibility of test results can be judged (Miles, 1963); but because in practice variability within laboratories rarely exceeds random variation, figures allowing for random sampling variation only are used to check the consistency of the results of the replicates of a single test or the compatibility of different tests in the same laboratory. Tables for all these purposes are incorporated in International Rules for Seed Testing (ISTA, 1966).

The tetrazolium test

In the laboratory germination test viability is measured by stimulating germination and judging the capacity of the resultant seedlings to produce plants. Although by controlling the environment to provide optimal conditions the time taken to complete the test is reduced to a minimum, it may nevertheless take several weeks, and the period may be prolonged if measures have to be introduced to overcome dormancy. Any deviation from ideal test conditions can affect results, either by depressing emergence, by influencing the development of the essential structures or by favouring the spread of microorganisms. These disadvantages are eliminated if viability can be measured by means of a biochemical assessment of the metabolic activity of the resting seed.

This is the principle of the tetrazolium test, developed by Lakon (1949), in which tetrazolium salts are used to indicate the activity of enzymes of the dehydrogenase group, which are responsible for reduction processes in living tissue. The chemical is imbibed by the seed as a colourless solution and is reduced by the enzymes to a red-coloured, stable, non-diffusible substance, the formazan (Bulat, 1961). In the absence of active enzymes dead tissues remain unstained, and the distribution of living and dead areas of the embryo can be studied. A 1 per cent aqueous solution of 2,3,5-triphenyl-tetrazolium chloride or bromide is used; pH should be between 6 and 7, since the reaction occurs satisfactorily only in neutral solutions; the chemical is light-sensitive and undue exposure to light must therefore be avoided.

The most satisfactory method of preparing the seeds depends on the species, and there are at present no internationally agreed procedures, except for certain tree seeds (ISTA, 1966). Techniques for very many species have been worked out and published by Lakon and Bulat at Stuttgart-Hohenheim since 1942 (Lindenbein, 1965) and an extensive list of references to these is given by Bulat (1969): they form the basis for most methods used or described subsequently.

Cereal seed may be soaked in water at 30°C for about 16 hours, after which the embryos are excised, together with a thin layer of endosperm; they are then submerged in tetrazolium chloride at 30°C for 24 hours, followed by rinsing in water and removal of the endosperm layer (MacKay and Flood, 1968) (Fig. 7.5). Alternatively, after over-

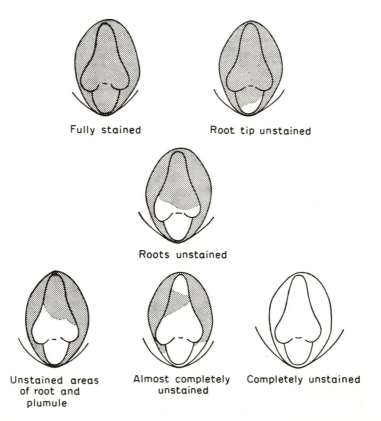

Fully stained   Root tip unstained

Roots unstained

Unstained areas   Almost completely   Completely unstained
of root and   unstained
plumule

FIGURE 7.5   Distribution of stained (live) and unstained (dead) areas in the rye embryo after treatment with tetrazolium chloride at different stages of storage. The two upper seeds are viable; the remainder are not. (From MacKay and Flood, 1968.)

night soaking in water the grains may be split longitudinally, bisecting the embryo, and soaked at 20°C for four hours (Cottrell, 1947, 1948). Although the latter method is somewhat quicker, complete evaluation of all parts of the embryo is impossible and defects in the scutellum and secondary roots may be missed. The larger seeded grasses can be soaked for 30 minutes in water at room temperature, after which the lemmas and paleas are removed and the caryopses soaked for a further five hours at 30°C; they are then halved transversely and the

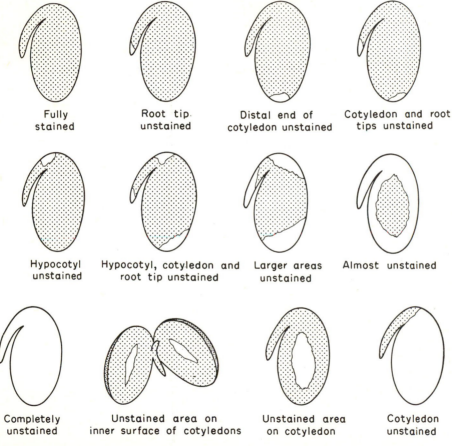

FIGURE 7.6 Distribution of stained (live) and unstained (dead) areas in the red clover embryo after treatment with tetrazolium chloride after different periods of storage. During deterioration there is usually an extension of dead tissue from points on the periphery corresponding with the corners of the seed. Seeds in the top row are viable; of the remainder only the seed showing the inner surface of the cotyledons unstained (bottom row) would germinate normally. (From MacKay and Flood, 1969.)

halves containing the embryos immersed in tetrazolium chloride solution for five hours (MacKay and Flood, 1969). Clover seed may be soaked in water for about 16 hours at 30°C; hard seeds are separated and the imbibed seeds immersed in tetrazolium chloride solution at 30°C for 24 hours, the seeds being assessed after rinsing in water and removal of the testas (MacKay and Flood, 1969) (Fig. 7.6). Brassica seed may be treated similarly, but the testas are removed before immersion in the tetrazolium chloride solution (MacKay and Flood, 1970).

Judgement of the ability of a seed to germinate is based on the degree to which the embryo is stained in the areas essential to growth. Complete staining of all parts is not necessarily essential for germination, but both the position and extent of necroses are critical. Thus a quite large unstained area at the distal end of the cotyledons of a clover embryo would not prevent germination, but a much smaller necrosis on the hypocotyl, by interfering with the movement of nutrients to the root, would result in the production of an abnormal seedling. Since the structure of the embryo varies in different species different systems of evaluation are essential for particular plant groups, and these have been described in the publications of the Hohenheim school (Bulat, 1969).

In the hands of a skilled technician the tetrazolium test can provide results agreeing closely with those of the laboratory germination test over a wide range of samples. But it fails to detect phytotoxicity caused by seed dressings, so that affected samples of no value for planting may be classed as of high viability. Heat injury caused by the use of excessive temperatures during artificial drying is also readily missed, although the presence of an unstained area at the centre of the scutellum is a valuable diagnostic feature of this condition in cereals (Wellington and Bradnock, 1964). The presence of living microorganisms, which are also stained by tetrazolium, may prove misleading if not recognised (Lakon, 1949).

Although the tetrazolium test cannot be regarded as a complete substitute for the laboratory germination test in all circumstances, it can be used in conjunction with it to provide additional information on the condition of a seed lot which may be valuable in assessing its potential planting or storage behaviour (MacKay, Tonkin and Flood, 1970; Moore, 1962) (See also pp. 96, and 228–229).

Other tests

While tetrazolium salts have proved the most successful chemical indicators of viability several other substances, such as indigo carmine and salts of selenium and tellurium, have also been used as vital stains (Moore, 1969). The level of viability in samples of seed of a number of species has been successfully deduced from X-ray photography. Solutions of salts of heavy metals, such as barium chloride, penetrate dead cells, but not living cells because of their semi-permeability: thus dead parts of the embryo and endosperm show up clearly as contrasted areas on X-ray photographs (Simak, Gustafsson and Granström, 1957). The seed is soaked at room temperature for 16 hours and then, after removal of excess water, transferred to a 20 or 30 per cent solution of barium chloride, usually for one hour (Kamra, 1964, 1966; Nakamura, 1968). After drying the seeds are radiographed using soft X-rays. The photographs are evaluated according to the extent to which the tissues have been impregnated; account has also to be taken of seeds which are without an embryo or have poorly developed embryos, since these would also be incapable of normal germination. An advantage of the X-ray method over the tetrazolium method is that evaluation can be confirmed by subsequent germination of the actual seeds examined, although some damage by the chemical has been reported (Nakamura, 1968) which might interfere with the validity of conclusions unless untreated seed was also tested. In certain tree seeds impregnation of live seeds by barium chloride has been observed, and for these the use of organic contrast agents, such as Urografin and Umbradil, has proved more reliable (Kamra, 1963).

Takayanagi and Murakami (1968) reported that the most noticeable biochemical characteristic for distinguishing between dead and viable seeds lay in the much greater exudation of sugars from the former when they were soaked aseptically in distilled water. They developed a test on this basis in which the germination capacity of rape, barley and rice is judged by the colour changes induced in urine sugar analysis papers by the exudates from either single seeds or a given weight of seeds (Takayanagi and Murakami, 1969a and b). The test is unsuitable for species in which the carbohydrate exuded is not primarily in the form of glucose, although other chemical analyses could be applied. However, a satisfactory correlation between exudate concentration and laboratory germination does not always hold good; Matthews and Bradnock (1967) found substantial differences in the exudation of soluble carbohydrates from samples of wrinkle-seeded peas of similar germination in the laboratory, although high exudation was associated with poor field emergence under early sowing conditions. This test is discussed further in Chapter 8 (pp. 232–233).

References to Chapter 7

ABDALLA, F. H., and ROBERTS, E. H., 1969. The effect of seed storage conditions on the growth and yield of barley, broad beans and peas. *Ann. Bot.*, **33**, 169–84.

ANDERSEN, A. M., 1950. The interpretation of normal and abnormal seedlings in some cereals. *Proc. int. Seed Test. Ass.*, **16**, 197–213.

ANDERSEN, A. M., 1954. A study of normal and abnormal seedlings of some small-seeded legumes. *Proc. Ass. off. Seed Analysts, N. Am.*, **44**, 188–201.

AUSTIN, R. B., 1963. Yield of onions from seed as affected by place and method of seed production. *J. hort. Sci.*, **38**, 277–285.

BALLARD, L. A. T., 1967. Effect of carbon dioxide on the germination of leguminous seeds. In *Physiology, Ecology and Biochemistry of Germination*, ed. H. Borriss, 209–19. Ernst-Moritz-Arndt-Universität, Greifswald.

BARTON, L. V., 1961. *Seed preservation and longevity*. Leonard Hill, London.

BARTON, L. V., 1965. Longevity in seeds and in the propagules of fungi. In *Encyclopedia of Plant Physiology*, ed. W. Ruhland, **15** (2), 1058–85. Springer-Verlag, Berlin.

BATTLE, J. P., and WHITTINGTON, W. J., 1969. The relation between inhibitory substances and variability in time to germination of sugar beet clusters. *J. agric. Sci., Camb.*, **73**, 337–46.

BEKENDAM, J., 1968. Germination of beet seed. *Proc. int. Seed Test. Ass.*, **33**, 308–13.

BEKENDAM, J., and BRUINSMA, J., 1965. The chemical breaking of dormancy of wheat seeds. *Proc. int. Seed Test. Ass.*, **30**, 869–86.

BEKENDAM, J., and BRUINSMA, J., 1966. The chemical breaking of dormancy of barley seeds. *Proc. int. Seed Test. Ass.*, **31**, 779–87.

BENTON, R. A., 1965. *Factors affecting water uptake by seed from various media*. Ph.D. thesis, University of Wales.

BLEASDALE, J. K. A., 1965. *The bed system of carrot growing*. STL 27. Ministry of Agriculture, Fisheries and Food, London.

BORTHWICK, H. A., 1965. Light effects with particular reference to seed germination. *Proc. int. Seed Test. Ass.*, **30**, 15–27.

BORTHWICK, H. A., HENDRICKS, S. B., PARKER, M. W., TOOLE, E. H., and TOOLE, V. K., 1952. A reversible photoreaction controlling seed germination. *Proc. natn. Acad. Sci. USA*, **38**, 662–66.

BORTHWICK, H. A., HENDRICKS, S. B., TOOLE, E. H., and TOOLE, V. K., 1954. Action of light on lettuce seed germination. *Bot. Gaz.*, **115**, 205–25.

BRANDT, F. O., 1964. Germat-germination testing equipment, a great help in germination. *Proc. int. Seed Test. Ass.*, **29**, 487–97.

BRETT, C. C., 1939. The production, handling, testing and diseases of seeds. *Ann. appl. Biol.*, **26**, 616–27.

BRETT, C. C., and DILLON WESTON, W. A. R., 1941. Seed disinfection. IV. Loss of vitality during storage of grain treated with organo-mercury seed disinfectants. *J. agric. Sci., Camb.*, **31**, 500–17.

BULAT, H., 1961. Reduktionsvorgänge in lebendem Gewebe, Formazane, Tetrazoliumsalze und ihre Bedeutung als Redoxindikatoren im ruhenden Samen. *Proc. int. Seed Test. Ass.*, **26**, 686–96.

BULAT, H., 1969. Keimlingsanomalien und ihre Feststellung am ruhenden Samen im topographischen Tetrazoliumverfahren. *Saatgut-Wirt.*, **21**, 575–79.

CHETRAM, R. S., and HEYDECKER, W., 1967. Moisture sensitivity, mechanical injury and gibberellin treatment of *Beta vulgaris* seeds. *Nature, Lond.*, **215**, 210–11.

CHING, T. M., and PIERPOINT, M., 1957. Evaluation of germinated seedlings of King Green beans by greenhouse planting. *Proc. Ass. off. Seed Analysts, N. Am.*, **47**, 122–25.

CHOWINGS, J. W., 1968. Testing the uniformity of seed lots. *J. natn. Inst. agric. Bot.*, **11**, 404–10.

CHOWINGS, J. W., 1970. Interaction of seed and substrate conditions in the laboratory germination of broad beans (*Vicia faba* L. var. *faba*). *Proc. int. Seed Test. Ass.*, **35**, 619–29.

CLARK, B. E., 1948. *Nature and causes of abnormalities in onion seed germination.* Memoir 282, Cornell University Agricultural Experiment Station, New York.

CLAYTON, L. C. W., and MACKAY, D. B., 1962. An automatic germination chamber. *Proc. int. Seed Test. Ass.*, **27**, 623–26.

COBB, R. D., and JONES, L. G., 1966. Development of a sensitive laboratory growth test to measure seed deterioration. *Proc. Ass. off. Seed Analysts, N. Am.*, **56**, 52–60.

COHEN, D., 1958. The mechanism of germination stimulation by alternating temperatures. *Bull. Res. Coun. Israel*, **6d**, 111–17.

COLBRY, V. L., 1965. Specifications for paper substrata. *Proc. int. Seed Test. Ass.*, **30**, 236–44.

COLLIS-GEORGE, N., and SANDS, J. E., 1961. Moisture conditions for testing germination. *Nature, Lond.*, **190**, 367.

COTTRELL, H. J., 1947. Tetrazolium salt as a seed germination indicator. *Nature, Lond.*, **159**, 748.

COTTRELL, H. J., 1948. Tetrazolium salt as a seed germination indicator. *Ann. appl. Biol.*, **35**, 123–31.

CROSIER, W., 1951. Blackening of seedlings of broad and velvet beans. *Proc. Ass. off. Seed Analysts, N. Am.*, **41**, 99–102.

CUDDY, T. F., 1959. Studies on the germination of sugar beet seed. *Proc. Ass. off. Seed Analysts, N. Am.*, **49**, 98–102.

ESSENBURG, J. F. W., and SCHOOREL, A. F., 1962. *Het verband tussen de Kiemkrachtsbepaling van zaaizaden in het laboratorium en de opkomst te velde.* Literatuuroverzicht No. 26. Centrum voor Landbouwpublikaties en Landbouwdokumentatie, Wageningen.

EVENARI, M., 1961. A survey of the work done in seed physiology by the Department of Botany, Hebrew University, Jerusalem (Israel). *Proc. int. Seed Test. Ass.*, **26**, 597–658.

EVENARI, M., 1965. Light and seed dormancy. In *Encyclopedia of Plant Physiology*, ed. W. Ruhland, **15** (2), 804–47. Springer-Verlag, Berlin.

FRITZ, T., 1966. Influence of temperature and parasitic fungi on germinating capacity of cereal seed. *Proc. int. Seed Test. Ass.*, **31**, 711–17.

GABER, S. D., and ROBERTS, E. H., 1969. Water-sensitivity in barley seeds. II. Association with micro-organism activity. *J. Inst. Brew.*, **75**, 303–14.

GADD, I., 1932. Undersökningar rörande förhållandet mellan grobarheten på laboratoriet och uppkomsten på fältet. *Meddn St. cent. Frökontrollanst.*, **7**, 87–133.

GRANT LIPP, A. E., and BALLARD, L. A. T., 1959. The breaking of seed dormancy of some legumes by carbon dioxide. *Aust. J. agric. Res.*, **10**, 495–99.

GRIFFITHS, D. J., and PEGLER, R. A. D., 1964. The effects of long-term storage on the viability of S.23 perennial ryegrass seed and on subsequent plant development. *J. Br. Grassld Soc.*, **19**, 183–90.

HARPER, J. L., 1959. The ecological significance of dormancy and its importance in weed control. *Proc. IVth int. Congr. Crop Prot.*, **1**, 415–20.

HARPER, J. L., and LANDRAGIN, P. A., 1955. The influence of the environment on seed and seedling mortality. IV. Soil temperature and maize grain mortality with especial reference to cold test procedure. *Pl. Soil*, **6**, 360–72.

HARPER, J. L., LANDRAGIN, P. A., and LUDWIG, J. W., 1955a. The influence of environment on seed and seedling mortality. I. The influence of time of planting on the germination of maize. *New Phytol.*, **54**, 107–18.

HARPER, J. L., LANDRAGIN, P. A., and LUDWIG, J. W., 1955b. The influence of environment on seed and seedling mortality. II. The pathogenic potential of the soil. *New Phytol.*, **54**, 119–31.

HARRINGTON, G. T., 1916. Agricultural value of impermeable seeds. *J. agric. Res.*, **6**, 761–96.

HEWETT, P. D., 1958. Effects of heat and loss of moisture on the dormancy of barley. *Nature, Lond.*, **181**, 424–25.

HEWETT, P. D., 1959. Effects of heat and loss of moisture on the dormancy of wheat, and some interactions with 'Mergamma D'. *Nature, Lond.*, **183**, 1600.

HEYDECKER, W., 1960. Can we measure seedling vigour? *Proc. int. Seed Test. Ass.*, **25**, 498–512.

HEYDECKER, W., 1962. From seed to seedling: factors affecting the establishment of vegetable crops. *Ann. appl. Biol.*, **50**, 622–27.

HEYDECKER, W., 1966. Clarity in recording germination data. *Nature, Lond.*, **210**, 753–54.

HEYDECKER, W., ORPHANOS, P. I., and CHETRAM, R. S., 1969. The importance of air supply during seed germination. *Proc. int. Seed Test. Ass.*, **34**, 297–304.

HITE, B. C., 1923. Effect of storage on the germination of bluegrass seed. *Proc. Ass. off. Seed Analysts, N. Am.*, **14–15**, 97.

HOLMES, G. D., 1951. Methods of testing the germination quality of forest tree seed, and the interpretation of results. *For. Abstr.*, **13**, 5–15.

Institute of Brewing, 1967. Recommended methods of analysis of barley, malt and adjuncts. *J. Inst. Brew.*, **73**, 233–45.

International Seed Testing Association, 1966. International rules for seed testing. *Proc. int. Seed Test. Ass.*, **31**, 1–152.

International Seed Testing Association, 1968. Interpretations of the international rules for seed testing, 1966. *Proc. int. Seed Test. Ass.*, **33**, 335–39.

JACOBSEN, I., 1910. Keimprüfung von Waldsamen. *Zentbl. ges. Forstw.*, **36**, 22–28.

JONES, L. G., and COBB, R. D., 1963. A technique for increasing the speed of laboratory germination testing. *Proc. Ass. off. Seed Analysts, N. Am.*, **53**, 144–60.

KÅHRE, L., KOLK, H., and FRITZ, T., 1965. Gibberellic acid for breaking of dormancy in cereal seed. *Proc. int. Seed Test. Ass.*, **30**, 887–91.

KÅHRE, L., and WIKLERT, P., 1965. Sand as a substrate for germination. *Proc. int. Seed Test. Ass.*, **30**, 245–50.

KAMRA, S. K., 1963. Studies on a suitable contrast agent for the X-ray radiography of Norway spruce seed (*Picea abies*). *Proc. int. Seed Test. Ass.*, **28**, 197–201.

KAMRA, S. K., 1964. Determination of germinability of cucumber seed with X-ray contrast method. *Proc. int. Seed Test. Ass.*, **29**, 519–34.

KAMRA, S. K., 1966. Determination of germinability of melon seed with X-ray contrast method. *Proc. int. Seed Test. Ass.*, **31**, 719–29.

KAMRA, S. K., 1968. Effect of different distances between water level and seed bed on Jacobsen apparatus on the germination of *Pinus silvestris* L. seed. *Stud. for. Suecica.*, **65**, 1–18.

KAMRA, S. K., 1969a. Studies on the effect of different distances between water level and seed bed on Jacobsen apparatus on the germination of *Picea abies* (L.) Karst. seed. *Sv. bot. Tidskr.*, **63**, 72–80.

KAMRA, S. K., 1969b. Further studies on the effect of different distances between water level and seed bed on Jacobsen apparatus on the germination of *Pinus silvestris* and *Picea abies* seed. *Sv. bot. Tidskr.*, **63**, 265–74.

KEARNS, V., and TOOLE, E. H., 1939. Relation of temperature and moisture content to longevity of Chewings fescue seed. *Tech. Bull. US Dep. Agric.*, No. 670.

KIETREIBER, M., 1962. Der *Septoria* – Befall von Weizenkornern (zur Methodik der Erkennung). *Proc. int. Seed Test. Ass.*, **27**, 843–55.

KIETREIBER, M., 1966. Atypische *Septoria nodorum* Symptome an Weizenkeimlingen (das Verhalten der Sorte Probus). *Proc. int. Seed Test. Ass.*, **31**, 179–86.

LAFFERTY, H. A., 1953. Abnormal germination in cereals. *Proc. int. Seed Test. Ass.*, **18**, 239–47.

LAKON, G., 1949. The topographical tetrazolium method for determining the germinating capacity of seeds. *Pl. Physiol., Lancaster*, **24**, 389–94.

LANG, A., 1965. Effects of some internal and external conditions on seed germination. In *Encyclopedia of Plant Physiology*, ed. W. Ruhland, **15** (2), 848–93. Springer-Verlag, Berlin.

LAWRENCE, W. J. C., and NEWELL, J., 1939. *Seed and potting composts*. Allen and Unwin, London.

LINDENBEIN, W., 1965. Tetrazolium testing. *Proc. int. Seed Test. Ass.*, **30**, 89–97.

MACKAY, D. B., 1961. The effect of pre-washing on the germination of sugar beet. *J. natn. Inst. agric. Bot.*, **9**, 99–103.

MACKAY, D. B., 1966. Seed testing – looking ahead after 50 years. *J. natn. Inst. agric. Bot.* **10** *Suppl.*, 42–46.

MACKAY, D. B., 1969. Report of the purity committee working group on crop seed definitions. *Proc. int. Seed Test. Ass.*, **34**, 551–62.

MACKAY, D. B., and FLOOD, R. J., 1968. Investigations in crop seed longevity. II. The viability of cereal seed stored in permeable and impermeable containers. *J. natn. Inst. agric. Bot.*, **11**, 378–403.

MACKAY, D. B., and FLOOD, R. J., 1969. Investigations in crop seed longevity. III. The viability of grass and clover seed stored in permeable and impermeable containers. *J. natn. Inst. agric. Bot.*, **11**, 521–46.

MACKAY, D. B., and FLOOD, R. J., 1970. Investigations in crop seed longevity. IV. The viability of brassica seed stored in permeable and impermeable containers. *J. natn. Inst. agric. Bot.*, **12**, 84–99.

MACKAY, D. B., and TONKIN, J. H. B., 1965. Studies in the laboratory germination and field emergence of sugar beet seed. *Proc. int. Seed Test. Ass.*, **30**, 661–76.

MACKAY, D. B., and TONKIN, J. H. B., 1967. Investigations in crop seed longevity. I. An analysis of long-term experiments, with special reference to the influence of species, cultivar, provenance and season. *J. natn. Inst. agric. Bot.*, **11**, 209–25.

MACKAY, D. B., TONKIN, J. H. B., and FLOOD, R. J., 1970. Experiments in crop seed storage at Cambridge. *Landw. Forsch.*, **24**, 189–96.

MANOHAR, M. S., and HEYDECKER, W., 1964. Effects of water potential on germination of pea seeds. *Nature, Lond.*, **202**, 22–24.

MARSCHALL, F., 1969. Eine neue Apparatur für die Keimprüfung nach der Jakobsenmethode. *Proc. int. Seed Test. Ass.*, **34**, 97–101.

MATTHEWS, S., and BRADNOCK, W. T., 1967. The detection of seed samples of wrinkle-seeded peas (*Pisum sativum* L.) of potentially low planting value. *Proc. int. Seed Test. Ass.*, **32**, 553–63.

MATTHEWS, S., and BRADNOCK, W. T., 1968. Relationship between seed exudation and field emergence in peas and French beans. *Hort. Res.*, **8**, 89–93.

MAYER, A. M., and POLJAKOFF-MAYBER, A., 1963. *The germination of seeds.* Pergamon Press, Oxford.

MILES, S. R., 1961. Germination variation and tolerances. *Proc. Ass. off. Seed Analysts, N. Am.*, **51**, 86–91.

MILES, S. R., 1962. Heterogeneity of seed lots. *Proc. int. Seed Test. Ass.*, **27**, 407–13.

MILES, S. R., 1963. Handbook of tolerances and of measures of precision for seed testing. *Proc. int. Seed Test. Ass.*, **28**, 525–686.

MOORE, R. P., 1962. Tetrazolium as a universally acceptable quality test of viable seed. *Proc. int. Seed Test. Ass.*, **27**, 795–805.

MOORE, R. P., 1969. History supporting tetrazolium seed testing. *Proc. int. Seed Test. Ass.*, **34**, 233–42.

MULLIN, J., 1968. Causes of variation in samples. *Proc. int. Seed Test. Ass.*, **33**, 235–39.

MUNN, M. T., 1950. *A method for testing the germinability of large seeds.* Bulletin 740. New York State Agricultural Experiment Station, Geneva, New York.

NAKAMURA, S., 1968. Determination of germinability of eggplant and pepper seeds with X-ray contrast method. *15th int. Seed Test. Congr.* Preprint 38.

NAKAMURA, S., WATANABE, S., and ICHIHARA, J., 1960. Effect of gibberellin on the germination of agricultural seeds. *Proc. int. Seed Test. Ass.*, **25**, 433–39.

NEERGAARD, P., 1965. Historical development and current practices in seed health testing. *Proc. int. Seed Test. Ass.*, **30**, 99–118.

NICHOLS, M. A., and HEYDECKER, W., 1968. Two approaches to the study of germination data. *Proc. int. Seed Test. Ass.*, **33**, 531–40.

NOBBE, F., 1876. *Handbuch der Samenkunde.* Wiegand, Hempel & Parey, Berlin.

Official Seed Testing Station for England and Wales, 1952. Fourteenth conference of seed analysts. *J. natn. Inst. agric. Bot.*, **6**, 257–74.

Official Seed Testing Station for England and Wales, 1964. Twenty-fifth conference of seed analysts. *J. natn. Inst. agric. Bot.*, **10**, 144–50.

Official Seed Testing Station for England and Wales, 1970. Thirtieth conference of seed analysts. *J. natn. Inst. agric. Bot.*, **12**, 207–13.

OOMEN, W. W. A., and KOPPE, R., 1969. Germination cabinets with day and night cycles. *Proc. int. Seed Test. Ass.*, **34**, 103–14.

ORPHANOS, P. I., and HEYDECKER, W., 1967. The danger of wet seedbeds to germination. *Rep. Sch. Agric. Univ. Nott.*, *1966–67*, 73–76.

OVERAA, P., 1960. On the value of hard seeds in red clover as judged by laboratory tests. *Proc. int. Seed Test. Ass.*, **25**, 422–31.

OVERAA, P., 1962. A new germination apparatus designed for alternating temperature and light exposure. *Proc. int. Seed Test. Ass.*, **27**, 742–47.

OWEN, E. B., 1956. *The storage of seeds for maintenance of viability.* Commonwealth Agricultural Bureaux, Farnham Royal, England.

PERRY, D. A., 1970. The relation of seed vigour to field establishment of garden pea cultivars. *J. agric. Sci., Camb.*, **74**, 343–48.

POLLOCK, J. R. A., 1962. The analytical examination of barley and malt. In *Barley and Malt*, ed. A. H. Cook, 399–430. Academic Press, New York and London.

REYNOLDS, J. D., 1955. Marsh spot of peas: a review of present knowledge. *J. Sci. Fd. Agric.*, **6**, 725–34.

ROBERTS, E. H., 1962. Dormancy in rice seed. III. The influence of temperature, moisture, and gaseous environment. *J. exp. Bot.*, **13**, 75–94.

ROBERTS, E. H., 1965. Dormancy in rice seed. IV. Varietal responses to storage and germination temperatures. *J. exp. Bot.*, **16**, 341–49.

ROBERTS, E. H., 1969. Seed dormancy and oxidation processes. *Symp. Soc. exp. Biol.*, **23**, 161–92.

SAUNDERS, C. B., 1923. *Methods of seed analysis.* National Institute of Agricultural Botany, Cambridge.

SEDGLEY, R. H., 1963. The importance of liquid-seed contact during the germination of *Medicago tribuloides* Desr. *Aust. J. agric. Res.*, **14**, 646–53.

SIMAK, M., GUSTAFSSON, A., and GRANSTRÖM, G., 1957. Die Röntgendiagnose in der Samenkontrolle. *Proc. int. Seed Test. Ass.*, **20**, 330–41.

SMITH, C. J., 1968. Seed dormancy in Sabi panicum. *15th int. Seed Test. Congr.* Preprint 24.

STAPLEDON, R. G., and ADAMS, M., 1919. The effect of drying on the germination of cereals. *J. Bd. Agric. Fish.*, **26**, 364–81.

STOKES, P., 1965. Temperature and seed dormancy. In *Encyclopedia of Plant Physiology*, ed. W. Ruhland, **15** (2), 746–803. Springer-Verlag, Berlin.

STOUT, M., and TOLMAN, B., 1941. Factors affecting the germination of sugar beet and other seeds, with special reference to the toxic effects of ammonia. *J. agric. Res.*, **63**, 687–713.

TAGER, J. M., and CLARKE, B., 1961. Replacement of an alternating temperature requirement for germination by gibberellic acid. *Nature, Lond.*, **192**, 83–84.

TAKAYANAGI, K., and MURAKAMI, K., 1968. Rapid germinability test with exudates from seed. *Nature, Lond.*, **218**, 493–94.

TAKAYANAGI, K., and MURAKAMI, K., 1969a. New method of seed viability test with exudates from seed. *Proc. int. Seed Test. Ass.*, **34**, 243–52.

TAKAYANAGI, K., and MURAKAMI, K., 1969b. Rapid method for testing seed viability by using urine sugar analysis paper. *Japan Agricultural Research Quarterly*, **4**, 39–45.

THOMSON, J. R., 1954. The effect of seed dressings containing an organo-mercurial and gamma BHC on germination tests of oats. *Emp. J. exp. Agric.*, **22**, 185–88.

THOMSON, J. R., 1962. A new seed germinator. *Proc. int. Seed Test. Ass.*, **27**, 675–78.

THOMSON, J. R., 1963a. New tolerances in seed testing. *J. natn. Inst. agric. Bot.*, **9**, 372–77.

THOMSON, J. R., 1963b. Seed dormancy. *Scott. Agric.*, **43**, 145–47.

THOMSON, J. R., 1965. Breaking dormancy in germination tests of *Trifolium* spp. *Proc. int. Seed Test. Ass.*, **30**, 905–9.

TIMSON, J., 1965. New method of recording germination data. *Nature, Lond.*, **207**, 216–17.

TOLMAN, B., and STOUT, M., 1940. Toxic effect on germinating sugar beet seed of water-soluble substances in the seed ball. *J. agric. Res.*, **61**, 817–30.

TOLMAN, B., and STOUT, M., 1944. Sheared sugar beet seed with special reference to normal and abnormal germination. *J. Am. Soc. Agron.*, **36**, 749–59.

TONKIN, J. H. B., 1969. Seedling evaluation: the use of soil tests. *Proc. int. Seed Test. Ass.*, **34**, 281–89.

United States Dept of Agriculture, 1952. Seed sampling and testing equipment. In *Testing agricultural and vegetable seeds, Agriculture Handbook No. 30*, 5–12. U.S.D.A., Washington.

VALLE, O., and MELA, T., 1965. Heikosti itävien kavätviljojen kylvösiemenarvosta. *Ann. Agric. Fenn.*, **4**, 121–33.

VERHEY, C., 1955. Germination equipment of the Seed Testing Station at Wageningen. *Proc. int. Seed Test. Ass.*, **20**, 5–28.

VERHEY, C., 1960. Is it still possible, with regard to modern views, to handle the conception 'germination energy'? *Proc. int. Seed Test. Ass.*, **25**, 391–97.

VERHEY, C., 1961. The influence of plumular damage on the yield of bush beans (*Phaseolus vulgaris* L.). *Proc. int. Seed Test. Ass.*, **26**, 162–69.

WELLINGTON, P. S., 1956. Effect of desiccation on the dormancy of barley. *Nature, Lond.*, **178**, 601.

WELLINGTON, P. S., 1957. Report of the committee on the effect of toxic substances in seed testing 1953–56. *Proc. int. Seed Test. Ass.*, **22**, 370–74.

WELLINGTON, P. S., 1965. Germinability and its assessment. *Proc. int. Seed Test. Ass.*, **30**, 73–88.

WELLINGTON, P. S., 1966. Seed production and seed testing. *J. R. agric. Soc.*, **127**, 164–86.

WELLINGTON, P. S., 1968. Seedling evaluation in germination tests on crop seeds; the essential structures of normal seedlings. *Proc. int. Seed Test. Ass.*, **33**, 299–307.

WELLINGTON, P. S., 1970. Handbook for seedling evaluation. *Proc. int. Seed Test. Ass.*, **35**, 449–597.

WELLINGTON, P. S., and BRADNOCK, W. T., 1964. Studies on the germination of cereals. 6. The effect of heat during artificial drying on germination and seedling development in barley. *J. natn. Inst. agric. Bot.*, **10**, 129–43.

WERNHAM, C. C., 1951. Cold testing of corn. *Prog. Rep. Pa. agric. Exp. Stn.* No. 47.

WHITTET, J. N., 1952. Essentials underlying selection of species for range and other dry-area zone reseeding. *Proc. 6th int. Grassld Congr.*, **1**, 521–25.

WITTE, B. O. H., 1928. On broken growths of leguminous plants, their causes, judgement and value. *Proc. 5th int. Seed Test. Congr.* International Institute of Agriculture, Rome.

WITTE, H., 1931. Some investigations on the germination of hard seeds of red clover, alsike clover and some other leguminous plants. *Proc. int. Seed Test. Ass.*, **3**, 135–47.

WITTE, H., 1934. Some international investigations regarding hard leguminous seeds and their value. *Proc. int. Seed Test. Ass.*, **6**, 279–310.

WITTE, H., 1938. New international investigations regarding the germination of hard leguminous seeds. *Proc. int. Seed Test. Ass.*, **10**, 93–121.

ZALESKI, A., 1957. Lucerne investigations. III. Effect of heat treatment on germination and field establishment of lucerne seed. *J. agric. Sci., Camb.*, **49**, 234–45.

CHAPTER 8

Vigour

W. Heydecker

In the scientific literature on seeds the word 'vitality' is taboo, and the word 'vigour' occurs almost exclusively in agronomic, rather than physiological, verbiage. Vigour is tantamount to the property of seeds which promotes a satisfactory performance; its antithesis has often been referred to as 'weakness' (de Tempe, 1961, 1964, 1966); and in this double sense it has become widely accepted. It is perhaps significant of the complexity and elusiveness of the term that the first documented use of its German near-equivalent was by Hiltner (Hiltner and Ihssen, 1911), a plant pathologist who showed that cereal seeds infected by *Fusarium* diseases were capable of germinating, but that the resulting seedlings were incapable of penetrating a 30–40 mm thick layer of Ziegelgrus (brick grit) of 2–3 mm particle diameter under specified conditions of temperature and moisture. He called this ability *Triebkraft*, and the term was later adopted, irrespective of the disease factor, to denote that seeds could not only germinate but also grow well. The term *Triebkraft* is, intentionally or not, a play on words, meaning both 'shoot strength' and 'driving force'. For a time this test was adopted in Germany (Eggebrecht, 1949) to indicate a quality of seeds beyond mere germinability; but it lost some of its appeal, and its result, expressed as a percentage of successful seedlings, was renamed *Ziegelgrus* value (Lindenbein and Bulat, 1955) when it became clear that, except for infected cereal seeds, the results obtained in brick grit were hardly ever worse, and sometimes better, than those from the official germination tests. The concept of *Triebkraft*, however, assumed a life of its own: it was translated into French and English as 'vigueur' and 'vigour', respectively. The original connotation was that of *seedling* vigour, the ability which enables a seedling upon germination to grow rapidly and well. Even here, however, objections can be raised: a cultivar may have been bred for dwarf growth; and the ultimate size of the plants may have little to do with their potential yield, either individually, where the product consists of fruits or seeds and not of the plant itself, or collectively, per unit area of ground, where close plant spacing can make up for small plant size. But within the same cultivar the seed lot* which produces plants with a faster growth rate is obviously the better one. The inherent vigour of the seedlings within a given lot is not usually uniform nor necessarily normally distributed; and moreover, individual seedlings may be more or less vigorous according to where they are growing: environmental factors greatly influence their performance (Laude and Cobb, 1969), and for a valid comparison between seedlings, whether of the same or different seed lots, growing conditions need to be strictly comparable. Both storability and versatile germinability depend on 'vigour factors' residing in the seed, but it is *seedling* vigour that is the ultimately most relevant expression of the seed quality. So what are the causes of differences in vigour in individual seeds, and how can this vigour be produced, maintained and improved at will?

Attempts to capture vigour

The vigour concept has been reviewed by (among others) Isely (1957), Moore (1962), Lindenbein and Bulat (1955), de Tempe (1961), Grahl (1965), Heydecker (1969), and in almost every report of the vigour Test Committee of the International Seed Testing Association (Schoorel, 1956, 1960; Heydecker, 1962, 1965, 1970b), but there is even now no agreement on a definition of vigour, and there is still no internationally accepted vigour test. There are, however, a number of so-called vigour tests which have been in operation on a more than experimental scale in certain seed testing establishments:

(a) The *Ziegelgrus* (brick grit) method which Hiltner and Ihssen (1911) devised to assess the ability of emerging cereal seedlings to penetrate a layer of ground brick. An adaptation by Fritz (1965) has been adopted in Sweden: here the obstacle to be penetrated is a layer of paper of a special type, overlain by sand.

(b) The cold test (e.g. Isely, 1950; Kietreiber, 1966), chiefly used for maize, to find out whether germinating seeds and emerging seedlings can survive suboptimal temperatures and the accompanying attacks of soil- and seed-borne microorganisms.

(c) The exhaustion test (Germ, 1960; Lindner, 1967) which measures the growth potential of seedlings before they begin to photosynthesise; this is again chiefly used on cereals.

(d) The electrical conductivity test (Matthews and Bradnock, 1967, 1968; Bradnock and Matthews, 1970), which assesses the

*For definition of 'seed lot' see Chapter 1 p. 11.

ability of cell membranes to prevent the leaching out of electrolytes.

(e) A more stringent interpretation of the tetrazolium test for viability (Lakon, 1945, 1950; Moore, 1962; Bulat, 1970) where living cells, i.e. those exhibiting dehydrogenase activity, are stained red but those which do not show this activity are not.

(f) The germination test iself (ISTA, 1966): this has tended to change into a vigour test through the adoption of the more rigorous standards of seedling evaluation (Wellington, 1970) which have in the course of time been accepted by the International Seed Testing Association as a realistic assessment of the capacity of seeds to give rise to good plants.

The diversity of these tests is a measure of the many shades of meaning which have, sooner or later, been attached to the term 'vigour'.

Schoorel (1956, 1960) and Isely (1957) combine to list the following conditions which influence the vigour of seeds: (a) weather during ripening and harvest; (b) post harvest treatments of the seeds (threshing, drying, cleaning and other operations); (c) duration and particularly conditions of storage; (d) the presence and activity of seed-borne microorganisms and possibly insects; (e) the wise or unwise use of chemical compounds ('fungicides', etc.); (f) genetic properties of the seeds.

The following definitions are a sample of definitions or circumscriptions of vigour.

According to Delouche and Caldwell (1960), seed vigour is, in a negative sense, generally thought of as 'something' not adequately measured or reflected by the standard germination test. On the positive side, no such precise definition is possible. Isely (1957) suggests that two ideas enter into most concepts of vigour: (1) rapidity of growth (e.g., Tables 8.1 and 8.2), and (2) non-susceptibility to unfavourable growing conditions, (e.g., Table 8.3), and he concludes: vigour is the sum total of all seed attributes which favour stand establishment under unfavourable field conditions, which in turn is primarily a function of the seed × microorganism interrelationship; a vigour test is an examination under specific environmental conditions to provide a means of detecting differences which are not discernible in an ordinary laboratory germination test. Ader (1965) states: 'vigour (*Triebkraft*) is the percentage of seeds able to produce . . . normally germinating seedlings, even though conditions are suboptimal.' (Note here that vigour 'equals' a percentage.) According to Bulat (1962), a test for the determination of vigour (*Triebkraft*) is a

TABLE 8.1   *Effects of storage conditions on seed deterioration of crimson clover seeds* (Trifolium incarnatum). (*After Ching and Schoolcraft, 1968.*)

Seed age	Original seed water content, %	Storage temperature, °C	Final seed water content, %	Germination, %	Seedling length,[a] mm	Conductivity,[b] µmho/g	amino acids µg/g dry seed	leached, %
new	5.7	—	5.7	99	84	48	513	0.5
10 yrs	6	3	4.4	98	83	70	299	23.4
10 yrs	12	3	11.7	98	76	89	377	23.8
10 yrs	6	variable	4.4	98	82	44	487	78.3
10 yrs	8	variable	6.9	99	74	51	498	22.1
10 yrs	6	22	4.4	94	77	49	430	28.4
10 yrs	8	22	6.8	86	55	78	368	37.0
10 yrs	6	38	4.5	89	67	57	375	38.7
10 yrs	8	38	6.8	0	—	74	650	69.1

[a] Difference required for significance (P = 0.05): 7 mm.
[b] Of water in which seeds have been steeped.

TABLE 8.2   *Effects of accelerated ageing treatment (45°C and saturated atmosphere) on cotton seeds* (Gossypium barbadense) *of the same original seed lot. (After Delouche, 1969.)*

Degree (time) of ageing	Slight	Moderate	Drastic
Germination, per cent	84	80	78
Germination in cold test, per cent	84	56	39
Seedling length after 5 days, mm	207	150	110
Free fatty acids in seeds, per cent	1.2	1.4	1.5
Relative yield of seed cotton, per cent	100	91	84

germination test under aggravated conditions, designed to discover and show the weakness of seeds. Neeb (1970) defines vigour (*Trieb-kraft*) – or rather one type of vigour – as the totality of properties contributing to the defence against, and successful resistance to, biotic and abiotic hazards during germination under suboptimal conditions . . . a gradual property, i.e. each normal seedling has a more or less high degree of vigour. Earlier, I had defined vigour as the ability to germinate and produce a stand in a suboptimal environment (Hey-decker, 1960); but later I suggested that vigour is a scientifically vague term which when applied to seeds is taken to denote that they are likely to . . . perform particularly well in the field, better than others

TABLE 8.3   *Effect of seed vigour on percentage seedling establishment : Field emergence of sugar beet* (Beta vulgaris) *sown on successive dates in Scotland, 1970. (With acknowledgments to Dr D. A. Perry for permission to use his unpublished data.)*

Seed lot	Sowing date			
	7th April	28th April	4th May	12th May
A	83	82	77	88
B	80	82	71	86
C	72	76	51	76
D	65	72	47	71
P for differences between seed lots	0.01	not significant	0.01	0.001
Mean daily temperature, °C	7.5	12.2	13.2	13.8
Total rainfall during the week after sowing, mm	28	33	40*	26

*Half of this fell during the 3 days immediately after sowing; no other sowing received so much so soon.

which may be equally satisfactory in the laboratory test (Heydecker, 1970a). (Note that the relevance of an adverse environment was no longer stressed.)

Vigour is viewed in a more positive sense by Nutile (1964) as the ability of the seeds to produce vigorous seedlings as compared to the maximum vigour attainable for the species under similar conditions, and by Woodstock (1969b) as that condition of active good health and natural robustness in seeds which, upon planting, permits germination to proceed rapidly and to completion under a wide range of environmental conditions. Grabe (1966) too prefers to reserve the term vigour for positive qualities of seedlings, such as the increased growth rate associated with hybridisation and seed size, but goes on to define 'poor seedling vigour' as 'a result of seed deterioration'. Germ (1960) defines vigour as the ability of seeds to produce seedlings well capable of increasing in length and volume while still dependent on their own reserves. Harrison (1966) – terse and strictly cytological: the vigour of seedlings can be assessed from the number of anaphases seen per root. Lindner (1967) – a tautological bullseye: vigour (*Triebkraft*) is the efficiency of seeds; likewise Delouche (1968): vigour is the physiological stamina of seeds. And, summing it up nicely, provided the seed is already a seedling, Qualls and Cooper (1969): seedling vigour is usually characterised by the weight of a seedling after a period of growth in a given environment ... the amount of reserves that are present, the rapidity with which they are mobilised, and the efficiency of their metabolism are all important aspects of seedling growth and development.

However defined, the phenomenon of vigour is of necessity influenced by many of the subjects dealt with elsewhere in this volume. And perhaps it is most fittingly described as the condition of a seed which is at the height of its potential powers, when all factors that may detract from its quality are absent and those that make up a 'good' seed are present in the right proportions, promising a satisfactory performance over a maximum range of environmental conditions.

Synoptically, vigour might be circumscribed by such terms as integrity, robustness, sturdiness, adaptability, flexibility, resilience – none of them scientifically respectable. Analysed, vigour results from the combination of all the following: adequate maturity at harvest; absence of physiological ageing, i.e. physiological deterioration or presence of necroses which are usually started by mechanical damage but increase in extent due to either microbial activity or auto-toxic reactions, or both; all making for a well-functioning enzymic synthesising biological apparatus.

A serious student of the vigour complex cannot be content with either the ecological (survival-under-adverse-conditions), compared with the inherent approach: it is the seed and not the environment that is the centre of attention; nor with the collective, compared with the individual, approach; and although a batch of seeds is often referred to as 'the seed' (*das Saatgut*), as if it were an organic whole, it is made up of individuals; and it is each one of these that is more or less vigorous.

Dormancy, lability, disease; or what lack of vigour is not

During and until shortly after maturation, most seeds go through a shorter or longer 'dormancy' period. But this dormancy is something essentially different from lack of vigour. Seeds may be dormant because they have water-impermeable coats or immature embryos or are prevented from germinating by soluble inhibitors. Moreover, recently matured seeds of many species are in a condition where, according to Vegis (1964) they will only germinate over a very narrow range of environmental conditions (frequently temperatures). Only as time passes this range widens and germination becomes more likely; and this is precisely the difference between this 'primary dormancy' and 'lack of vigour': dormancy can be 'broken' and thereupon the performance of the seed improves; low vigour inevitably leads to lower vigour.

Seeds which will germinate under certain, but not under other, apparently little different, conditions, have, in Lakon's widely accepted terminology (Lakon, 1950), a *labile* germination tendency (*Keimtendenz*). Unfortunately, this lability can be caused either by partial dormancy or by ageing, and sometimes it is not immediately obvious whether a seed is labile because it is too young or because it is too old to have a 'stable germination tendency'. Seeds like those of *Beta vulgaris* (Battle and Whittington, 1969; Heydecker and Chetram, 1971), whose surrounding dispersal unit may contain water-soluble germination and growth inhibitors, may present a different problem – given marginal moisture, these chemicals may either prevent germination or, if not, inhibit growth. In such cases washing, followed by drying, before sowing will enable the seedlings to grow more 'vigorously'.

Seed-borne sources of disease present a further problem. They can severely endanger seedling and plant vigour, but whether, how much and when they are going to attack depends on their nature and on the ecological balance of all the microorganisms in and around the seeds (Thomson, 1970).

Many disease organisms can be eradicated at source (Maude, Vizor and Schuring, 1969): and with luck or care, diseased seeds may still produce vigorous seedlings. Conversely, the absence of a disease organism does not guarantee the vigour of a seed.

What vigour is: the physiological approach

THE SIZE OF SEEDS

Seeds are easily separated by diameter, weight or density (apparent specific gravity). For distinctions in vigour, it is the weight and the density, rather than the volume of seeds that are important, and the 'thousand seed weight' of seed lots is in fact often given as part of the sales information. There are two possible but incongrous agronomic reasons for this: sowing rates are more logically adjusted by number than by weight; and larger seeds are often considered better seeds (see pp. 127–132).

Often larger seeds, with more 'initial capital', do have at least an intitial advantage over smaller ones. Nobbe (1876) quotes results obtained with peas (Table 8.4). Schachl (1970), working with red clover, and Scaife and Jones (1970), who studied lettuce, present supporting evidence. Hewston (1964) found that many vegetable species almost invariably produce larger seedlings when grown from

TABLE 8.4 *Effect of 1000-seed weight of peas* (Pisum sativum) *on establishment and yield.* (*After Nobbe, 1876, as quoted by Schachl, 1970.*)

Seed size	Number of seeds per 1000 g	Number of plants established from 1000 g seeds	Relative yield (mean = 100)
Small	4129	3611	78
Medium	2820	2689	109
Large	2040	1914	113

larger seeds, but that in the field such differences usually disappear within the first few weeks of growth. Oelke, Ball, Wick and Miller (1969) showed that both size and density of rice grains are important; but grain dry weight levels off during maturation before the lowest water percentage is reached, i.e. density still increases; and the seedlings with the best growth rate come from the grains with the lowest water content (13 per cent).

Similarly, Williams, Black and Donald (1968) have shown that,

TABLE 8.5 *Effect of age and size of wheat seeds* (Triticum vulgare, *cv. 'Elmar'*) *on their performance.* (*After Kittock and Law, 1968.*)

Seed age, years		Germination, %	Emergence from soil, %	Shoot dry weight per plant, mg[a]	'Speed'[b] of emergence (higher figures mean greater speed)
1		97	77	108	28.5
3		96	55	107	26.4
6		98	39	103	20.8
Seed diameter, n/64 inch	Mean seed weight, mg				
7	37	95	32	35	13.6
6	28	97	33	25	13.5
5	22	92	19	24	12.9

[a] 56 days from sowing for seed age, 21 days for seed size.

[b] Speed of emergence $= \Sigma \dfrac{n}{D} \times \dfrac{100}{\Sigma n}$, where D = days from sowing and n (for seed age) = daily stand, and n (for seed size) = daily emergence. Formula adapted from original to separate time to emergence from percentage emerged.

TABLE 8.6 *Effect of size of carrot seeds* (Daucus carota) *on germination, seedling emergence and yield of roots. (After Austin and Longden, 1967.)*

Seed diameter, mm	Seed weight, mg	Per cent germination (mean of 3 treatments)[a]	Per cent emergence above ground (mean of 3 treatments)[a]	Root weight, g,[b] after	
				15 weeks	24 weeks
1.00–1.25	0.67	48	18	1.73	3.67
1.25–1.50	0.91	69	37	1.85	3.72
1.50–1.75	1.24	83	49	1.95	3.58
1.75–2.00	1.51	85	45	1.99	3.72

[a]1962.
[b]1964.

within a cultivar, and with nutrients and moisture non-limiting, the early effect of seed weight on growth is essentially linear for a considerable time. But from a certain stage on, the volume of the maturing seed may not increase and may even shrink as water is lost; its fresh weight may remain constant, whilst its dry weight will still increase. Therefore high density is probably a better guide to maturity than is size.

But seed volume can matter, if other things are equal; for instance Kittock and Law (1968) demonstrated this on wheat (Table 8.5), Houghton (1969) with cauliflower, and Austin and Longden (1967) found that within any one properly matured sample of carrot seeds there was a distinct advantage of large over small diameter seeds, both in germination percentage and more particularly in seedling emergence (Table 8.6), and the roots grown from large seeds remained heavier for the first four months of growth. However, to some extent seed size may be a reflection of seed maturity: in Austin and Longden's work the time of harvest influenced the performance of all, but particularly of the small carrot seeds (Table 8.7). Though total fresh weight hardly changed throughout the harvesting period, the quantity and nature of the seed reserves clearly did improve. The number of

TABLE 8.7 *Per cent germination of carrot seeds* (Daucus carota) *harvested on different dates during 1962 in England. (After Austin and Longden, 1967.)*

Seed diameter, mm	5th Sept.	12th Sept.	20th Sept.
1.00–1.25	34	45	66
1.75–2.00	82	84	90

TABLE 8.8   *Effect on root growth of treatment of cotton seeds* (Gossypium barbadense) *with vitamins. (After Ovcharov, 1969, quoting Kalieva.)*

	Length of main root, mm			Number of lateral roots		
Weight of 1000 seeds	104g	114g	129g	104g	114g	129g
Treatment						
Water	146	186	211	2	4	5
Nicotinic acid, 100 ppm	157	183	258	8	10	11
Vitamin B₆, 100 ppm	163	224	272	8	14	8

large seeds able to germinate increased by only 10 per cent when the harvest was delayed by a fortnight, whilst with small ones it doubled, and these seedlings had a better relative growth rate than those from seeds of equal volume harvested prematurely.

More recently Austin, Longden and Hutchinson (1969) found that when carrot embryos were allowed to enlarge by letting the seeds imbibe carefully controlled quantities of water, and this was followed by drying back, this advancement resulted in earlier emergence, larger seedlings after unit time and superior ultimate yield of edible roots from the seeds so treated. The authors suggest that increase in embryo size resulting from the treatment may be analogous to the extended ripening process which in a favourable climate takes place on the mother plant. But additional metabolically advancing effects of the pre-treatment (Henckel, 1967) are not ruled out and will be discussed later.

Size, however, is only one relevant factor contributed by the mother plant. Ovcharov (1969) for instance, whilst reporting striking differences in the biosynthetic activity of seedlings from seeds of different weights, e.g. with cotton (Table 8.8), also showed that the position of the seed on the mother plant or even on the same inflorescence, e.g. of maize (Table 8.9), can greatly affect its biochemical constitution. Whalley, McKell and Green (1966), working on the competitive ability of different grass species, found, as did McDaniel (1969), that the nature of the species and even the cultivar largely overrides the effect of seed size, and comparisons across these divisions are largely meaningless. But within a cultivar large seeds have two advantages: the initial growth rate of their seedlings is higher, and if forced to continue to grow in the dark they are capable of growing to a larger size (Fig. 8.1). Whalley *et al.* conclude that the essential aspects of 'growth vigour' of dark-grown seedlings are (1) rapidity of germination, (2) rate of extension of root and shoot, and (3) overall efficiency

TABLE 8.9 *Effect of the position of seeds in the inflorescence on their chemical composition.* (*After Ovcharov, 1969.*)

Protein concentration in wheat caryopses

Spikelet zone	Position of caryopsis	Protein, %
Lower	Outer	16.6
Middle	Outer	19.1
Middle	Inner	16.4

Concentration of vitamins in different parts of corn cobs (µg/g)

Part of cob	Vitamin B₂	Nicotinic acid
Upper	5.0	22.6
Middle	3.3	28.9

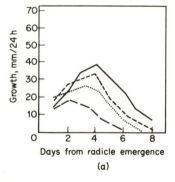 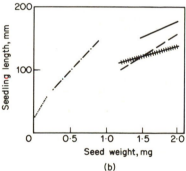

(a)   (b)

FIGURE 8.1   (a) Effect of seed weight on the heterotrophic growth rate of seedlings.
———— large seeded *Phalaris caerulescens*
– – – – small seeded *Phalaris caerulescens*
·········· large seeded *Oryzopsis miliacea*
— — small seeded *Oryzopsis miliacea*
   (b) Regression lines of seedling length 14 days after radicle emergence on seed weight.
· · · · *Schismus arabicus*
— —· *Oryzopsis miliacea*
———— *Phalaris caerulescens*
ⵙⵙⵙⵙⵙ *Phalaris tuberosa* 'hirtiglumis SCN 872'
— — *Phalaris tuberosa* 'hirtiglumis SCN 850'
(After Whalley, McKell, and Green, 1966.)

of conversion of endosperm reserves to seedling length. But beyond this they point out that between the initial heterotrophic subterranean growth phase during which a seedling lives completely on its seed reserves and the final autotrophic, i.e. photosynthesising, phase there occurs an important transition phase: here, despite incipient photosynthesis, an appreciable amount of reserves is still intact and is in fact drawn upon, to boost the growth of the seedling above ground.

Like Whalley *et al.*, Voigt and Brown (1969), whilst acknowledging the possible importance of seed size for seedling vigour, report success with breeding for seedling vigour regardless of seed size.

McDaniel (1969) analyses the effects of seed size in barley (*Hordeum vulgare*) on seedling vigour, expressed as seedling size after three days. He suggests that mitochondrial efficiency, at least as a link in the metabolic chain, is responsible for the vigour of seedling growth; and since the specific mitochondrial activity remains constant over a wide range of seed sizes, within a given cultivar the larger seeds produce the more vigorous seedlings. However, the mitochondrial activity, and hence the seedling vigour, may differ between equal-sized seeds of two different cultivars; more particularly, F_1 hybrids exhibit a greater activity and hence produce more vigorous seedlings than their parents. The size of seeds is therefore an indication of their vigour only if they are strictly comparable in all respects.

HYBRID VIGOUR

It is still not known why seeds derived from crossing inbred lines often exhibit heterosis or hybrid vigour, but the results of Göring (1967), working with maize (*Zea mays*), are of interest: in his studies parents and hybrids did not differ in their rate of oxygen uptake *per unit fresh weight*; but the cells of the hybrid were significantly smaller, with a significantly lower, i.e. more economical, respiration rate *per unit cell number*, than those of the parents.

The history of the seed

MATURITY

The importance of the seed maturity has already been mentioned. Klein (private communication, 1969) terms the final dehydration stage of maturation 'organised disorganisation' and emphasises its physiological importance. Perry (1969a) stresses that when seeds are harvested in the immature state or have prematurely experienced fierce natural drying conditions, such as 'atmospheric drought' produced, say, by a hot wind, this could seriously weaken a prospective plant; and that this immaturity is likely to be a more important cause than faulty storage of the wide differences in vigour of some quite young lots of pea seeds. Perry and Harrison (1970) consider that under wet seed-bed conditions it is the poorly matured pea seeds that are predisposed to damage by the sudden inrush of water. 'Death probably occurs because the seeds . . . fail to achieve the sub-cellular coordination necessary for the successful resumption of the active state.' Indeed, in Perry and Harrison's view the much-stressed effect of inimical microorganisms, though it has to be reckoned with, may well play a secondary role under such circumstances. In a similar vein Rampton and Lee (1969) also showed that despite identical ultimate seed weights rapid artificial drying at 35°C of *Dactylis glomerata* caryopses can dramatically reduce germination capacity as compared with gradual drying in the windrow.

Khan and Laude (1969), working with barley, go further: they report that, depending on its timing, two hours' heat stress (about 50°C applied to the parent plant) could either improve (through an effect on an inhibitor) or delay and depress (probably through an effect on seed membranes) the emergence of the seedlings, and they suggest that quite a brief period of environmental stress during seed maturation may well explain differences noted in the germinability of seeds produced in successive years at the same location or in the same year at different locations.

GEOGRAPHICAL FACTORS

Other examples exist of the profound effects of the environment during seed production. In particular, the geographic location of a seed crop can influence the composition and thereby the vigour of seeds. According to Marshall (1969), in some regions the ratio of

TABLE 8.10 *Effect of the location of the mother crop on the composition of 'fully ripe' wheat caryopses grown in Southern and Northern Siberia. (After Galatschalowa and Marussina, 1967.)*

Year of production	Region where grown	Per cent of dry matter		
		Total sugars	Starch	Protein-N
1961	South	2.21	63.9	2.72
	North	2.82	58.3	1.64
1962	South	1.99	58.4	2.80
	North	3.50	51.8	2.13

phosphate and potassium to nitrogen in the endosperm of 'mature' oat grains is consistently high enough to impart cold-hardiness to the resulting seedlings, but in others it is not. Pollock and Toole (1966) had analogous results with lima beans. Galatschalowa and Marussina (1967) reported similar findings: wheat grains produced in Northern Siberia do not compare well for yielding ability with those produced in Southern Siberia, which are higher in nutrient concentration when both groups are as ripe as they will ever grow (Table 8.10). Put differently, in certain locations the maturest seeds are still immature.

THE EFFECT OF OVER-DRYING ON GERMINATION

Even when the seed crop is home and dry it is not out of danger. Although it was shown in Chapter 2 that the lower the moisture content the better the preservation of the seed, it was also pointed out that there is evidence of damage in some cases if desiccation is too severe. Nutile (1964) has shown that rapid soaking, instead of slow imbibition, increases the appearance of this severe desiccation damage. Before affecting the germination *percentage* or the normal growth of seedlings, over-drying increases the time to germination; in more extreme cases it causes abnormalities in germinating seedlings which are, be it noted, indistinguishable from those due to over-moist storage conditions: this suggests that storage microorganisms are not a major cause of deterioration in moist storage. Moreover, when even after years of storage (of sorghum seeds at least) careful rehydration precedes germination conditions this can 'reverse' the damage from over-desiccation. (See also pp. 37–39.)

THE IMPORTANCE OF A GOOD START

The efficiency of the events which are peculiar to the very outset of the germination process is largely responsible for the well being and efficiency of the resulting seedling. Because living and dead seeds *appear* to take up water at similar rates during the early stages of imbibition the initial phases of germination are often thought of as 'purely passive' and preliminary to any physiological activity. But there is evidence of enormous effects of the environment at the very start of imbibition. The work by Pollock and Toole (1966) and by Woodstock and Pollock (1965) on Lima bean seeds provides a striking example. These seeds are susceptible to chilling over a period limited to the earliest stages of rehydration. Even when followed by favourable temperatures, a temperature of 15°C or less during the first hour of imbibition significantly depresses the subsequent rate of respiration of their embryonic axes and the growth rate of the seedlings. Woodstock and Pollock conclude that the early period of imbibition is critical because at this time respiration has to be rapid enough to supply sufficient energy for the *orderly* rehydration and stretching of the membranes within the embryonic axis. Anaerobic conditions at these critical sites lead to irreversible membrane damage and hence to impairment of growth and to attacks by microorganisms, encouraged by the leakage of substances through the membranes.

Referring to the classical work by Kidd and West (1918–1919), Woodstock and Pollock find that their own results are an example of 'physiological predetermination'. But their predetermining treatment, and similarly those reported by Highkin and Lang (1966) and by Durrant (1962) who both worked on the effects of the history of the parent generation, all had *detrimental* effects: the 'predetermination' here really amounts to predeterioration: the physiological potential of the seeds has in some way been impaired at the outset. Conversely, Pollock and Toole (1966) found that provided the germination temperature is high enough at the very start, lima bean seedlings become subsequently much more resistant to chilling. Similarly, Orphanos and Heydecker (1968) found that a brief initial aerobic imbibition period can sometimes prevent the permanent soaking damage to which French bean (*Phaseolus vulgaris*) seeds are known to be highly susceptible (See also pp. 38–39 and 159–162.).

Manifestations and measurements of vigour

It is often assumed that the percentage germination of batches of seeds in the laboratory is closely reflected by their field performance and, in fact, quite a good correlation between fitness and germination percentage often exists (Abdalla and Roberts, 1969) (see Chapter 10). Nevertheless, often seed lots of a given cultivar with similar laboratory germination percentages do not become established equally well, and vigour tests are being sought which do not so much forecast the actual establishment in the field (how could they? what field?) but which give a greater estimate of the relative capacity of different seed lots to cope with field conditions. Generally speaking the lots which have become established best will also tend, on an individual plant basis, to grow and yield best afterwards: the early vigour persists (See also pp. 193–195.).

YOUTH, VIGOUR AND ENERGY

It is sometimes said that vigorous seeds are 'physiologically young'. Faulty storage conditions, too moist and/or too warm, can 'age' even a chronologically young seed enormously fast (see Chapter 2).

According to Grabe (1966), quoting Zeleny (1954), many detrimental physiological changes (also listed by Abdul-Baki, 1969, see page 226) occur as seeds age in storage, as a result of which susceptibility to seed-rotting organisms increases and the speed of germination and the seedling growth rate decrease; but – very important – germination percentage begins to be reduced only after all these changes have occurred to some extent. Thus according to Grabe (1964) and Delouche, Rushing and Baskin (1967), the sequence of deterioration is usually, so to speak, back to front: yield declines first, growth second, and then the ability of the seed to get established in the field, and germination capacity last, so that deterioration may set in well before it is recognisable from the results of a germination test, or even of a modified germination test under environmental stress. In contrast to these conclusions, however, the results of Abdalla and Roberts on peas, beans and barley, described in Chapter 10, suggest that little deterioration of final yield occurs before the seed deterioration has been reflected in some loss of germination capacity. Furthermore in their experience declines in rates of seedling growth occur before detectable losses in final yields.

SPEED OF GERMINATION

Vigorous seeds, except when dormant, will normally be expected to germinate rapidly; but seed-bed conditions may not permit germination immediately after sowing, and in this case a vigorous seed is capable of surviving until conditions improve and then still goes on to produce a vigorous and healthy seedling and a good crop. (Here we have a combination of the two, only tenuously related, aspects of vigour.) But even in an environment that favours germination, the speed of germination is not always an essential component of seedling vigour. For instance, seeds which have matured particularly well (Hepton, 1957) may imbibe, and therefore germinate, more slowly than less fortunate ones; furthermore, seeds infected with seed-borne pathogens may occasionally be stimulated into earlier germination than healthier ones (Nancy Montgomery, personal communication). Again, seeds may be slow to germinate because they are still partly dormant. For these, among other, plausible reasons the term 'germination energy' has entirely disappeared from the vocabulary of seed technologists. But for a short period, terminated by a frontal attack by Verhey (1960), it had the definite meaning of *time to germination*, or rather the percentage of seeds capable of germinating within, for the species in question, a short time from the start of the germination test in the laboratory.

And indeed, when the germination performance of the seed lot is compared with *its own performance at an earlier stage in an identical environment*, a delay in germination and often also a greater scatter in the time to germination of individual seeds, is an expression of deterioration, which moreover is easy to measure. Delouche (1969) who finds that germination speed declines well ahead of germination percentage makes a plea for the reinstatement of germination speed (or energy) as a meaningful parameter. In agreement with this, Broniewski (1967) found that the intensity of cell division during the early phase of germination declines well ahead of the germination percentage. Wanjura, Hudspeth and Bilbro (1969), like Whalley, McKell and Green (1966) and Delouche (1968), stress the relevance of early emergence of the seedlings above ground. They find that this is much more dependent on seed quality than on sowing depth and that the ultimate yield is better correlated with early emergence than with most other parameters. Attempts to express germination speed and percentage in one combined index number are well meant but confusing (Heydecker, 1966).

It is worth noting that for agronomic purposes a logical bridge between germination and vigour tests is formed by seedling evaluation, whether in the shape of the exhaustion test (Germ, 1960) or the 'early growth' test which is now widely favoured (e.g. Perry, 1969b, Tonkin, 1969) and which has been accepted into the International Rules for Seed Testing (Wellington, 1970). Here three components are judged: (a) the rapidity of germination, (b) the growth rate after germination, and (c) the integrity and normality of the seedlings.

PHYSIOLOGICAL AND BIOCHEMICAL TESTS FOR VIGOUR

Cooper and Qualls (1969) and Qualls and Cooper (1969) have pointed out that measurements of seedling vigour can be made *via* the relative growth rate once the seedlings are photosynthesising, but that the qualities required while they are not yet above ground may be very different from those needed for growth in the light. For the preceding, heterotrophic, phase, the capacity for rapid and continued extension growth by the seedling is a useful index of yield potential, at least of birdsfoot trefoil (*Lotus corniculatus*) (Table 8.11). A test of the potential endurance of seedlings growing in the dark is more easily standardised than is one involving light. Germ (1960) and, following him, Lindner (1967) have developed the *Keimrollen* (upright rolled paper towel) germination test into the so-called exhaustion test in which seeds are mounted in rolls of filter paper and placed in the dark at a favourable temperature for a number of days, after

TABLE 8.11   *Relationship of performance early in life to yield of four cultivars of Birdsfoot Trefoil* (Lotus corniculatus). (*After Qualls and Cooper, 1968.*)

Cultivar	Emergence from soil after 3 days at 26.7°C, per cent of final emergence		'Speed' of elongation in the dark		Yield after 4 weeks, mg per plant in growth room		Yield at full bloom, g per m of row in field	
Empire	10	c	24.9	b	32.6	c	125	c
Viking	26	b	25.4	b	50.5	b	225	b
Tana	22	b	25.6	b	52.0	b	259	b
Leo	51	a	27.2	a	65.0	a	316	a

In each column, values not carrying the same letter differ significantly ($P = 0.05$). Seed respiration rates of the four cultivars during germination did not differ significantly.

which the length and health of the resulting seedlings are measured.

Attention has however recently been focused more and more upon tests which are not growth tests at all. Here measurements are made within a day or even within hours of the beginning of incubation, or even on the unimbibed seed. In a literature survey on biochemical changes known to be associated with the reduction in vigour or germinability, Abdul-Baki (1969) lists four categories: (1) a decline in metabolic acitivity and its manifestations (e.g. reduced respiration, slower seedling growth, and lower germination); (2) an increase in the total activity of certain enzymes (e.g. phytase, proteases, phosphatases); (3) a decrease in the activity of other, chiefly respiratory, enzymes (e.g. catalase, peroxidase, dehydrogeneses, cytochrome oxidases, glutamic acid decarboxylase); and (4) an increase in membrane permeability and thereupon greater leakage of sugars, amino acids and inorganic solutes from the seed, to which should be added the increase in free fatty acids mentioned by Zeleny (1954).

Biochemical tests for the vigour status of seeds can thus be divided into three related groups: they may show the unsound areas of seeds which are faulty; they may use the response of an essential enzyme as an index of general vitality; or they may monitor a more complex process vital for the seed in the early stages of germination. As Woodstock (1969a) says, such a process should ideally be both essential and precisely measurable and, if intended for use on a wide advisory scale, it should also be easy, rapid and cheap to carry out.

Seeds can in fact be tested before they are imbibed by infiltrating them with barium chloride and X-raying them (Simak and Kamra, 1963; Kamra, 1964, 1966; Banerjee and Singh, 1969). This shows at the same time to what extent the embryo fills the seed and whether any barium chloride has entered; it can only do so if there are spaces which are not intact and this provides a measure of the integrity of individual seeds.

A different test which is also made on unimbibed seeds, but here usually on a bulk of homogenised seeds (which never adds to clarity), is the glutamic acid decarboxylase (GADA) test used so far mainly on cereals (Grabe, 1964) (Fig. 8.2; Table 8.12): here the activity of a single enzyme is tested by measuring the response of the homogenate to the addition of glutamic acid: a vigorous evolution of carbon dioxide indicates vigorous seeds, although the actual function of the enzyme in the live seeds is far from clear. The reaction has to be calibrated for each cultivar but Grabe has shown that once this has been done the result can forecast the growth and potential yield, rather than germinability (which may still be deceptively high) of the seed lot tested. James (1968), however, points out that with French bean seeds the axis is the most vital organ, and this only contributes 9 per cent (in poor seeds) to 13 per cent (in good seeds) of the GADA test activity and that the cotyledons which contribute the rest can be highly active even when the seeds have deteriorated.

Moore (1969) has traced the history of tests employing vital stains

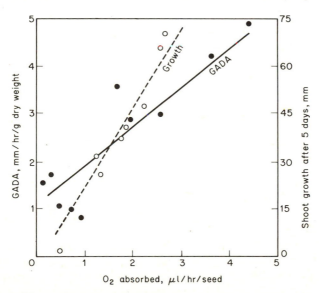

FIGURE 8.2   The relationship between the Q_{O_2} of maize seeds in air at 25°C after 2–6 hours and: • glutamic acid decarboxylase activity (ml Brodie's solution/hour/g 'dry' weight of seeds at 30°C); ○ shoot growth after 5 days. (After Woodstock and Grabe, 1967.)

TABLE 8.12   *Differences between and within cultivars of maize* (Zea mays). (*After Grabe, 1966.*)
Two differently aged lots of each of five cultivars. The two ages (a and b) are the same for all cultivars.

Cultivar	Age	Germina-tion, %	Cold test, %	Root length, mm	GADA	Days of adverse storage required to kill 50 per cent of the seeds
		(1)	(2)	(3)	(4)	
1	a	99	96	195	214	34
1	b	77	93	150	152	14
2	a	99	91	147	203	34
2	b	98	94	166	135	20
3	a	98	96	177	175	32
3	b	97	92	140	91	6
4	a	100	95	214	251	30
4	b	98	97	156	177	10
5	a	99	94	195	155	24
5	b	96	91	156	98	10

(1) Laboratory germination at 27–31°C.
(2) 7 days at 10°C followed by laboratory germination at 27–31°C.
(3) After 5 days.
(4) Glutamic Acid Decarboxylase Activity: evolution of CO_2 after 30 min from mixture of standard quantities of ground seeds and glutamic acid.
Note that only the GADA test forecasts differences in storability; but cultivars need individual calibration.

(a misnomer, incidentally, the seeds are usually killed in the process). These include the selenium, the resazurin, recently the tellurium (Asakawa, 1970), but chiefly the tetrazolium test (e.g. Lakon, 1945; Moore, 1962; Bulat, 1970) which is in fact used in an ever-growing number of laboratories although considerable training is required in both its technique and its interpretation. Each of these tests is based on the response of the applied reagent to one enzyme or enzyme complex, the tetrazolium test on the reduction of colourless triphenyl tetrazolium chloride (TTC or TZ) to red formazan by dehydro-genases, a group of oxidising enzymes assumed to persist only in living cells. Live areas become stained regardless of whether the seed is 'dormant' or not; this has the advantage of demonstrating what Lakon (1950) has termed *Keimpotenz* (potential ability to germinate) but may be misleading concerning the *Keimkraft* (ability to germinate at the time of the test). Though not the originator of this test, Lakon was probably the first to see its many possibilities. He named his version the 'topographic tetrazolium test', since it enables the investigator to *map* areas of live and dead cells in individual seeds. It is the location and not necessarily the size of these areas that is important. It was Lakon who coined the phrase that seeds can be 'half-dead' and who showed that the germination is not, as often suggested, an on-off process but that the ability of seeds which are partially dead to produce seedlings depends largely on favourable environmental circumstances. He suggested the term *labile* (in contrast to *stable*) for their *Keimtendenz* (tendency to germinate) as mentioned earlier.

Following Lakon, Bulat (1970) has developed the tetrazolium test so expertly that the differences in results between assessments of the germinability of non-dormant seeds *via* the tetrazolium and the germination tests in expert hands are usually negligible. A more stringent assessment of the TTC test than is used for germination capacity might suggest itself as a test for 'vigour', i.e. integrity of seeds, but the experts are not agreed on the degree of stringency to use: Schubert (1967) suggests the division of seeds into three categories: entirely stained, entirely unstained, and all others. Some of the third group might be capable of germinating and growing well under some but not all conditions. By contrast Moore (1962) suggests five categories. It is all a question of where to draw arbitrary lines: an unenviable task.

What is much more relevant, Moore has developed his expertise in interpreting results of the TTC method to such an extent that he can use it to discover causes of injury to seeds: necroses may be due to weather before harvest, mechanical damage of the seeds during extraction or preparation (see Chapter 4), or to ageing in store, and each shows different TTC stain symptoms. Moore (1970) has further demonstrated that initially small necroses extend during storage in the presence of xerophytic microorganisms which may start as saprophytes but may end up as parasites (Christensen and López, 1963).

More recently, Woodstock (1966), Woodstock and Grabe (1967) (Table 8.13), Woodstock and Feeley (1965) (Table 8.14) and Kittock and Law (1968) have shown that respiration, both expressed as oxygen uptake and as respiratory quotient during the first few hours of incubation, whatever its place in the causal chain, is a sufficiently vital and sufficiently changeable process to be a good index not only of the germinability but also of the seedling growth rate to be expected. Woodstock (1969a) also suggests that the incorporation of radioactive leucine by germinating seeds might furnish a good estimate of their capacity for protein synthesis. However, evident discrepancies be-

TABLE 8.13   *Effects of seed moisture during 4 years' cold storage on parameters of 'vigour' of maize* (Zea mays) *grains.* (*After Woodstock and Grabe, 1967.*)

Seed water content %	Q_{O_2} in air (1)	Anaero-biosis quotient (2)	GADA (3)	Germina-tion, %	Root length, mm (4)	Shoot length, mm (4)
9	5.0	1.9	4.4	97	93	48
11	2.6	2.6	3.1	93	77	44
13	1.7	3.2	3.7	95	75	44
15	0.1	22.0	1.6	91	43	20

(1) 4 hours at 25°C: μl per seed.
(2) Q_{O_2} at 100 per cent O_2 compared with Q_{O_2} in air (computed from authors' data).
(3) Glutamic acid decarboxylase activity; see Table 8.12.
(4) Mean of 3, 4 and 5 days.

TABLE 8.14   *Freezing injury to immature grains of maize* (Zea mays cv. 'Delkalb 441'). (*After Woodstock and Feeley, 1964.*)

Duration of freezing, hours	Q_{O_2} during first two hours of imbibition	Shoot length after 4 days, mm
0	11.5	66
0.5	7.0	17
1	5.1	4
2	5.4	0

tween results from respiration measurements and leucine incorporation caused Woodstock to call for caution in the interpretation of either as an index of vigour. In this connection, Kittock and Law (1968) have found differences between the respiration rates of seeds of different cultivars of a given species. This means that only the respiration rates of seeds of the same or closely related lines can be meaningfully compared as indices of vigour.

EFFECTS OF AGEING ON VIGOUR

It is clear that the overall metabolic efficiency of seeds declines with age. This phenomenon has been studied by Abu-Shakra and Ching (1967). A biochemical and electron microscopic examination of dark-grown axes of germinating soya bean seeds revealed that deterioration ('ageing') causes a visible increase in the density of the matrix of mitochondrial pellets. And whilst in ageing seeds the oxygen uptake per unit weight of mitochondrial nitrogen rose to 110–140 per cent, the P/O ratio, representing phosphorylative efficiency, sank to 40–70 per cent of that of new seeds, which suggests that the respiratory activity of older seeds becomes uncoupled (see p. 266).

Anderson (1970), working with barley, warned against the assumption that in deteriorating seeds all responses change at a similar rate: in his experiments, the rate of carbon dioxide evolution of old seeds during germination was up to twice as high as that of new ones, whereas the amylase activity after 36 hours was only about two thirds. But old seeds did not differ from new in percentage germination or in rate of osygen uptake. However, *accelerated* ageing (at 45°C and 100 per cent relative humidity) caused a more rapid drop in germination capacity in old than in young seeds.

Abdul-Baki (1969) goes a step further. He traces the *utilisation* (in contrast to absorption) of externally supplied glucose by wheat and barley grains of different physiological ages and finds that this provides an index of deterioration well ahead of measurements of seedling shoot growth, respiration and germination, processes which lag behind in sensitivity in increasing order (Fig. 8.3). This decline in synthetic ability is therefore an early danger signal which indicates that seeds are beginning to 'go downhill'.

All these biochemical methods have a scientifically more respectable content than the short-cuts of the growth-response school, even though the need to reconcile discrepancies in their results provides a

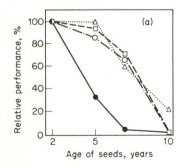 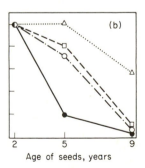

FIGURE 8.3   Effect of age of seed on vigour parameters (expressed in relation to two-year-old seeds whose performance = 100 per cent).
●———● Glucose utilisation
□— — —□ Germination, per cent
○—·—·—○ Shoot growth
△·············△ Oxygen uptake
(a) Barley 'Wisconsin X–691–1'; (b) Wheat 'Vermilion'. (After Abdul-Baki, 1969.)

challenge for studying their interrelations more closely. Of the methods, Abdul-Baki's has the advantage that it employs a complex process (analysed in detail by himself) and gives therefore perhaps a particularly convincing estimate of a seeds' synthetic ability. This, together with the analysis of mitochondrial metabolism (McDaniel, 1969; Abu-Shakra and Ching, 1967) and nucleic acid metabolism (Chen, Sarid and Katchalski, 1968) go further than any previous attempt to unravel the cause of vigour in seeds.

MEMBRANE INTEGRITY

It has been postulated that most phenomena of waning vigour in seeds are associated with, if not due to, a weakening of cell membranes (see the section on *Damage to Cytoplasmic Organelles* in Chapter 9, p. 265). The consequent leakage of sugars and electrolytes, among other water-soluble contents, from the cells, has at least two effects: a deterioration of the metabolic and transport efficiency, and the encouragement of microorganisms by the leachate exuded, to colonise first the spermosphere around the seed in the soil and the seed surface and ultimately any attackable places within the leaky seed itself. Even in store, before sowing, leakage of metabolites would encourage the surface microflora which is always present on seeds. Takayanagi and Murakami (1968) base a successful *viability* test on urine sugar (glucose) analysis paper which, by staining green, detects leakage of glucose out of imbibed seeds of, e.g. *Brassica napus*. However, Abdul-Baki and Anderson (1970) give a number of reasons why the leaching of sugars from seeds is not a reliable index of their *vigour*: thus, both natural and accelerated ageing depress viability, but although natural ageing increases the leaching of sugars from seeds accelerated ageing does not; whilst mechanical injury, which causes increased leaching of sugars, may not decrease viability. Though leaching of sugars may well be an indication of increased membrane permeability, what may matter more is the rate of utilisation of the remaining sugars, which is not measured by this technique. Also, often only 10–20 per cent of the sugars are present in the embryo itself.

The conductivity test (Matthews and Bradnock, 1967, 1968; Bradnock and Matthews, 1970) on the other hand has some promising features (Table 8.1), especially if its interpretation is safeguarded by applying it exclusively to seed samples with high germination percentages (otherwise the dead seeds might obscure the contribution of the live ones). Here, seeds are steeped in a measured quantity of water for a standardised time; the 'steep-water' is then well stirred and decanted and its electric conductivity measured. The weaker the membranes, the larger the quantity of electrolytes leached from the seeds, and the greater the conductivity of the steep-water. The measurement is rapid, precise and easy to carry out, and there is some evidence that the leakage of electrolytes is representative of membrane quality in the general; but once again, individual cultivars may require separate calibration. Though normally used collectively on a group of seeds for advisory purposes, conductivity measurements have been made successfully on individual pea seeds for research into effects of adverse climatic conditions (Perry and Harrison, 1970).

A further test for intactness of cell membranes developed by Effmann and Specht (1967) is the acid fuchsin test, which has the advantage that it is carried out on individual seeds and is topographic.

None of these tests is concerned with causes, but only with the phenomenon, of leakage. Koostra and Harrington (1969), by contrast, have shown that a decline in polar lipids, chiefly owing to their oxidation, may be the immediate cause of leaky cell membranes in seeds (Table 8.15). Considering the nature of the leachate and its possible causes, Ching and Schoolcraft (1968) point out that a dramatic increase in leachable amino acids in seeds of crimson clover stored at high humidities and/or temperatures (Table 8.1) reflects an increase in protease activity *during storage*, which is probably associated with the degradation of 'organellar membranes, nucleoproteins, ribosomes and enzymes', whereas larger quantities of inorganic phosphate in the leachate indicate a rise in phytase activity: this, though not affecting viability, results in distinctly depressed seedling growth.

TABLE 8.15   *Effects of accelerated ageing (at 38°C in saturated air) on seeds of cucumber* (Cucumis sativus cv. 'Ashley'). (*After Koostra and Harrington, 1969.*)

Seeds	Seed weight g/100 seeds	Radicles emerged, %	Seedlings complete, %	Fresh weight per seedling, mg	Total lipid, %	Polar lipid, %	µg P in 300 g polar lipid	% P at origin	% P as LPC[a]	% P as PC[b]
Not aged	2.85	99	98	173	28.6	1.4	5.16	3.8	2.9	61.0
Aged 2 wks	2.15	81	71	89	27.8	1.1	1.83	9.5	2.4	56.8
Aged 4 wks	1.99	2	0	0	13.4	0.5	0.96	20.5	19.0	33.1

[a]LPC = lysophosphatydil choline   } P in phospholipids after thin layer chromato-
[b]PC = phosphatydil choline   } graphy.

CYTOLOGICAL DETERIORATION

It is well known that, but less known why, cytological deterioration goes hand in hand with ageing and loss of vigour. This subject is dealt with in some detail in Chapter 9 but it is important to emphasise here that the cytological changes associated with seed deterioration may well have effects on vigour as well as on viability.

Vigour as a response to adversity: the ecological approach

There are two views within the ecological vigour camp: one school of thought, with plenty of vision though perhaps not enough evidence, claims that the response of seeds to *any* stress will be representative of their ability to cope with any other stress, whether they are steeped in hot water or a solution of, say, ammonium chloride (Table 8.16), stored in a hot and/or humid atmosphere (Fig. 8.4) for the purpose of 'accelerated ageing', exposed to atomic radiation, or subjected to partial drought, excessive wet or massive attacks by pathogens. The other school, more down to earth, considers that one might just as well test the performance of seeds under the specific conditions – including soil conditions – which they are likely to meet.

Accordingly there are two groups of so-called stress tests: (a) in the 'general' group seeds are given a treatment – any treatment – that is likely to impair their vitality and then placed under conditions favourable to germination; (b) in the 'specific' tests the responses of seeds are tested to adverse seed-bed conditions under which they may be capable of germinating and growing provided they are 'vigorous'.

Webster and Dexter (1961) demonstrated that any maltreatment, such as mechanical injury, high storage temperature or high storage

TABLE 8.16   *Response of different seed lots of lucerne* (Medicago sativa) *to accelerated ageing, to pre-soaking in ammonium chloride and to normal storage. (After Delouche et al., 1967.)*

Seed lot	Germination, %				
	Initial	After accelerated ageing (40°C, saturated atmosphere) for		After 1.5 hours in 2 per cent NH₄Cl at 40°C	After open storage for 15 months
		4 days	6 days		
A	90	85	87	90	86
B	90	84	76	87	71
C	89	70	70	75	73
D	89	69	54	59	55

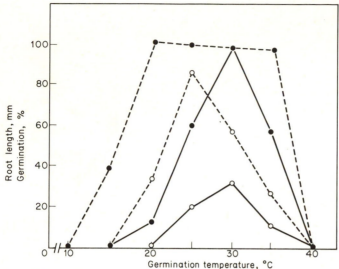

FIGURE 8.4   Effect of heat treatment on the performance of maize.
— — —   Germination, per cent
—————   Root length, mm
  •   No heat treatment
  ○   Four days' heat treatment
(After Woodstock, 1968b.)

humidity does indeed affect percentage germination, rapidity of germination and rate of seedling growth. But these do not deteriorate at the same time or in the same order in all species and it is suggested that for different types of seeds the characteristic that begins to decline first should be taken as the index of vigour. Mark and McKee (1968), comparing a large number of treatments and of criteria, came to similar conclusions.

Since vigour is measurable on a large number of manifestations and in response to a large variety of environmental conditions, Woodstock (1969b) suggests the use of a subscript for the environmental variable (the environmental range vector) and a superscript for the attribute to be measured (the intensity vector) to express the various aspects of seed vigour (SV). For instance, SV_t^{Germ} = seed vigour expressed as germination percentage over a range of temperatures, and $SV_{soil\ H_2O}^{Growth}$ = seed vigour expressed as seedling growth over a range of soil moisture levels.

THE COLD TEST

One of the treatments where the two groups coalesce is the cold test where seeds, usually of warm season crops, particularly maize, are

TABLE 8.17   *Cold test on maize in upright filter paper rolls (germination, per cent).*
(After Kietreiber, 1966.)

Cultivar	Laboratory germination (25°C)	Proportion soil: sand in cold test		
		0:4	1:3	4:0
A	98	95	77	60
B	96	97	93	91
C	92	90	84	79
D	88	82	47	36
E	65	35	1	1

sown in trays filled with soil (Svien and Isely, 1955) or placed in paper towels lined with soil (Kietreiber, 1966) (Table 8.17) and kept in the cold so that they cannot germinate and are exposed to the activity of inimical soil microorganisms for a substantial period; they are then transferred to a temperature favourable for germination, in order to find out about their capacity for survival and possibly also their 'residual' growth potential. But, and this is a vital but, soil conditions are notorious irreproducible (Svien and Isely, 1955), even if temperature and moisture are controlled, and therefore results of cold tests should never be regarded as absolute values, although they may be useful when a number of similar seed lots are sown side by side and ranked according to their relative performance. It may also be noted

TABLE 8.18   *Effect of high-temperature storage (30°C at 57 per cent relative humidity) on maize seeds (Zea mays). (After Delouche, 1968, and Gill, 1969.)*

Storage period, months	Laboratory germination at 30°C, % (1)		Cold test germination, % (2)	Field emergence, % (3)		Respiration (4)		GADA (5)	Growth, mm in 3 days (6)		Relative grain yield (7)
	4 days	7 days		6 days	13 days	O₂	RQ		root	shoot	
0	94	95	88	88	92	9.9	1.8	145	118	39	100
5	88	95	74	77	86	7.4	1.9	97	86	26	77
10	58	92	21	54	79	6.7	2.2	78	60	14	54

(1) Normal seedlings at least 50 mm long.
(2) Soil and sand mixture at '70 per cent saturation'; 10°C for 7 days, then 30°C for 6 days.
(3) At least 25 mm above ground.
(4) O₂ = μl O₂ absorbed per seed per hour, measured after 6 hours at 25°C.
    RQ = respiratory quotient, CO₂ produced/O₂ absorbed.
(5) Glutamic acid decarboxylase activity: CO₂ released in 30 minutes by 30 ground up seeds mixed with glutamic acid; expressed as mm height of column of Brodie's solution.
(6) At 30°C (in upright rolled paper towel).
(7) Adjusted for equivalent populations.

that the deterioration which results from the imposition of the cold test may vary from imperceptible to drastic, so subtle are the factors involved (Tables 8.2, 8.12, 8.17, 8.18).

ACCELERATED AGEING

One group of the preliminary treatments, however, has an additional value: seeds artificially aged can foreshadow the 'storage vigour' of seeds lots, i.e. their ability or otherwise to survive well when stored for any length of time. A few weeks' or even days' storage in adverse conditions – a hot and/or moisture-laden atmosphere – can work wonders of deterioration, even when expressed simply as a decline in percentage germination (see Chapter 2). However, like Grabe (1965), Delouche (1968, 1969) and his school (Aponte, 1970; Baskin, 1970; Byrd, 1970; Castro, 1970; Chang, 1970; Chen, 1970; Gill, 1970; Islam, 1967; Sittisroung, 1970) have raised their sights beyond the counting of seedlings and have employed many other measures of deterioration (Table 8.18). Largely by using different storage treatments but chiefly 'accelerated ageing' at 40–45°C in a saturated atmosphere and measuring various biochemical and physiological parameters during and after seed germination in the laboratory and attributes of the resulting crops in the field, they have come to the following conclusions: the vigour (i.e. performance potential) of a seed is at its highest at maturation. From then on it deteriorates 'inexorably, continuously and irreversibly'. The rate of deterioration is largely determined by genetic factors, i.e. species and cultivar, but also by treatment and environment before and certainly during storage, which accounts for differences between seed lots and even individual seeds within the same lot. Loss of germination is clearly an important indication of loss of vigour but it is, equally clearly, the last relevant indication, the final catastrophe. Many important detrimental changes take place before seeds lose their ability to germinate. The unravelling of their causal connections requires much further work, but from the evidence to date the sequence of deterioration is usually approximately as follows (Delouche, 1969):

(1) Degradation of cellular membranes and subsequent loss of control of permeability;
(2) impairment of energy-yielding and biosynthetic mechanisms, and consequently
(3) reduced respiration and biosynthesis;
(4) slower germination and slower heterotrophic seedling growth;
(5) reduced storage potential;

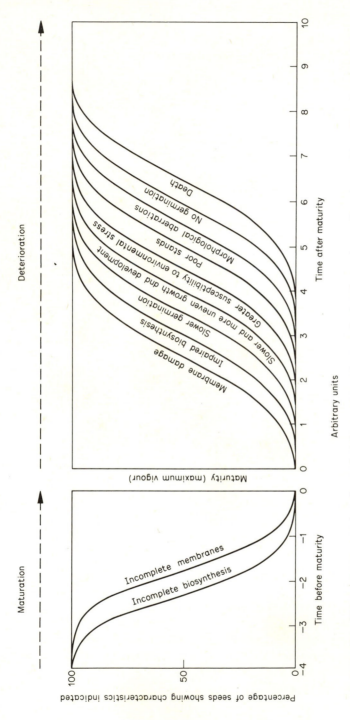

FIGURE 8.5 Diagram representing the typical time-course of events in the change in quality of a seed lot. The time scale is in arbitrary units but, very roughly, the units during maturation would be of the order of a day; whereas during deterioration the units would be of the order of a day during typical accelerated ageing treatments, or the order of a year for 'good' storage conditions. But the time scale will vary according to the principles described in Chapter 2. The curves are meant to represent cumulative normal distributions. Although there is evidence that such curves are appropriate to the death curve (Chapter 2), there is as yet no evidence as to the actual distribution of any other criterion. In addition to the criteria indicated in the diagram, nuclear damage also occurs before death, but this subject is dealt with in detail in Chapter 9. (With acknowledgments to J. C. Delouche for the suggested order of events.)

  (6) slower growth and development of the autotrophic plant;

  (7) less uniformity in growth and development amongst plants in the population;

  (8) increased suceptibility to environmental stresses (including microorganisms);

  (9) reduced stand-producing potential;

 (10) increased percentage of morphologically abnormal seedlings and, lastly,

 (11) loss of germinability.

Figure 8.5 is an attempt to represent what happens to a population of seeds during ageing. It assumes that accelerated ageing, caused by exposing seeds to an unfavourable hot and moist atmosphere, is truly representative of more normal ageing, i.e. that it is a true time-lapse process. This assumption is in agreement with the arguments presented in Chapter 2, with the possible exception of seeds stored under *extremely* severe conditions. Accordingly accelerated ageing provides a powerful research tool which makes it possible to study over a conveniently short period the processes of deterioration, their sequence and relationships. And most important from the practical point of view, this *time machine* can forecast the order in which different seed lots are likely to deteriorate in store, which they inevitably will, though seedsmen are doing their best to make this a slow-motion process. Delouche uses the decline in germination percentage, the rate at which a population slides to its *ultimate* doom, as the criterion of deterioration of seed lots through accelerated ageing, no doubt in deference to the ease of this measurement and the probability that it will be accepted as a true measure of value in the seed trade for a long time to come. Nevertheless his chief contribution is the demonstration that loss of viability, though the most blatantly obvious result, is in fact the *last* stage of deterioration, preceded by many much more subtle changes. It is clear that accelerated ageing is not itself yet another alternative vigour test but a technique through which all other tests can acquire an additional dimension (Table 8.19).

How then is the ability of seed lots to endure in store to be measured? Of the agronomic characteristics, root growth during the early stages is often a reliable index of the future potential of the seed. In fact, Schachl (1949) has suggested this as a basis for a vigour test. Time to germination (i.e. as discussed earlier, the no longer fashionable concept of 'germination energy'), is also a property affected long before germination percentage. But a single 'first count' at an arbitrary early date is open to many objections, and an integrating method,

TABLE 8.19   *Vigour tests which may show effects of accelerated ageing.*

Biochemical tests

Necroses	Respiration
Tetrazolium stain (TTC)	O₂ consumption
X-ray examination	RQ
	Mitochondrial activity
	Glucose utilisation
	Protein synthesis (¹⁴C-leucine
	incorporation)
	GADA

Micro-organisms	Membrane integrity
Turbidity	Conductivity
Plating	Sugar leakage
TTC	

Performance tests

Straight (physiological and agronomic)	Under stress (agronomic)*
Percentage germination	Germination and emergence
Time to germination	Under sub-optimal conditions,
Root growth rate in the dark	e.g.:
Shoot growth rate in the dark	Low or high pH
Relative growth rate in light	Cold or hot soil
Yield of total dry matter	Wet or dry soil
Yield of crop	Pathogen-laden soil
	Compacted soil
	After soaking in:
	Cold water
	Hot water
	Salt solution (e.g. NH₄Cl)

*All these are, more or less, relevant to field performance but do not actually *predict* it.

though more laborious, is really required. The coefficient suggested by Kotowski (1926) may be useful:

$$\text{CVG} = \frac{\Sigma n \times 100}{\Sigma(Dn)} \qquad (8.1)$$

where *n* is the number of seeds germinated on separate days and *D* is the number of each day after day zero, the day of sowing. CVG is the 'coefficient of velocity of germination', the reciprocal of the *average time to germination.* The maximum CVG value is 100 if all germinative seeds germinate on day one.

It is suggested that measurements of germination 'speed' (CVG), germination uniformity (Coefficient of Variation of the individual times or velocities of the seeds to germinate) and, to a lesser extent, germination percentage, *before and after accelerated ageing* will provide both a forecast of 'storage vigour' and an estimate of the likely relative performance of a seed lot, compared with a standard lot or with a large number of lots of the same kind.

Invigoration of seeds

Is it possible to improve the vigour of seeds?

(a) RESTORING AND MAINTAINING HEALTH

Fungicidal and insecticidal seed treatments have considerable advantage over similar treatments applied in the field: they are easier to administer, more economical of material, but most important, they are prophylactic at the stage of life where the plant is quantitatively most susceptible since a small injury could do a maximum of damage, both because of its size and because it has its whole life still before it. In this connexion it has frequently been shown that fungicidal seed treatment can often, though not always, largely prevent the potential loss of viability to be expected when seeds have to be sown in soil under conditions which are suboptimal for germination, as exemplified by the cold test.

Protective seed treatments are in fact widely applied as a matter of routine, as an insurance policy so to speak. Treatments by dusting have been largely superseded by slurry applications and more recently (Maude, Vizor and Schurling, 1969) by a combined hot water plus chemical treatment in which the chemical is capable of entering the seed and eradicating the pathogen, whilst the seed is not harmed in any way. This treatment is designed to eliminate even deep-seated seed-borne pathogens; but the soaking-and-drying may itself have an additional physiological effect, see (b) below. On the other hand, surface seed dressings to protect the seeds from soil-borne attacks may well improve in efficiency in future by the more sophisticated use of the practice of seed pelleting: here the seeds are coated in small spheres of initially inert materials, tough when dry but water-unstable, for purposes of precision drilling; much larger quantities of fungicidal, and any other, substances can be applied to individual seeds whose natural surface is small by incorporating the chemical in the pelleting material. But all this, though important, is only one aspect of invigoration.

(b) THE ADVANCEMENT OF SEEDS

According to Henckel (syn. Genkel) (1967), and his school, seeds can be 'hardened' – trained so that the seedlings are able to withstand drought, heat, frost and salinity better than before – by a basically simple treatment involving repeated partial imbibition followed by drying back. He attributes this to a far-reaching activation of the biochemical potential of the embryonic plants through forcing them to become adapted to the imposed drought (see also May, Milthorpe and Milthorpe, 1962). Austin, Longden and Hutchinson (1969) and Hegarty (1970) obtained beneficial results from such a treatment applied to carrot seeds, although they attribute the success to an advancement by pre-sowing enlargement of the embryo, as discussed earlier under *Seed size,* rather than true hardening against inimical conditions; Longden (personal communication) obtained similar results for sugar beet and Hafeez and Hudson (1967) obtained similar results with radish seeds (*Raphanus sativus*), as did Hegarty (1970) with sweetcorn (*Zea mays*). T. Sivanayagam (personal communication, 1970) had no response from maize grains pre-imbibed and dried back three times; but seeds of *Capsicum annum* responded to the same treatment by more uniformly early germination; and in addition, if stored after treatment instead of being sown immediately, they survived better than untreated seeds.

CHEMICAL ADJUVANTS

For true 'invigoration', however, one must turn to the application of physiologically active chemicals. Woodstock (1969a) reports lasting benefit from preliminary imbibition of combined 2 per cent solutions of KNO_3 and KH_2PO_4, followed by drying back, to tomato and pepper seeds. This treatment increased the subsequent speed and completeness of germination and the rate of respiration of the germinating seeds. It is not clear, however, to what extent the effect is due to the preliminary soaking and drying back treatment by which the chemicals were administered – which might support the results of Henckel or Austin *et al.* – and to what extent to the chemicals themselves.

Salim and Todd (1968) applied large numbers of inorganic chemicals at different concentrations to seeds of a diverse crop species and concluded that the seed, the chemical and the concentration had all to be right for success. No one recipe was applicable to all seeds and often no effect was obtained, but occasionally a treatment was mildly successful.

TABLE 8.20 *Effect after 4 days on root growth of treatment of maize* (Zea mays) *grains with 100 ppm biotin.* (*After Ovcharov, 1969.*)

Source: part of cob	Length of roots, mm Grains treated with		Number of roots Grains treated with	
	water	biotin	water	biotin
Top	182	183	2.6	3.1
Middle	188	206	2.1	3.2
Bottom	155	220	2.5	3.2

However, where it is a question of supplying a micronutrient element in short supply (e.g. Reisenauer, 1963) success can sometimes be spectacular (see Chapter 5), and the question of incorporating major as well as minor elements in seed pelleting materials is at present under consideration. Against the potential benefits must be set the hazard of mineral toxicity to which the young seedling might be exposed should the soil dry out.

Work by Dexter and Miyamoto (1959) and Miyamoto and Dexter (1960) is of interest here: they showed that the addition of colloidal substances, especially alginate, accelerated water uptake and seedling emergence of sugar beet from dryish soils; and furthermore, addition of inorganic or organic nutrients increased the weight of the emerging seedlings. A more direct and promising way of 'invigorating' seeds is through the adequate nutrition, especially regarding phosphorus, of their mother plants (see Chapter 5).

THE ROLE OF APPLIED GROWTH FACTORS

Ovcharov (1969) applied vitamin B_2, vitamin B_6, or similar compounds to seeds before sowing and states that this can improve their performance during the vital early stages of seedling growth (Table 8.20). These chemicals are not applied, as are some other organic substances, e.g. gibberellic acid, in order to break dormancy: they are administered in the hope that they may remove 'biochemical bottlenecks' by supplying metabolites which are essential or beneficial to germination and early growth the innate supply of which may however be inadequate during the earliest stages of germination. A successful curative treatment of beetroot seeds with gibberellin against the effects of sub-microscopic damage sustained through abrasion ('rubbing') of the seed clusters has been reported by Chetram and Heydecker (1967). But what is often wanted is the improvement of seeds which are already of reasonably high quality, rather than the rehabilitation of the sub-standard.

Most attempts so far have however been 'empirical', hit and miss rather than based on a detailed knowledge of the biochemical processes of germination. A possible exception is the use of ascorbic acid by Chinoy (1967), based on the knowledge that ascorbic acid does not exist in seeds before germination but has to be synthesised *de novo*. Similarly, beneficial carry-over effects from cytokinin applications to seed crops of Brussels sprouts on seed germination and seedling growth have been reported (Thomas and Comber, 1969). Although many stimulatory effects are short-lived or have undesirable side effects, the prospects are exciting, especially since Meyer and Mayer (1971) have developed a technique for incorporating chemicals into dry seeds with organic solvents.

Summary

CAUSES OF DIFFERENCES IN SEED VIGOUR

There are a number of distinct causes for low vigour in seeds:

(i) Genetic: certain cultivars are more susceptible than others to adverse environmental conditions, and certain cultivars are less capable of growing rapidly than others. On the other hand, the heterosis exhibited by hybrid cultivars results in resistance to adverse conditions, possibly as a result of their ability to grow rapidly, which itself may be due to some extent to the high efficiency of their mitochondrial metabolism.

(ii) Physiological: the 'physiological state' of the seeds which can be suboptimal for two reasons: immaturity at harvest and deterioration during storage.

(iii) Morphological: within a cultivar the smaller seeds often produce the less vigorous seedlings than the larger seeds.

(iv) Cytological: the production, with ageing, of chromosome aberrations resulting, possibly, from the production of auto-mutagens.

(v) Mechanical: gross breakages or the creation of necroses which may spread through physiological mechanisms (production of auto-toxins) or through microbial activity.

(vi) Microbial: the burden of fungi and/or bacteria which accumulate on and in the seed during its maturation may endanger its performance under appropriate storage or field conditions by heating in store, by direct attack, including the invasion and enlargement of necroses, or by competition for oxygen. (Heydecker and Chetram, 1971).

EFFECTS OF SEED VIGOUR

Vigour of seeds expresses itself in, broadly, four ways:

(i) Survival intact in the non-active state: a vigorous seed is one that remains vigorous.

(ii) Survival upon sowing in the field: a vigorous seed resists or out-strips its attackers.

(iii) Ability to establish a plant: a vigorous seed has plenty of all the right reserves, and it uses them during the heterotrophic and transition phases of growth.

(iv) Ability to grow well: a vigorous seed results in a seedling that grows vigorously during the autotrophic phase of growth.

THE NATURE OF VIGOUR

(i) Vigour differs fundamentally from 'absence of dormancy'.

(ii) Seeds in any one seed lot are not divisible into 'good' and 'bad' but form a continuously variable population.

(iii) To speak of lack of vigour in seeds may mean two things: *either* that seeds which are capable of germinating under reasonably favourable conditions will not necessarily germinate under unfavourable ones, *or* that some seeds which do germinate will produce 'worse' plants, i.e. plants with a lower performance potential, than others *growing under identical conditions.*

(iv) Overall, vigorous seeds and their seedlings are capable of doing well over a wide range of conditions, but

(v) methods exist by which it is possible to map, so to speak, the position of each individual seed tested on its inevitable downhill path in respect of certain vital characteristics which one might term 'components of vigour', and which are themselves the resultants of complex, largely interacting processes.

(vi) A plant cannot be better than the seed from which it has grown, and any decrease in seed vigour generally accompanies the resulting plant throughout its life (although cases are known in which early lack of vigour tends to disappear with time – see Chapter 10). To base an estimate of the quality of a seed lot merely on its laboratory germination percentage, the least sensitive index of its quality, is therefore a short-sighted short cut which may not be very meaningful.

(vii) More meaningful comparative forecasts of plant performance could be made if they were based on a more sensitive and more representative measurable characteristic of seeds or germinating seedlings.

(viii) There are several promising biochemical tests of vigour (Table 8.19). A distinction should be drawn between tests which are simple and rapid enough for use on large numbers of commercial samples and those which are more refined and may well serve as research tools into the nature of seed vigour, especially where their respective results do not quite agree.

(ix) No seed is perfect, and invigoration (and not only re-invigoration) of seeds may be possible at some future time. However, at present too little is known about the biochemical mechanisms involved, and attempts at invigoration have been largely empirical, though some reported successes may well point the way to further progress in both explaining and improving the vigour of seeds.

References to Chapter 8

ABDALLA, F. H., and ROBERTS, E. H., 1969. The effect of seed storage conditions on the growth and yield of barley, broad beans, and peas. *Ann. Bot.*, **33**, 169–84.

ABDUL-BAKI, A. A., 1969. Relationship of glucose metabolism to germinability and vigour in barley and wheat seeds. *Crop Sci.*, **9**, 732–37.

ABDUL-BAKI, A. A., and ANDERSON, J. D., 1970. Viability and leaching of sugars from germinating barley. *Crop. Sci.*, **10**, 31–34.

ABU-SHAKRA, S. S., and CHING, T. M., 1967. Mitochondrial activity in germinating new and old soybean seed. *Crop Sci.*, **7**, 115–17.

ADER, F., 1965. Zur Definition eines einheitlich anwendbaren Begriffs der Triebkraft. *Proc. int. Seed Test. Ass.*, **30**, 1005–12.

ANDERSON, J. D., 1970. Physiological and biochemical differences in deteriorating barley seed. *Crop Sci.*, **10**, 36–39.

APONTE, A., 1970. *Quality of sesame seed* (Sesamum indicum L.) *influenced by storage conditions and artificial aging*. M.S. thesis, Mississippi State University, Miss.

ASAKAWA, S., 1970. Some proposals to amend the international rules for seed testing. *Proc. int. Seed Test. Ass.*, **35**, 641–43.

AUSTIN, R. B., and LONGDEN, P. C., 1967. Some effects of seed size and maturity on the yield of carrot crops. *J. hort. Sci.*, **42**, 339–53.

AUSTIN, R. B., LONGDEN, P. C., and HUTCHINSON, J., 1969. Some effects of 'hardening' carrot seed. *Ann. Bot.*, **33**, 883–95.

BANERJEE, S. K., and SINGH, A., 1969. Radiographic detection of seed characteristics in some horticultural crops. *Indian J. agric. Sci.*, **39**, 27–31.

BASKIN, C. C., 1970. *Relation of certain physiological properties of peanut* (Arachis hypogaea L.) *seed to field performance and storability*. Ph.D. thesis, Mississippi State University, Miss.

BATTLE, J. P., and WHITTINGTON, W. J., 1969. The relation between inhibitory substances and variability in time to germination of sugar beet clusters. *J. agric. Sci.*, **73**, 337–46.

BRADNOCK, W. T., and MATTHEWS, S., 1970. Assessing field emergence potential of wrinkled-seeded peas. *Hort. Res.*, **10**, 50–58.

BRONIEWSKI, S., 1967. Die Intensität der Zellteilung in früher Keimphase als Kriterium der Vitalität von Samen. In *Physiologie, Ökologie und Biochemie der Keimung, Vol. 1*, ed. H. Borriss, 55–64. Ernst-Moritz-Arndt-Universität, Greifswald.

BULAT, H., 1962. Probleme der Triebkraftbestimmung. *Saatgutwirtsch.*, 1962, 305–7.

BULAT, H., 1970. Das topographische Tetrazoliumverfahren in der Saatgutprüfung. In *Hundert Jahre Saatgutprüfung*, ed. F. Ader, 95–103. Sauerländer, Frankfurt am Main.

BYRD, H. W., 1970. *Effect of deterioration in soybean* (Glycine max) *seed on storability and field performance.* Ph.D. thesis, Mississippi State University, Miss.

CASTRO, L. A. B. DE, 1970. *Some factors influencing the yield and quality of carrot* (Daucus carota L.) *seed.* M.S. thesis, Mississippi State University, Miss.

CHANG, S. S., 1970. *Physiological study of differences in quality and longevity among seed of two inbred lines of corn and the hybrid.* M.S. thesis, Mississippi State University, Miss.

CHEN, C. C., 1970. *Influence of physiological quality of seed on emergence, growth and yield of some vegetable crops.* M.S. thesis, Mississippi State University, Miss.

CHEN, D., SARID, S., and KATCHALSKI, E., 1968. Studies on the nature of messenger RNA in germinating wheat embryos. *Proc. natl. Acad. Sci. US*, **60**, 902–9.

CHETRAM, R. S., and HEYDECKER, W., 1967. Moisture sensitivity, mechanical injury and gibberellin treatment of *Beta vulgaris* seeds. *Nature, Lond.*, **215**, 210–11.

CHING, T. M., and SCHOOLCRAFT, I., 1968. Physiological and chemical differences in aged seeds. *Crop Sci.*, **8**, 407–9.

CHINOY, J. J., 1967. Role of ascorbic acid in crop production. *Poona agric. Coll. Mag.*, **57**, 1–6.

CHRISTENSEN, C. M., and LÓPEZ, F. L. C., 1963. Pathology of stored seeds. *Proc. int. Seed Test. Ass.*, **28**, 701–11.

COOPER, C. S., and QUALLS, M., 1969. Seedling vigor evaluation of four birdsfoot trefoil varieties grown under two temperature regimes. *Crop Sci.*, **9**, 756–57.

DELOUCHE, J. C., 1968. Physiology of seed storage. *23rd Corn and Sorghum Res. Conf., Amer. Seed Trade Ass.*, 83–90.

DELOUCHE, J. C., 1969. Planting Seed Quality. Journal paper No. 1721, Mississippi Agric. Exp. Sta., Mississippi State University. *Proc. 1969 Beltwide Cotton Production-Mechanization Conf., New Orleans, La.*, 16–18.

DELOUCHE, J. C., and CALDWELL, W. P., 1960. Seed vigor and vigor tests. *Proc. Assoc. off. Seed Anal.*, **50**, 124–29.

DELOUCHE, J. C., RUSHING, T. T., and BASKIN, C. C., 1967. *Predicting the relative storability of crop seed lots.* Rep. to Amer. Seed Res. Foundation, Seed Technology Lab., Mississippi State University, Miss.

DEXTER, S. T., and MYAMOTO, T., 1959. Amelioration of water uptake and germination of sugarbeet seedballs by surface coatings of hydrophilic colloids. *Agron J.*, **51**, 388–89.

DURRANT, A., 1962. The environmental induction of heritable change in *Linum*. *Heredity*, **17**, 27–61.

EFFMAN, H., and SPECHT, G., 1967. Bestimmung der Lebensfähigkeit der Samen von Gramineen mit der Säurefuchsinmethode unter Anwendung der Sequenz-analyse. *Proc. int. Seed Test. Ass.*, **32**, 27–47.

EGGEBRECHT, H., 1949. *Die Untersuchung von Saatgut; Methodenbuch*. **5**, 23–25, 27. Neumann, Hamburg.

FRITZ, T., 1965. Germination and vigour tests of cereal seed. *Proc. int. Seed Test. Ass.*, **30**, 923–27.

GALATSCHALOWA, S. N., and MARUSSINA, T. M., 1967. Biochemie der Reifung und Saatgutqualität von Weizenkaryopsen in der Wald – und Steppenzone von Westsibirien. In *Physiologie, Biochemie und Ökologie der Keimung, Vol 2*, ed. H. Borriss, 991–97. Ernst-Moritz-Arndt-Universität, Greifswald.

GERM, H., 1960. Methodology of the vigour test for wheat, rye and barley in rolled filter paper. *Proc. int. Seed Test. Ass.*, **25**, 515–18.

GILL, N. S., 1970. *Deterioration of corn* (Zea mays *L.*) *seed during storage*. Ph.D. thesis, Mississippi State University, Miss.

GÖRING, H., 1967. Der Gaswechsel keimender Karyopsen von Inzuchtlinien und Hybriden von *Zea mays* L. In *Physiologie, Ökologie und Biochemie der Keimung, Vol. 2*, ed. H. Borriss. Ernst-Moritz-Arndt-Universität, Griefswald.

GRABE, D. F., 1964. Glutamic acid decarboxylase activity as a measure of seedling vigor. *Proc. Assoc. off. Seed Anal.*, **54**, 100–9.

GRABE, D. F., 1965. Prediction of relative storability of corn seed lots. *Proc. Assoc. off. Seed Anal.*, **55**, 92–96.

GRABE, D. F., 1966. Significance of seedling vigor in corn. *21st Ann. Hybrid Corn Ind.-Res. Conf.*, 39–44.

GRAHL, A., 1965. Die Triebkraft des Saatgutes. *Kali-Briefe, Fachgebiet,* 39. Folge.

HAFEEZ, A. A. T., and HUDSON, J. P., 1967. Effect of 'hardening' radish seeds. *Nature, Lond.*, **216**, 688.

HARRISON, B. J., 1966. Seed deterioration in relation to storage conditions and its influence upon germination, chromosomal damage and plant performance, *J. natn. Inst. agric. Bot.*, **10**, 644–63.

HEGARTY, T. W., 1970. The possibilities of increasing field establishment by seed hardening. *Hort. Res.*, **10**, 59–64.

HENCKEL, P. A., 1967. Über die Determination neuer physiologischer Eigenschaften bei keimenden Samen, Vol. 1. In *Physiologie, Ökologie und Biochemie der Keimung*, ed. H. Borriss, 79–85. Ernst-Moritz-Arndt-Universität, Greifswald.

HEPTON, A., 1957. *Studies on the germination of* Brassica oleracea *var.* botrytis (Linn.) *with special reference to temperature relationships*. B.Sc.(Hons.) dissertation, University of Nottingham.

HEWSTON, L. J., 1964. Effect of seed size on crop performance. *Rep. natn. Veg. Res. Sta., Wellesbourne, 1963*, 45–46.

HEYDECKER, W., 1960. Can we measure seedling vigour? *Proc. int. Seed Test. Ass.*, **25**, 498–512.

HEYDECKER, W., 1962. Report on the activities of the Seedling Vigour Test Committee. *Proc. int. Seed Test. Ass.*, **27**, 211–19.

HEYDECKER, W., 1965. Report of the Vigour Test Committee. *Proc. int. Seed Test. Ass.*, **30**, 369–80.

HEYDECKER, W., 1966. Clarity in recording germination data. *Nature, Lond.*, **210**, 753–54.

HEYDECKER, W., 1969. The 'vigour' of seeds – a review. *Proc. int. Seed Test. Ass.*, **34**, 201–19.

HEYDECKER, W., 1970a. Samentriebkraft und Saatbett. In *Hundert Jahre Saatgut-prüfung*, ed. F. Ader, 88–94. Sauerländer, Frankfurt am Main.

HEYDECKER, W., 1970b. Report of the Vigour Test Committee, 1965–68. *Proc. int. Seed Test. Ass.*, **34**, 751–74.

HEYDECKER, W., and CHETRAM, R. J., 1971. Water relations of beetroot seed germination. I. Microbial factors, with special reference to laboratory germination. *Ann. Bot.*, **35**, 17–29.

HIGHKIN, H. R., and LANG, A., 1966. Residual effect of germination temperature on the growth of peas. *Planta*, **68**, 94–98.

HILTNER, L., and IHSSEN, G., 1911. Uber das schlechte Auflaufen und die Aus-winterung des Getreides infolge Befalls durch *Fusarium*. *Landwirtsch. Jb. Bayern*, **1**, 20–60, 231–78, 315–62.

HOUGHTON, B. H., 1970. Winter cauliflower – seedbed density. *15th Rep. Rosewarne expl. Hort. Sta.*, *1969*, 63–64.

ISELY, D., 1950. The cold test for corn. *Proc. int. Seed Test. Ass.*, **16**, 299–311.

ISELY, D., 1957. Vigor tests. *Proc. Assoc. off. Seed Anal.*, **47**, 176–82.

ISLAM, A. J. M. A., 1967. *Comparison of methods for evaluating deterioration in rice seed*. M.S. thesis, Mississippi State University, Miss.

ISTA (International Seed Testing Association), 1966. International rules for seed testing. *Proc. int. Seed Test. Ass.*, **31**, 1–152.

JAMES, E. 1968. Limitations of glutamic acid decarboxylase activity for estimating viability in beans (*Phaseolus vulgaris* L.). *Crop Sci.*, **8**, 1403–4.

KAMRA, S. K., 1964. The use of X-rays in seed testing. *Proc. int. Seed Test. Ass.*, **29**, 71–79.

KAMRA, S. K., 1966. Determination of germinability of melon seed with X-ray contrast method. *Proc. int. Seed Test. Ass.*, **31**, 719–30.

KHAN, R. A., and LAUDE, H. M., 1969. Influence of heat stress during seed matura-tion on germinability of barley seed at harvest. *Crop Sci.*, **9**, 55–58.

KIDD, F., and WEST, C., 1918–19. Physiological predetermination: the influence of the physiological conditions of seed upon the course of subsequent growth and upon yield, I–V. *Ann. appl. Biol.*, **5**, 1–10, 111–42, 157–70, 220–51; **6**, 1–26.

KIETREIBER, M., 1966. Der Erde-Keimrollentest, eine raum- und zeitsparende Kaltprüfungsmethode für Mais. *Bodenkultur* **17**. *Sonderh.*, 56–59.

KITTOCK, D. L., and LAW, A. G., 1968. Relationship of seedling vigor to respiration and tetrazolium chloride reduction by germinating wheat seeds. *Agron. J.*, **60**, 286–88.

KOOSTRA, P. T., and HARRINGTON, J. F., 1969. Biochemical effects of age on mem-branal lipids of *Cucumis sativus* L. seed. *Proc. int. Seed Test. Ass.*, **34**, 329–40.

KOTOWSKI, F., 1926. Temperature relations to germination of vegetable seeds. *Proc. Amer. Soc. Hort. Sci.*, **23**, 176–84.

LAKON, G., 1945. The topographical tetrazolium method for determining the ger-mination capacity of seeds. *Pl. Physiol.*, **24**, 389–94.

LAKON, G., 1950. Die 'Triebkraft' der Samen und ihre Feststellung nach dem topographischen Tetrazoliumverfahren. *Saatgurwirtsch.*, *1950*, 37–39.

LAUDE, H. M., and COBB, R. D., 1969. Germination temperature in relation to growth performance evaluation. *Proc. int. Seed Test. Ass.*, **34**, 291–95.

LINDENBEIN, W., and BULAT, H., 1955. 'Triebkraft', Ziegelgruswert und Tetra-zoliumwert. *Saatgutwirtsch.*, **7**, 315–17.

LINDNER, H., 1967. Möglichkeiten und Grenzen der 'Triebkraft' von Getreide mit dem Keimrollentest. In *Physiologie, Ökologie und Biochemie der Keimung, Vol. 2*, ed. H. Borriss, 965–67. Ernst-Mortiz-Arndt-Universität, Greifswald.

MCDANIEL, R. G., 1969. Relationships of seed weight, seedling vigor and mitochondrial metabolism in barley. *Crop Sci.*, **9**, 823–27.

MARK, J. L., and MCKEE, G. W., 1968. Relationships between five laboratory stress tests, seed vigor, field emergence and seedling establishment in reed Canary grass. *Agron. J.*, **60**, 71–76.

MARSHALL, H. G., 1969. Effect of seed source and seedling age on freezing resistance of winter oats. *Crop Sci.*, **9**, 202–5.

MATTHEWS, S., and BRADNOCK, W. T., 1967. The detection of seed samples of wrinkle-seeded peas (*Pisum sativum* L.) of potentially low planting value. *Proc. int. Seed Test. Ass.*, **32**, 553–63.

MATTHEWS, S., and BRADNOCK, W. T., 1968. Relationships between seed exudation and field emergence in peas and French beans. *Hort. Res.*, **8**, 89–93.

MAUDE, R. B., VIZOR, A. S., and SCHURING, C. G., 1969. The control of fungal seed-borne diseases by means of a thiram seed soak. *Ann. appl. Biol.*, **64**, 245–57.

MAY, L. H., MILTHORPE, E. J., and MILTHORPE, F. L., 1962. Pre-sowing hardening of plants to drought. *Field Crop Abstr.*, **15**, 93–98.

MEYER, H., and MAYER, A. M., 1971. Permeation of dry seeds with chemicals: use of dichloromethane. *Science*, **171**, 583–84.

MIYAMOTO, T., and DEXTER, S. T., 1960. Acceleration of early growth of sugar beet seedlings by coating of seedbeds with hydrophilic colloids and nutrients. *Agron. J.*, **52**, 269–71.

MOORE, R. P., 1962. Tetrazolium as a universally acceptable quality test of viable seed. *Proc. int. Seed Test. Ass.*, **27**, 795–805.

MOORE, R. P., 1969. History supporting tetrazolium seed testing. *Proc. int. Seed Test. Ass.*, **34**, 233–42.

MOORE, R. P., 1970. Tetrazolium for diagnosing causes for disturbances in seed quality. In *Hundert Jahre Saatgutprüfung, 1869–1969*, ed. F. Ader, 104–9. Sauerländer, Frankfurt am Main.

NEEB, O., 1970. Keimfähigkeit, Triebkraft und Feldaufgang bei Zuckerrübensaatgut. In *Hundert Jahre Saatgutprüfung, 1869–1969*, ed. F. Ader, 76–82. Sauerländer, Frankfurt am Main.

NOBBE, F., 1876. *Handbuch der Samenkunde*. Wiegandt-Hempel-Parey, Berlin.

NUTILE, G. E., 1964. Effect of desiccation on viability of seeds. *Crop Sci.*, **4**, 325–28.

OELKE, E. A., BALL, R. B., WICK, C. M., and MILLER, M. D., 1969. Influence of grain moisture at harvest on seed yield, quality and seedling vigor of rice. *Crop Sci.*, **9**, 144–47.

ORPHANOS, P. I., and HEYDECKER, W., 1968. On the nature of the soaking injury of *Phaseolus vulgaris* seeds. *J. exp. Bot.*, **19**, 770–84.

OVCHAROV, K. E., 1969. The physiology of different quality seeds. *Proc. int. Seed Test Ass.*, **34**, 305–13.

PERRY, D. A., 1969a. Seed vigour in peas (*Pisum sativum* L.). *Proc. int. Seed Test. Ass.*, **34**, 221–32.

PERRY, D. A., 1969b. A vigour test for peas based on seedling evaluation. *Proc. int. Seed Test. Ass.*, **34**, 265–70.

PERRY, D. A., and HARRISON, J. G., 1970. The deleterious effect of water and low temperature on germination of pea seed. *J. exp. Bot.*, **21**, 504–12.

POLLOCK, B. M., and TOOLE, V. K., 1966. Imbibition period as a critical temperature sensitive stage in germination of lima bean seeds. *Pl. Physiol.*, **41**, 221–29.

QUALLS, M., and COOPER, C. S., 1969. Germination, growth and respiration rates
of birdsfoot trefoil at three temperatures during the early non-photosynthetic
stage of development. *Crop Sci.*, **9**, 758–60.

RAMPTON, H. H., and LEE, W. O., 1969. Effects of windrow curing *vs.* quick drying
on pre-harvest development of orchard grass (*Dactylis glomerata* L.) seeds. *Agron.
J.*, **61**, 483–84.

REISENAUER, H. M., 1963. Relative efficiency of seed-and-soil-applied molybdenum
fertilizer. *Agron. J.*, **55**, 459–60.

SALIM, M. H., and TODD, G. W., 1968. Seed soaking as a pre-sowing drought hardening treatment in wheat and barley seedlings. *Agron. J.*, **60**, 179–82.

SCAIFE, M. A., and JONES, D., 1970. Effect of seed weight on lettuce growth. *J. hort.
Sci.*, **45**, 299–302.

SCHACHL, M., 1949. Die Wurzelbildtriebmethode und inhre Anwendung zur Beurteilung von Leguminosensaatgut. *Festschrift 50 Jahre Landw. – Chem. Bundes-
versuchsanstalt in Linz*, 99–116.

SCHACHL, M., 1970. Der Einfluss der Korngrösse bei Rotkleesamen auf Ertrag und
Qualität. In *Hundert Jahre Saatgutprüfung, 1869–1969*, ed. F. Ader, 116–22.
Sauerländer, Frankfurt am Main.

SCHOOREL, A. F., 1956. Report of the activities of the committee on seedling vigour.
Proc. int. Seed Test. Ass., **21**, 282–86.

SCHOOREL, A. F., 1960. Report on the activities of the Vigour Test Committee.
Proc. int. Seed Test. Ass., **25**, 519–24.

SCHUBERT, J., 1967. Grundlagen und Möglichkeiten der Saatgutbeurteilung nach
dem topographischen Tetrazoliumverfahren. In *Physiologie, Ökologie und Biochemie der Keimung, Vol. 2*, ed. H. Borriss, 933–46. Ernst-Moritz-Arndt-Universität, Greifswald.

SIMAK, M., and KAMRA, S. K., 1963. Comparative studies on Scots pine seed
germinability with tetrazolium and X-ray contrast methods. *Proc. int. Seed Test.
Ass.*, **28**, 3–18.

SITTISROUNG, P., 1970. *Deterioration of rice* (Oryza sativa) *seed in storage and its
influence on field performance.* Ph.D. thesis, Mississippi State University, Miss.

SVIEN, A., and ISELY, D., 1955. *Factors affecting the germination of corn in the cold
test.* Journal paper No. J–2792, Iowa Agric. Exp. Sta., Ames, Iowa.

TAKAYANAGI, K., and MURAKAMI, K., 1968. Rapid germinability test with exudates
from seeds. *Nature, Lond.*, **218**, 493–94.

DE TEMPE, J., 1961. Seed weakness. *Proc. int. Seed Test. Ass.*, **22**, 3–11.

DE TEMPE, J., 1964. *Proeven over zaadzwakte I, Die zaadcontrole in 1963–64*, 88–94.

DE TEMPE, J., 1966. *Proeven over zaadzwakte II, Die zaadcontrole in 1964–65*, 68–71.

THOMSON, J. R., 1970. Health as a factor in seed quality. *Proc. int. Seed Test. Ass.*,
35, 9–17.

THOMAS, T. H., and COMBER, M., 1969. Plant hormones. In *Rep. natn. Veg. Res.
Stn., Wellesbourne, 1968*, 63.

TONKIN, J. H. B., 1969. Seedling evaluation: the use of soil tests. *Proc. int. Seed
Test. Ass.*, **34**, 281–89.

VEGIS, A., 1964. Dormancy in higher plants. *Ann. Rev. Pl. Physiol.*, **15**, 184–224.

VERHEY, C., 1960. Is it still possible, with regard to modern views, to handle the conception, germination energy? *Proc. int. Seed Test. Ass.*, **25**, 391–97.

VOIGT, P. W., and BROWN, H. W., 1969. Phenotypic recurrent selection for seedling vigor in side-oats Grama, *Bouteloua curtipendula* (Michx.) Torr. *Crop Sci.*, **9**, 664–66.

WANJURA, D. F., HUDSPETH, E. B., jr., and BILBRO, J. D., jr., 1969. Emergence time, seed quality, and planting depth effects on yield. *Agron. J.*, **61**, 63–65.

WEBSTER, L. V., and DEXTER, S. T., 1961. Effects of physiological quality of seeds on total germination, rapidity of germination, and seedling vigor. *Agron. J.*, **53**, 297–99.

WELLINGTON, P. S., 1970. Handbook for Seedling Evaluation. *Proc. int. Seed Test Ass.*, **35**, 449–597.

WHALLEY, D. B., MCKELL, C. M., and GREEN, L. R., 1966. Seedling vigor and the early non-photosynthetic stage of seedling growth in grass. *Crop Sci.*, **6**, 147–50.

WILLIAMS, W. A., BLACK, J. N., and DONALD, C. M., 1968. Effect of seed weight on the vegetative growth of competing annual Trifoliums. *Crop Sci.*, **8**, 660–63.

WOODSTOCK, L. W., and FEELEY, J., 1965. Early seedling growth and initial respiration rate as potential indicators of seed vigor in corn. *Proc. Ass. Off. Seed Anal.*, **55**, 131–39.

WOODSTOCK, L. W., and POLLOCK, B. M., 1965. Physiological predetermination: Imbibition, respiration, and growth of lima bean seeds. *Science*, **150**, 1031–32.

WOODSTOCK, L. W., 1966. A respiration test for corn seed vigor. *Proc. Ass. Off. Seed Anal.*, **56**, 95–98.

WOODSTOCK, L. W., and GRABE, D. F., 1967. Relationships between seed respiration during imbibition and subsequent seedling growth in *Zea mays* L. *Pl. Physiol.*, **42**, 1071–76.

WOODSTOCK, L. W., 1969a. Biochemical tests for seed vigor. *Proc. int. Seed Test. Ass.*, **34**, 253–64.

WOODSTOCK, L. W., 1969b. Seedling growth as a measure of seed vigor. *Proc. int. Seed Test. Ass.*, **34**, 273–80.

ZELENY, L., 1954. Chemical, physical and nutritive changes during storage. In *Storage of Cereal Grains and their Products*, eds J. A. Anderson and A. W. Alcock, 46–76. Am. Ass. Cereal Chemists, St Paul, Minn.

CHAPTER 9

Cytological, Genetical, and Metabolic Changes Associated with Loss of Viability

E. H. Roberts

Early observations on chromosomal damage and genetic mutations

It has long been known that plants produced from old seeds are more likely to produce sports than plants from new seeds. Kostoff (1935) considered the earliest relevant observation to be that of De Vries (1901) who noticed that 5-year old seeds of *Oenothera* gave a considerably higher proportion of phenotypic variants than fresh ones. De Vries explained this phenomenon by postulating a longer period of viability for those seeds whose hereditary constitution had been altered. Nilsson (1931), who also worked on the same species, observed that as the percentage germination fell with age, so the frequency of mutant plants increased. He explained this in a similar way to De Vries by suggesting a greater longevity of the mutant forms. In other words these workers thought that the mutant genotypes were contained in the seed population from the beginning: the period of storage simply selected out these mutant forms.

Historical hindsight makes these observations seem peculiarly perverse since a great deal of subsequent work has shown that this type of explanation is untenable: there is no doubt now that aberrations are actually produced during ageing. The chromosome damage occurs and accumulates in the seeds during storage. The evidence comes mainly from three types of observation: (1) examination of the frequency of chromosome abnormalities, particularly in the first mitoses in root tips of germinating A_1 seeds,* (2) examination of the frequency of pollen abortion produced by A_1 plants, and (3) the examination of phenotypic mutations in A_2 and A_3 plants.

---

*The convention adopted in this chapter is to designate seeds which have received an ageing treatment and plants which have developed from them the A_1 generation; subsequent generations are designated A_2, A_3, etc.

Several important independent lines of work came to fruition in 1933. Navashin (1933a and b) working on seeds of *Crepis tectorum* and Peto (1933) who investigated seeds of maize (*Zea mays*) appear to have been the first to have observed that high frequencies of visible chromosome aberrations occur in roots produced from old seeds. At the same time Cartledge and Blakeslee (1933, 1934) showed that old seeds of *Datura* gave increased rates of pollen-abortion mutations. Further work by this group (Blakeslee and Avery, 1934; Avery and Blakeslee, 1936) showed that high rates of phenotypic mutations were also induced by ageing in seeds; and further work on other species of *Crepis* by Navashin and Gerassimova (1935, 1936a and b) confirmed the earlier work on the increase in chromosome aberrations that could be detected by cytological observation on young roots. A number of other workers have subsequently reported increased frequencies of chromosome aberrations with increase in age of seed from a wide range of species: e.g. in onion (*Allium cepa*) (Nichols, 1941; Sax and Sax, 1964); in spring onion (*Allium fistulosum*) (Kato, 1951); in peas (*Pisum sativum*) (D'Amato, 1951); in durum wheat (*Triticum durum*) and common wheat (*T. aestivum*), barley (*Hordeum distichon*), rye (*Secale cereale*) and peas (*Pisum sativum*) (Gunthardt, Smith, Haferkamp and Nilan, 1953); in lettuce (*Lactuca sativa*) (Harrison and McLeish, 1954; Harrison, 1966); and in maize (Berjak, 1968b).

In a great deal of the early work visible chromosome damage, or the symptoms of chromosome damage – pollen abortion or phenotypic changes in the progeny – tended to be related to age. However, it gradually became apparent that chronological age of the seed is not the only factor involved in the production of chromosome aberrations. Evidence that factors such as temperature and moisture during storage are important came not only from cytological work on *Crepis* (Navashin and Gerassimova, 1936a and b) but also from investigations on pollen abortion in *Datura* by Cartledge, Barton and Blakeslee (1936) who concluded that 'in general the mutation rate increased with increased temperature, with increased moisture content, and with increased duration of treatment'.

The importance of temperature is also suggested by the many investigations which have shown that heat treatments induce chromosome breakage. What is not always realised though is that any temperature above absolute zero can be considered as a heat treatment. Peto (1933) reported that treatment of barley seeds at 95°C for 25 minutes or at 40°C at high humidity for 30 days resulted in the appearance of chromosome abnormalities. A series of papers by Navashin and Shkvarnikov (Navashin and Shkvarnikov, 1933a and b;

Shkvarnikov and Navashin, 1934, 1935; Shkvarnikov, 1936, 1939) on wheat and *Crepis* showed that treatment of fresh seeds with temperatures of 50–60°C for 20 days had a comparable effect on the production of chromosome aberrations with that of ageing at room temperature for 6–7 years. It was also reported that increasing the relative humidity of the air also raised the frequency of chromosome aberrations, particularly at high temperatures. In contrast with this work, however, Smith (1943, 1946) reported that exposing seeds of cereals to temperatures of 50–70°C for 5–15 days or 80°C for 45–80 minutes had little, if any, effect upon the frequency of chromosome aberrations. Apart from Smith's work, however, all the evidence above supports the view that chromsome damage in seeds during storage is accelerated by the combined effects of temperature, moisture content and time.

In addition to temperature and humidity, another factor in the normal storage environment which can contribute to nuclear damage is oxygen. Some of the early evidence on this factor, however, was ambiguous. Shkvarnikov (1935, 1939) reported that storage of fresh seeds for 20 days in an atmosphere consisting entirely of carbon dioxide, nitrogen, or oxygen increased the rate of mutation; furthermore, he claimed that an increase *or* reduction in the oxygen content of the air increased mutation rate. More recently, Moutschen-Dahmen, Moutschen and Ehrenberg (1959) studied the effect of oxygen applied at 30–60 atmospheres on broad bean (*Vicia faba*) seeds treated for 3–28 days (at 20–22°C). A high frequency of breaks was induced which was shown to be due to oxygen since similar treatments with nitrogen had practically no effect. Jackson (1959) showed that the breakage of onion seed chromosomes is increased by storage in increased oxygen concentrations.

This brief review indicates that the production of chromosome damage during the ageing of seeds probably depends on the integration with time of the collective effects of temperature, moisture and oxygen. These are the major factors which control the viability period of seeds (Chapter 2); consequently it is possible that there could be some connexion between the accumulation of chromosome damage and loss of viability.

The correlation between loss of viability and the accumulation of chromosome damage

The fact that the pattern of loss of viability in seeds can now be closely defined (Chapter 2) has enabled a fresh approach to be made

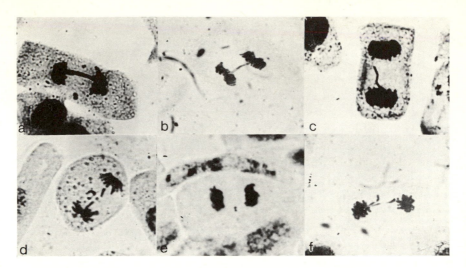

FIGURE 9.1 Chromosome aberrations induced by ageing treatments in barley seeds, observed at anaphase during the first cell divisions of the radicle meristem. (a) two parallel bridges and two dot fragments; (b) two bridges twisted round each other; (c) lagging chromosome; (d) two lagging chromosomes and a fragment; (e) two dot fragments; (f) bridge and lagging chromosome? (a, c, d, e × 640; b, f × 575.)

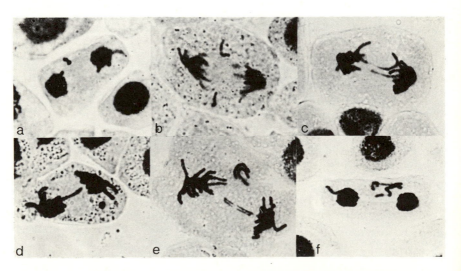

FIGURE 9.2 Chromosome aberrations induced by ageing treatments in broad bean seeds, observed at anaphase during the first cell divisions of the radicle meristem. (a) two dot fragments; (b) acentric fragments; (c) two bridges; (d) a single bridge and dot fragment; (e) two parallel bridges and two fragments; (f) two broken bridges? (× 640.)

to investigating the relationship between loss of viability and the accumulation of chromosome damage. Some investigations have been carried out on peas, beans and barley in which the seeds were treated with various combinations of temperature and moisture content to produce different rates of loss of viability; observations were then made at intervals on the reduction of viability and the accumulation of chromosome damage (Roberts, Abdalla and Owen, 1967; Abdalla and Roberts, 1968). The cytological examinations were made on the first meristematic divisions in the root tips of samples of the surviving seeds which had been set to germinate. Some preliminary observations were made on the various abnormalities occurring at all phases of mitosis but the majority of chromosome breakages are most easily detectable at anaphase (see Figs. 9.1–9.3). For this reason calculations of the frequency of aberrant cells were made on each treatment at each sampling time on observations of 300 anaphase figures from about 10 seeds. Those cells showing bridges or fragments were scored as aberrant and this fraction was expressed as a percentage of all cells observed.

Examples of the results obtained with broad beans are shown in Fig. 9.4 which also shows the corresponding survival curves for the treatments. (Very similarly shaped curves were obtained with barley

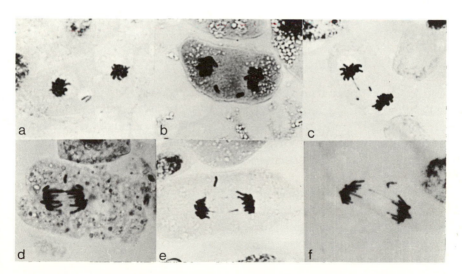

FIGURE 9.3   Chromosome aberrations induced by ageing treatments in pea seeds. (a) single fragment with secondary constriction; (b) two fragments; (c) broken bridge? and two small fragments; (d) several bridges and fragments; (e) two bridges and a fragment; (f) single bridge and fragment (× 640).

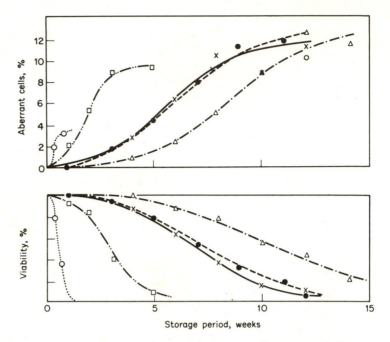

FIGURE 9.4   Upper graph: the increase in the mean frequency of aberrant cells in the surviving population of broad bean seeds stored under various combinations of temperature and moisture content. Lower graph: survival curves for the same treatments. Storage conditions: ∘ · · · · · ∘ 45°C, 18 per cent m.c.; •— — • 45°C, 11 per cent m.c.; □ – · · – · · □ 35°C, 18 per cent m.c.; x —— x 35°C, 15.3 per cent m.c.; △ – · – · △ 25°C, 18 per cent m.c. (Redrawn from Roberts, Abdalla and Owen, 1967.)

and peas except that in these cases the maximum levels of aberrant cells were about 4 and 8 per cent respectively.) It is immediately obvious that there is an increase in chromosome damage with increase in period of treatment and that, if a treatment leads to rapid loss of viability, it also leads to a rapid accumulation of aberrations. In Fig. 9.5 the data have been plotted in a different fashion and it becomes apparent that, for all treatments except one, for any given percentage seed survival, the mean frequency of aberrant cells in the survivors is the same. Thus irrespective of the rate of loss of viability, or which environmental factor was chiefly responsible, the proportion of aberrant cells in the survivors is a function of the percentage viability. Very similar results were obtained with both peas and barley (Abdalla and Roberts, 1968). As discussed in greater detail elsewhere (Roberts *et al.*, 1967; Abdalla and Roberts, 1968), the asymptotic nature of these curves could be explained on the basis that a seed with more than

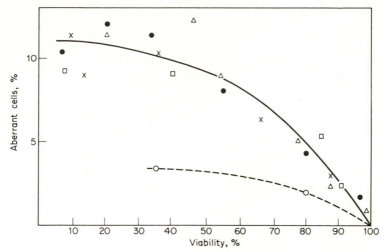

FIGURE 9.5   The relationship between percentage viability and the mean frequency of aberrant cells in the surviving seeds of broad bean. The symbols for the points are the same as in Fig. 9.4. The unbroken curve represents the relationship typical of all treatments except the most severe treatment (at 45°C and 18 per cent m.c.) which is represented by a broken line. (Redrawn from Roberts, Abdalla and Owen, 1967.)

a critical amount of chromosome damage becomes non-viable and thus automatically exluded from the sample for cytological examination.

The exception to this generalisation in all three species – peas, beans and barley – was the most severe treatment (45°C in combination with 18 per cent moisture content) in which the mean viability period was less than a week. In this case, for any given percentage viability, the frequency of aberrant cells found in the survivors is less than for all other treatments. This suggests the possibility that some additional deleterious factor may operate under the most severe conditions which is not reflected in chromosome breakage. Harrison (1966) has observed what appears to be a similar phenomenon in lettuce seed. He reported that a treatment of seeds with 10 per cent moisture content at 40°C for 2–5 days reduced germination to within the range 65 to 6 per cent. These samples showed a very much smaller accumulation of anaphase abnormalities than seed of similar germination percentages which had lost viability slowly (2–5 years at about 6 per cent moisture content and 18°C). It must be concluded, then, that although there is an excellent correlation between loss of viability and the accumulation of chromosome damage, this relationship may break down under severe storage conditions when loss of viability

TABLE 9.1   *The effect of oxygen pressure on the viability and induction of chromosome aberrations in broad bean seeds stored at 25°C. (From Roberts, Abdalla and Owen, 1967.)*

Partial pressure of oxygen, % of 1 atm.	Stored at 18% moisture content for 35 days		Stored at 27% moisture content for 3 days	
	Viability, %	Aberrant cells in survivors, %	Viability, %	Aberrant cells in survivors, %
0	70	6.0	95	0
21	15	11.3	20	3.3
100	20	10.7	10	3.0

may be greater than would be expected from the amount of chromosome damage which has occurred after any given time.

Experiments in which the rate of loss of viability was increased by increasing the partial pressure of oxygen also supports the thesis that loss of viability is closely correlated with frequency of aberrant cells in the surviving embryos, providing that the other conditions of storage are not extreme. Table 9.1 shows the percentage seed viability and mean percentage of aberrant cells in the surviving population after 35 days under moderate storage conditions or 3 days under severe storage conditions. It can be seen that, in both cases, raising the oxygen concentration from 0 to 21 per cent increased the rate of loss of viability and, at the same time, increased the frequency of aberrant cells. Under less severe conditions the numerical relationship between percentage germination and frequency of aberrant cells is very similar to that obtained under the majority of storage conditions but, under the severe storage conditions, as found previously, the seeds tend to die before the typical amount of chromosome breakage has accumulated. Again these results emphasise that under most conditions, irrespective of the environmental factors which lead to loss of viability, there is a very close relationship between chromosome damage and loss of viability.

In some work carried out by Harrison and McLeish (Harrison and McLeish, 1954; Harrison, 1966) on lettuce seeds stored in various gases, it is evident that the curve showing the relationship between percentage viability and the frequency of abnormal anaphases is essentially the same as that described here for peas, beans and barley. The maximum frequency of aberrations was, however, much higher: the proportion of abnormal anaphases per root approached 90 per cent in lettuce seeds when the viability dropped to 50 per cent.

Harrison and McLeish also investigated onion seeds and found a low incidence of chromosome breakage at all levels of germination, from which they concluded that 'a decrease in germination is not necessarily followed by an increase in the frequency of abnormal cells'. However, the points plotted for onion only include three levels of germination, and the highest of these represents only 76 per cent viability; at this and the lower levels of germination the frequency of abnormal anaphases varied between 10 and 15 per cent. Although these frequencies are not high in relation to lettuce, they are in line with the values obtained in peas and beans and higher than that obtained in barley (Abdalla and Roberts, 1968). Consequently, in view of the shortage of points for onion and the lack of information on frequency of aberrations at high germination values, there may be some doubt about the suggested lack of correlation between viability and frequency of aberrant cells in onion. This doubt is reinforced by the work of Nichols (1941) and Sax and Sax (1964) who reported that the frequency of aberrant cells in onion increased with storage period.

In support of the cytological examinations made on peas, beans and barley, an examination was also made of the relationship between loss of viability produced by different combinations of temperature and moisture content on the induction of mutations during storage (Abdalla and Roberts, 1969). Two techniques were employed. The first depended on the assumption that a number of recessive mutations may be lethal and express themselves when unmasked in the haploid condition: accordingly an examination was made of pollen abortion in A_1 plants. The second method depended on allowing A_1 plants to self-pollinate and examining the progeny for recessive chlorophyll mutants (which are easy to recognise – see Frontispiece) in the A_2 generation. The different combinations of time, temperature and moisture content used in the various treatments were designed to reduce viability to about 50 per cent and the survivors of these treatments were compared with the control seeds which were stored under good conditions (dry in cold storage) so their original high viability was maintained. On the hypothesis that genetic damage is a function of loss of viability these treatments would be expected to give the same number of mutants. In fact the investigation was not sufficiently large to test this rigorously, but the results, which are summarised in Tables 9.2 and 9.3 are at least not incompatible with this hypothesis.

On the basis of these results it seems reasonable to assume that under most storage conditions percentage viability is a good index of chromosome breakage and gene mutation in the surviving seeds;

TABLE 9.2   *The effect of seed storage conditions on the frequency of pollen abortion in the A_1 plants. (From Abdalla and Roberts, 1969.)*

Species	Treatment			Viability, %	Viable pollen, %
	Temperature, °C	Moisture %	Period of storage, days		
Barley	control (cold storage at low moisture content)			100	100.00
	45	12.6	21	37	98.81*
	25	17.8	63	40	98.88*
Broad beans	control (cold storage at low moisture content)			90	99.33
	45	11.7	49	37	93.60*
	25	18.3	70	32	92.40*
Peas	control (cold storage at low moisture content)			100	99.81
	45	12.4	47	45	96.26*
	25	18.2	126	40	97.20*

* Significantly different from control (P = 0.01).

this is so irrespective of how rapidly viability is lost or which factor in the storage environment – temperature, moisture content or oxygen – could be considered as the main deleterious factor. In these studies mutations in relatively few genes were investigated: only those which affect pollen abortion and a few which result in chlorophyll deficiencies were examined. Undoubtedly these represent a very small proportion of the total gene complement. Yet by storing seeds under conditions which lead to a 50 per cent loss of viability, the mutation rate of these chlorophyll genes was increased from some extremely low level (apparently nil) to between 1 and 3 per cent. It seems reasonable to assume, therefore, that the increase in the total number of all types of mutant genes induced in the surviving seed population would be very high.

Although the evidence described above shows that gene mutations are induced under storage conditions which lead to a loss of viability of about 50 per cent, there is very little direct evidence that heritable mutations are induced under storage conditions which lead to only small losses of viability. Nevertheless, we know that for a given loss of viability, however small, a given amount of visible chromosome damage may be predicted in the surviving seeds (Fig. 9.4 and 9.5). Furthermore there is a great deal of evidence that visible chromosome damage is normally linked with invisible nuclear changes, i.e. with gene mutation. For example, in barley seeds Caldecott (1961) has

TABLE 9.3 *Frequency of A_1 seeds containing mutant chlorophyll genes as determined by phenotypic segregation in the A_2 generation. (From Abdalla and Roberts, 1969.)*

Species	Treatment			Viability of A_1 seeds, %	No. of A_1 plants observed	Proportion of A_1 plants segregating mutants, %[a]
	Temp., °C	Moist. cont., %	Period, days			
Barley	Control			100	180	0
	25	18.0	54	50	170	1.76
	35	18.0	12	40	142	2.82[b]
	45	12.1	17	47	179	1.12 (7.26)[b]
Broad beans	Control			90	152	0
	25	18.5	45	48	150	2.67[b]
	35	18.5	48	48	139	0
	45	11.5	53	53	160	0
Peas	Control			100	170	0
	25	18.0	100	54	160	3.13[b]
	35	18.0	24	53	153	0
	45	12.3	45	53	130	1.54

[a] Mutants observed included the following (photographs of the phenotypes are shown in the Frontispiece):

Barley { Albina (single recessive)
{ Striata (genetical status obscure: did not persist after A_2)

Beans { Xantha (single recessive)
{ Maculata (double recessive)
{ Greenish-yellow (single recessive)
{ Chlorina (double recessive)

Peas { Xantha (probably single recessive)
{ Maculata (double recessive)
{ Yellowish edges (probably single recessive)

The figure in brackets for barley includes the figures for the enigmatic striata mutant.

[b] Significant increase over control (P = 0.05).

shown that the frequency of chlorophyll mutations induced by X-rays under anaerobic conditions is linearly proportional to the frequency of chromosome aberrations. Thus in absence of evidence to the contrary, we may assume that percentage viability may be taken as a good index of the heritable genetic mutations which have been induced in the surviving population.

These results are of practical importance for those concerned with the preservation of genetic seed stocks and the implications for the design of seed-storage systems have been considered in Chapter 2. One further minor implication is that deliberately 'poor' storage conditions could be used as a simple method for inducing mutations in breeding work.

Apart from the work using 'normal' storage conditions described so far, a correlation has also been observed between viability and frequency of chromosome damage induced by artificial means. For example, there is a great deal of evidence, some of which will be discussed later, which shows that ionising radiations increase chromosome damage (as indicated by breakage or point mutations) and at the same time decrease viability, and it is particularly interesting to note that Sax and Sax (1964) have shown that the effects of 'normal' ageing and X-irradiation are additive in their effects on the induction of chromosome aberrations. A different type of example is provided by Dodson and Yu (1962) who treated sorghum, wheat and oats with centrifugal force for 3 or 6 hours. They showed that over the range 5,000–25,000g there is a negative linear relationship between centrifugal force and percentage viability, whereas previously they had shown (Yu and Dodson, 1961) that the induction of chromosomal aberrations in barley is proportional to the centrifugal force and the time for which it is applied.

NON-HERITABLE MORPHOLOGICAL ABNORMALITIES

In addition to the recessive mutations which, of course, do not appear until the A_2 generation, very often 'old' seeds produce distinctly abnormal plants in the A_1 generation. In peas and beans it was found that any storage treatment which caused a significant loss of viability resulted in some abnormal plants in the A_1 generation. The frequency did not seem to relate to the percentage viability, but the abnormal plants were found with a frequency of 5 per cent in peas as compared with 0 per cent in the control seeds and the corresponding figures in broad beans were 10 and 2 per cent (Abdalla and Roberts, 1969). In most cases the abnormal plants set viable seed but in no case were the abnormalities inherited, consequently their cause is obscure. It seems most probable that they are not due to chromosomal changes but to some kind of cytoplasmic or physiological change.

There would be no point in detailing these A_1 morphological abnormalities but briefly they included anomalous leaf shapes, chlorophyll-deficient spots (often associated with thickening of the leaves) and abnormal branching. Cartledge, Barton and Blakeslee (1936) found similar abnormalities in plants produced from 'aged' *Datura* seeds. The peculiar morphological abnormality of 'pinched root tips' has been observed to occur at high frequency in seedlings from aged cotton seeds (*Gossypium hirsutum*) (Hunter and Presley, 1963). Harrison (1966) reported the occurrence of dwarf and chlorotic A_1 plants of lettuce; the proportions of each varied with storage treatment and their genetical status is not known. Examples of various other abnormalities found at the seedling stage which are probably not genetical are given in Chapter 7.

Damage to cytoplasmic organelles

Although there has long been evidence of an association between loss of viability and the accumulation of nuclear damage, evidence that damage also occurs to the cytoplasmic organelles has only just begun to emerge. This is due to the fact that it has only been possible to obtain direct evidence on possible damage to the extra-nuclear organelles since the development of the use of the electron microscope in biology.

Abu-Shakra and Ching (1967) examined the mitochondrial activity of two samples of soya beans (*Glycine max*); one sample was a few months old and the other three years old. Both had been stored 'dry' at room temperature. Although the second sample was much older than the first, the germination of both was said to be 'in the high 80s'; the older sample, however, showed a much slower rate of seedling growth. Metabolic investigations were carried out on mitochondria isolated from seedling axes four days after the seeds had been set to germinate. Although the mitochondria from both samples showed about the same rate of oxygen uptake, the amount of inorganic phosphate esterified by the mitochondria from the fresh seeds was more than twice as much as those from old: the P/O ratios were 3.0 and 1.4 respectively. Thus there is evidence that the mitochondria from the old seeds tend to be endogenously uncoupled. Electron micrographs of the mitochondrial pellets isolated from the new seeds showed that they contained intact and swollen mitochondria as well as plastids and vesicles. In the old material the mitochondria appeared to have dilated or inflated cristae, a coagulated matrix, and some of the mitochondria did not have an intact outer membrane. Abu-Shakra and Ching suggested that these morphological features did not necessarily explain their inefficiency since functional mitochondria isolated from other plant material show similar characteristics. Nevertheless the mitochondria from the old seeds were specially distinguishable by their dense matrix. They also reported that the number of mitochondria per seedling was apparently lower in old as compared with new tissue; however, it is difficult to know without further information whether the difference in numbers is a function of the age of the seed or the rate of growth of the seedling. (The possible implications of these findings as they affect seedling vigour are further discussed in Chapter 8, p. 230.)

Abu-Shakra and Ching (1967) suggested that the *in situ* condition should be examined in order to discern possible morphological differences with greater validity. This has now been done in maize seeds by Berjak and Villiers (Berjak, 1968b; Berjak and Villiers, 1970, 1971, and in preparation) who have also examined changes in all the other detectable cell organelles in a series of intensive observations. In these studies unaged seeds of maize were compared with another treatment in which the seeds were stored at 14 per cent moisture content at a temperature of 40°C for periods up to 34 days. Under these conditions the mean viability period was about 18 days and viability was lost completely by 30 days. Root squashes confirmed that chromosome aberrations accumulated in a manner very similar to that already described for peas, beans and barley; furthermore the curve showing the relationship between percentage viability and percentage of aberrant cells in the survivors was of essentially the same type as that shown by Abdalla and Roberts (1968). In maize the percentage of aberrant anaphase configurations rose to an asymptote of about 9 per cent. But apart from these light-microscope observations on the nucleus, most of Berjak's observations were made on electron micrographs of the cytoplasmic organelles. For these observations the seeds were imbibed at 25°C and sampled after 12, 24 and 48 hours. Although some observations were made on the whole of the root tips, detailed examinations were restricted to the root cap since this enabled a study to be made of two distinct ageing processes. The first process is the organised senescence which occurs in the root-cap cells as they move out from the cap initials: this is an event occurring during growth necessary to the functional efficiency of the root cap which, Berjak's observations suggest, involves hypersecretory dictyosomal activity and the release of lysosomal enzymes. It is suggested that this type of ageing is genetically programmed. In this book, however, we are more concerned with the second ageing process which can be observed in root-cap cells, i.e. the unprogrammed changes which occur in the stored seeds.

Ultrastructural studies on the root-caps from seeds which had been 'aged' for up to 18–20 days revealed a wide range of degenerative changes. In different seeds it was observed that these changes occurred

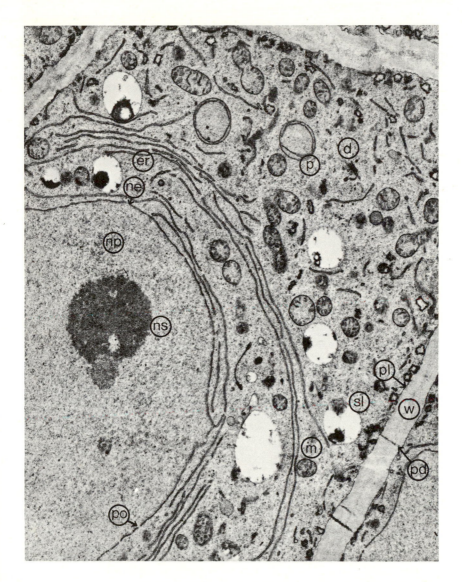

FIGURE 9.6　General view of a portion of a root cap cell of an unaged embryo of maize in the zone of differentiation, 48 hours after the start of imbibition (×11,000).

d = dictyosome; er = endoplasmic reticulum; m = mitochondrion; ne = nuclear envelope; np = nucleoplasm; ns = nucleolus; p = plastid; pl = plasmalemma; pd = plasmodesma; po = pore; sl = second-phase lysosome; w = cell wall.

with various degrees of severity and on this basis the embryos were designated as Type 1, Type 2 and Type 3 in ascending order of severity of damage. All the organelles seemed to be affected by 'ageing' but the damage in the Type 1 embryos was largely reversible in that the abnormalities apparent after 12 hours of imbibition had often disappeared by 24 or 48 hours – presumably as the result of the operation of repair mechanisms. The damage in Type 2 and Type 3 embryos was, however, mostly irreversible. Type 3 embryos, and probably Type 2, are almost certainly non-viable.

In unaged material the typical ultrastructural features of cap cells were observed and some of these are shown in Figs. 9.6–9.9 as a basis for comparison with Figs. 9.10–9.17 which show aged material. What follows now is a brief resumé of the main changes; more details of the complexities involved are summarised in Table 9.4.

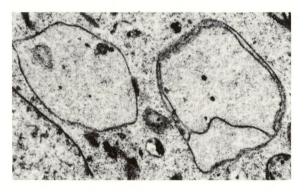

FIGURE 9.7 Relatively large proplastid typical of the cells of the mature zone of cap cells of unaged maize embryos 12 hours after the start of imbibition (× 16,000).

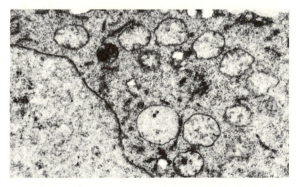

FIGURE 9.8 Mitochondria in cap initial of an unaged maize embryo 12 hours after the start of imbibition (× 16,000).

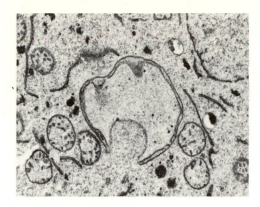

FIGURE 9.9   Mitochondria in a mature cap cell of an unaged maize embryo
12 hours after the start of imbibition (×11,500).

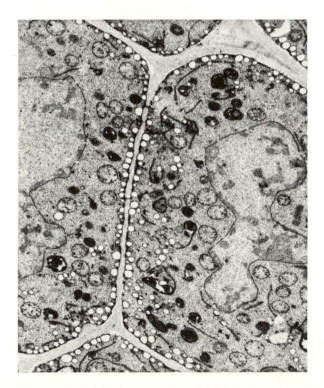

FIGURE 9.10   Lobed nuclei in the zone of cap initials in an aged maize embryo
(×7,300).

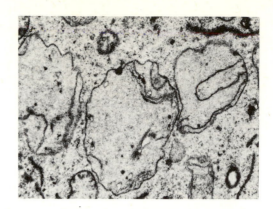

FIGURE 9.11   Plastid in cap cell in zone of differentiation of a Type 1 aged maize embryo, 12 hours after the start of imbibition (× 13,700).

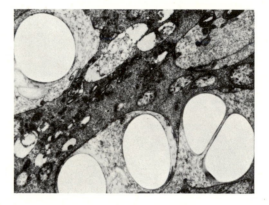

FIGURE 9.12   Swollen plastids in a cap cell of a Type 3 aged maize embryo, 12 hours after the start of imbibition (× 8,000).
Note that the density of the contents is less than that found in unaged material.

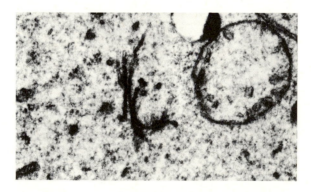

FIGURE 9.13   'Unstacked' dictysome, typical of aged maize embryos, 12 hours after the start of imbibition (× 33,000).

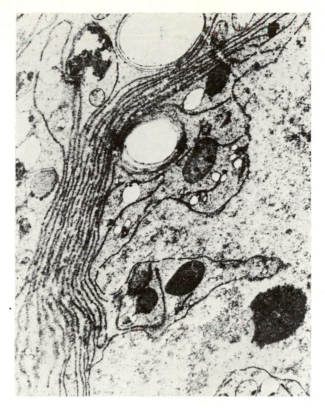

FIGURE 9.14   Proliferation of endoplasmic reticulum in the perinuclear region of a root cap cell in a Type 2 aged embryo, 48 hours after the start of imbibition (× 8,200).

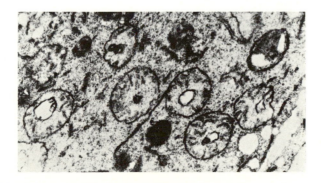

FIGURE 9.15   Damaged mitochondria in a mature cap cell of a Type 1 aged maize embryo (× 16,000).
Note peculiar internal membrane-bounded space.

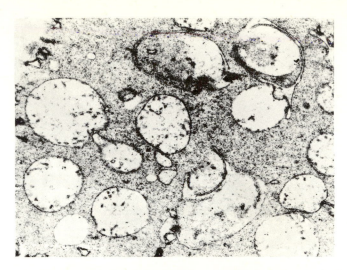

FIGURE 9.16 Swollen and disorganised mitochondria in a cap cell of a Type 3 aged maize embryo 24 hours after the start of imbibition (× 22,000).

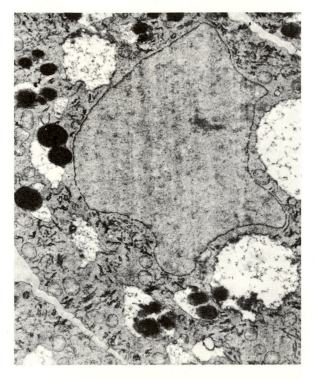

FIGURE 9.17 Cap cell of aged maize embryo in the zone of division 48 hours after the start of imbibition, showing the occurrence of second-phase lysosomes (× 7,200).

TABLE 9.4 *Ultrastructural changes in root-cap cells of maize seed 'aged' at 14 per cent moisture content and 40°C. (From Berjak, 1968.)*

	Unaged seeds	Aged seeds		
		Type 1 embryos	Type 2 embryos	Type 3 embryos
		Protoplasts relatively well-organised. Organelles show a measure of degenerative change, but largely recover normal structure after 48 hr imbibition.	Disorientation of protoplasts: organelles crowded round nucleus leaving areas of cytoplasm virtually devoid of organelles. Organelles show a measure of degeneration. Cells probably non-viable. Increased incorporation of ^{3}H-Thymidine into non-meristematic zones (cf. Figs. 9.18 and 9.19).	Deterioration of protoplasts in an advanced state. Cells non-viable. No recovery from damage observed after 12 hr imbibition.
Nuclei	Cap initials have spherical or oval nuclei with no lobing; mature cap cells show some lobing. Chromatin does not stain with potassium permanganate. Mean cross-sectional diameter about 74 nm (Fig. 9.6).	Lobing of nuclei which persists (for 48 hr, at least). Chromatin stains with potassium permanganate (Fig. 9.10).	Severe lobing of nuclei (no recognisable central core). Chromatin stains with potassium permanganate. Condition deteriorates with increasing period of imbibition.	Extreme lobing of nuclei. Chromatin stains with potassium permanganate. Nucleoli not visible.
Mitochondria	Regular profiles. Normal internal structure. Roughly circular after 12 hr imbibition with few, short cristae. Mean diameter in cap initials: 410 nm increasing to 460 nm after 24 hr imbibition. Mean diameter in mature cells:	Irregular profiles. Unusual internal structure present in mature cells: internal membrane-bounded space, probably representing disorganisation of inner membrane (Fig. 9.15). Occasional elongated cristae appear after	Abnormalities as in Type 1 embryos. Little development of cristae, except for a few abnormal elongated cristae. Abnormalities still persist after 48 hr, i.e. no sign of repair or replication (cf. plastids).	Swollen: mean diameter 702 nm after 12 hr reaching 860 nm after 24 hr imbibition. Marked reduction in density of matrix. Symptoms still persist after 48 hr (cf. Type 2 embryos) (Fig. 9.16).

TABLE 9.4 *Ultrastructural changes in root-cap cells of maize seed 'aged' at 14 per cent moisture content and 40°C. (From Berjak, 1968.)*

	Unaged seeds	Aged seeds		
		Type 1 embryos	Type 2 embryos	Type 3 embryos
Plastids	660 nm decreasing to 540 nm after 24 hr imbibition (Figs. 9.6, 9.8, 9.9).	about 24 hr imbibition. Anomalies tend to disappear within 48 hr imbibition, presumably because of repair and replication.		
	Three types present up to 12 hr imbibition: (a) plastid initials – undifferentiated, circular profile, mean diameter 510 nm; (b) proplastids – mean diameter 710 nm with invaginations of inner membrane; these increase to 1,100 nm in the mature zone and become completely disorganised in the outermost layer of cells of the cap; (c) small quantities of amyloplasts, starch lost after a few hours imbibition (Figs. 9.6, 9.7).	After only 6 days of the ageing treatment many plastids in all cell zones had distorted internal profiles (Fig. 9.11). Internal membranes of proplastids distorted. Damage disappears after 24 hr imbibition.	All plastids distorted. Damage not reversed after 48 hr. Inner membrane damage becomes more marked with period of imbibition.	Mean plastid diameter in all cell zones is 1,900 nm. Contents less dense (Fig. 9.12). Amyloplasts atypically retain starch contents after 48 hr imbibition.
Dictyosomes	Comprised of two or three loosely associated cisternae with small associated vesicles. No cisternae visible in outermost senescing cells, but	Less frequent and more disorganised. Tend to become 'unstacked' and lose cisternae (Fig. 9.13). No large vesicles. Most of this damage is	None evident.	None evident.

274

the vesicles still persist. Mean cisternal lengths in zones of differentiation and maturity: 650 nm and 850 nm respectively (Fig. 9.6).	reversed after 48 hr imbibition, except that the cisternal lengths remain shorter – 362 nm and 580 nm in the zones of differentiation and maturity respectively.		
Ribosomes	Disaggregated monosomes for up to 4 hr after the start of imbibition. After 6 hr polysomes form in all but the outermost senescing cells.	In the cap cells the ribosomes of seeds aged for 18–20 days are still largely dis-aggregated as monosomes after 12 hr imbibition. It is suggested that this is the result of the slowing down of cellular processes rather than the destruction of long-lived messenger-RNA. Polysome formation occurs in all cap cells within 24 hr of the start of imbibition.	Ribosomes largely dis-aggregated after 12 hr imbibition but polysomes are evident after 24 hr.
			Monosomes only.
Lysosomes	Thought to originate from dilation of the ER in cap initial cells. Only occur as fully formed, first-phase lysosomes, mean diameter 550 nm, in mature cells. Closely associated with ER. In the outermost layer of cells of cap, lysosomes are in the second development stage – swollen and bounding membrane lifted (Fig. 9.6).	Precocious development: fully formed, first-phase lysosomes found in young cap cells. Otherwise membranes appear normal. Mean size 600 nm in initials and in zone of division and 720 nm in zones of differentiation and maturity. Second-phase lysosomes found in all cellular zones of cap after 18–20 days' ageing (Fig. 9.17).	As Type 1 embryos, but crowded in perinuclear region. Lysosomal vacuoles occur containing what appear to be remnants of other organelles.
			Membranes not closely applied to contents. May show membrane rupture.

TABLE 9.4 *Ultrastructural changes in root-cap cells of maize seed 'aged' at 14 per cent moisture content and 40°C.* (*From Berjak, 1968.*)

	Unaged seeds	Aged seeds		
		Type 1 embryos	Type 2 embryos	Type 3 embryos
Endoplasmic reticulum	Sparse with short profiles in the cap initials. In the zone of division the profiles are longer but not orientated. In the zone of differentiation there are long profiles parallel with each other and the nuclear envelope (Fig. 9.6). In the mature zone the ER is relatively sparse. Senescing cells of the outermost layer have short profiles associated with second-phase lysosomes.	After 12 days' ageing the short, scattered profiles typical of initials in the younger material persist in cell zones of division and differentiation. In mature cells there is an increase in length of some profiles. After 24 hr imbibition, the profiles in the zones of differentiation and maturity are atypically long. After 48 hr the ER pattern is similar to unaged material.	After 24 hr imbibition profiles are sparse and short, scattered in the cytoplasm particularly near the nucleus in all cap cell types. After 48 hr imbibition banks of long parallel profiles are found encircling the perinuclear region, and the ends of the profiles are often distended (Fig. 9.14).	Only a few short distended profiles are present. It is suggested that this organelle has almost completely degenerated during storage.

With increasing age of the seeds there is an increasing tendency for nuclear lobing to occur (Fig. 9.10). At the same time the chromatin becomes more susceptible to staining with potassium permanganate.

Plastids are amongst the first organs to show changes and these are evident in the distortion of the inner and outer membranes followed by swelling with a concommitant decrease in density of the matrix (Figs. 9.11, 9.12).

The early degeneration of dictyosomes involves 'unstacking' of the cisternae (Fig. 9.13) followed by their apparent loss.

Distortion of the profiles of the endoplasmic reticulum (ER) takes place: the irregularities are first visible in cap cells as an increase in the diameter of the cisternal lumen; with further ageing the ER profiles become longer and thinner, often showing considerable hypertrophy (Fig. 9.14); finally in Type 3 embryos only short distended profiles remain.

In unaged material polysome formation was evident 12 hours after the start of imbibition but in aged embryos the ribosomes were still largely disaggregated as monosomes at this stage of imbibition. This change is thought to result more from a slowing down of metabolism than from a destruction of messenger RNA. In Type 3 embryos only monosomes are found at any stage.

Together with the plastids, the mitochondria are amongst the first organelles to suffer age-related changes. The profiles become irregular and, in addition, there is apparent disorganisation of the inner membrane particularly in cells of the mature zone; and often a peculiar internal membrane-bounded space arises (Fig. 9.15). In Type 3 embryos there are signs of extreme degeneration: the mitochondria then have little internal structure and the earlier membrane distortion appears reversed since the organelles have become swollen – probably because of loss of transport control, with a concommitant reduction in matrix density (Fig. 9.16).

In intermediate stages of ageing, precocious development of lysosomes is observed so that second-phase lysosomes are found in the zones of division and differentiation (Fig. 9.17);* quite often rem-

*Lysosomes, it is suggested (Berjak, 1968a, 1968b), originate from dilation of the cisternae of the ER in cells of the zone of differentiation, but in unaged material they only occur as fully formed first-phase lysosomes, 550 nm in diameter, with dense contents in the mature cells where they are found in close association with the ER. In the outermost cell layer of the cap the lysosomes increase in size and the bounding membrane tends to lift away from the dense contents; they are then called secondary-phase lysosomes. It is also suggested that the plant cell vacuole may be homologous with the secondary-phase lysosome.

nants of the mitochondria and plastids are found within the lysosomal vacuoles. In Type 3 embryos, lysosomes swell and boundary membranes often appear to be incomplete; as the period of imbibition increases there is evidence of general organelle break-down and few intact lysosomes are visible. An observation consistent with this disappearance is that, at this stage, tests show that acid phosphatase becomes widespread in the protoplasm.

Practically all the degenerative changes in the ultrastructure outlined here could be the result of a general deterioration in the lipo-protein membranes of the cells. If these have not deteriorated too far, as in Type 1 embryos, then apparently some repair is possible. But if the deterioration has gone beyond some critical point, then the cell becomes irreversibly disorganised.

Berjak and Villiers suggest that although the membrane damage is general, the symptoms of mitochondrial damage are among the first to appear and therefore it may be that it is this damage which is the most critical. They suggest (Berjak and Villiers, 1971) that the temporary hypertrophy of the ER which can occur during the germination of aged seeds may be a symptom of mechanisms which compensate for the mitochondrial damage. This suggestion is based on the fact that there is some evidence that the ER may have some connexion with respiratory metabolism since proliferation of ER has been observed in plant cells at low oxygen tensions and in yeast strains devoid of mitochondria. If mitochondrial damage is an early event in seed deterioration, then it might be expected that as seeds lose viability so there would be some decrease in the respiration rates of seeds when they are set to germinate; Throneberry and Smith (1955) reported such a decrease in maize, but the correlation with loss of viability was by no means perfect. On the other hand it could be agreed that if compensating alternatives to mitochondrial respiration exist, then a deterioration in mitochondrial respiration may be obscured in simple respiratory measurements. However, unless compensatory mechanisms do exist, observations on old and new barley and wheat seeds showing little difference in oxygen uptake (Abdul-Baki, 1969; Anderson, 1970) are difficult to reconcile with Berjak's suggestion that mitochondrial damage may be the major factor which prevents germination in old seeds.

The loss of membrane integrity with ageing might be thought to explain the fact that aged seeds show a greater tendency to leak metabolites into the germination medium than fresh seeds do. This phenomenon has been known for some time but one of the more comprehensive investigations into it was carried out by Ching and Schoolcraft

(1968) who compared seeds of ryegrass and red clover which had been stored under various conditions for ten years. They showed that there was some correlation between loss of viability and the conductance of the leachates and the concentration of sugars and amino acids in the leachates from the clover seeds – although the correlation was by no means perfect. In the case of ryegrass, however, there was little correlation with conductance measurements or sugar concentration, although the correlation with concentration of amino acids in the leachates was somewhat better. In Chapter 7 it is pointed out that Takayangi and Murakami (1968a and b) developed simple viability tests based on the concentration of sugar in the leachates. But the validity of such tests, at least in barley, has since been questioned by Abdul-Baki and Anderson (1970) who found that the correlation broke down in rapidly aged seeds. Furthermore, what is even more relevant to the present discussion, they suggested that leachable glucose was related to the internal concentration of glucose and thus it may not reflect membrane integrity.

Autoradiographic experiments were also undertaken by Berjak and Villiers to investigate the pattern of ^{3}H-thymidine incorporation during a 4-hour incubation period at the end of the 48-hour period of imbibition. These experiments showed that incorporation, indicating DNA replication, is confined to the nuclei of the meristematic zone of the cap (Fig. 9.18). The proportion of labelled nuclei per cap

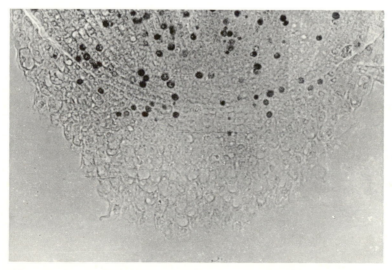

FIGURE 9.18   Autoradiograph showing that the incorporation of ^{3}H-Thymidine is limited to nuclei in the meristematic zone of cap cells in unaged embryos of maize after 48 hours from the start of imbibition ($\times$210).

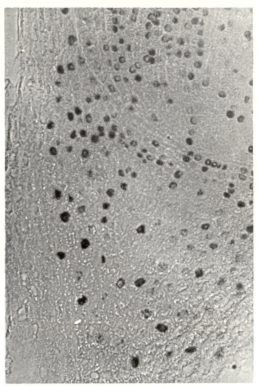

FIGURE 9.19   Autoradiograph showing that in maize embryos which have
received intermediate periods of ageing (10–14 days at 40°C and 14 per cent
moisture content), the incorporation of ³H-Thymidine occurs in non-meristematic
as well as meristematic zones of the cap after 48 hours from the start of
imbibition (× 440).

was found to be 6 per cent. However, in embryos which had been
subjected to intermediate periods of ageing (10–14 days), the proportion of labelled nuclei increased to about 8.5 per cent. This increased
proportion was due, at least in part, to incorporation of the label into
zones which are normally non-meristematic (Fig. 9.19). With further
periods of ageing (18–20 days) the proportion of cap cells with labelled
nuclei dropped to less than 1 per cent; and the incorporation then
appeared random and seldom occurred in the zones of initials or
division. Berjak and Villiers (1971) offer two possible explanations of
this behaviour. First it could indicate that the meristematic cells are
more susceptible to damage during ageing, and are replaced from
normally quiescent cells which must be assumed to be less prone to
damage. This would be analogous to the phenomenon that cells in
the quiescent zone of roots undergo division after X-irradiation, and

replenish the meristem cells which have been damaged (Clowes, 1969). Secondly it could indicate that the repression mechanisms in non-meristematic cells which inhibit DNA replication may have been damaged.

From the discussion so far it will be seen that there is now a great deal of evidence which shows that there is an increase of damage to practically all cell organelles – both in the nucleus and in the cytoplasm – before a seed loses viability. In the case of the nuclear damage, with the exception of very severe storage conditions, the amount of damage which occurs correlates well with the amount of viability which has been lost, irrespective of how rapidly viability has been lost or which are the environmental factors mainly responsible. These generalisations apply to all three species on which detailed investigations have been carried out. There is now an urgent need to extend the work on cytoplasmic organelles to other species and to investigate the changes which occur under different rates and different conditions of ageing. Such investigations would help to shed further light on which changes are the most critical.

Anderson, Baker and Worthington (1970) have published a brief paper on the ultrastructural changes which occur during storage to the embryonic axes of wheat embryos. They compared wheat seeds infected with the natural seed-borne mycoflora placed in storage at 25°C and 75 per cent relative humidity for 26 weeks with control seeds kept at −20°C. Various abnormalities were noted including changes which were interpreted as the coalescence of lipid bodies (or spherosomes), withdrawal of the plasmalemma from the cell wall, and rupture of the plasmalemma. They suggested that these changes were the result of fungal activity but, since the presence and absence of fungal activity was confounded with storage environment, their interpretation cannot yet be accepted with any certainty: these changes could be the result of a more direct effect of the storage environment on the cell organelles. There is obviously a need for further work to distinguish between the direct effects of environment and the possible indirect effects of the mycoflora on damage to cell organelles.

Although this review is primarily concerned with orthodox seeds in which viability is increased by decreasing moisture content to low levels, it should be mentioned that in seeds in which the optimum moisture content appears to be fairly high, the cytological factors associated with loss of viability may be quite different. For example, Genkel and Shih-Hsū (1958) have associated the loss of viability in coffee with break-down of the plasmodesmata.

Viability theories

In addition to the genetical and cytological changes which occur during seed storage there have been large numbers of reports on miscellaneous metabolic age-related differences between young and old seeds. It is not proposed to attempt a complete catalogue of these. In most cases although age-associated changes have been shown they do not often correlate well with percentage loss of viability. In this chapter it is proposed to deal only with those changes which have been implicated in some of the theories proposed to account for loss of viability. For further information other reviews may be consulted (particularly Crocker and Barton, 1953, and Barton, 1961).

We still do not know why seeds lose viability but there is no shortage of theories. In Table 9.5 I have attempted a classification of the major suggestions which have been made. It will be seen that the viability theories can be conveniently divided into two groups: the first group suggests that the activities which cause death are extrinsic to the seed; the second group includes theories suggesting that death is brought about as the result of events occurring within the seed. To some extent the two types may overlap. For example, one theory that has been fairly widely supported is that loss of viability is the result of nuclear damage, but this could be postulated to be the result of either internal or external events.

IONISING RADIATION

It has sometimes been suggested (e.g. Went and Muntz, 1949) that loss of viability under normal storage conditions could be caused by the harmful effects of background ionising radiation. This idea is based on the well-known fact that increased doses of applied radiation can accelerate loss of viability in seeds (e.g. Nilan and Gunthart, 1956; Read, 1959; Sax and Sax, 1962, 1964) as well as accelerate ageing in other organisms (e.g. Comfort, 1964; Grosch, 1965; Lindop and Sacher, 1966; Curtis, 1967). Furthermore ionising radiations induce mutations and chromosome abnormalities which, as we have seen, are closely correlated with loss of viability in seeds. Nevertheless there are at least three good arguments why background radiation cannot be an important factor under normal storage conditions.

First there is a great deal of evidence that ionising radiations cause greater damage to diploids in terms of loss of viability and decreased seedling growth (Stadler, 1929; Fröier, Gelin and Gustafsson, 1941; Palenzona, 1961; Matsumura and Nezu, 1961) or in the induction of recessive mutant phenotypes (Krishnaswami, 1968) as compared

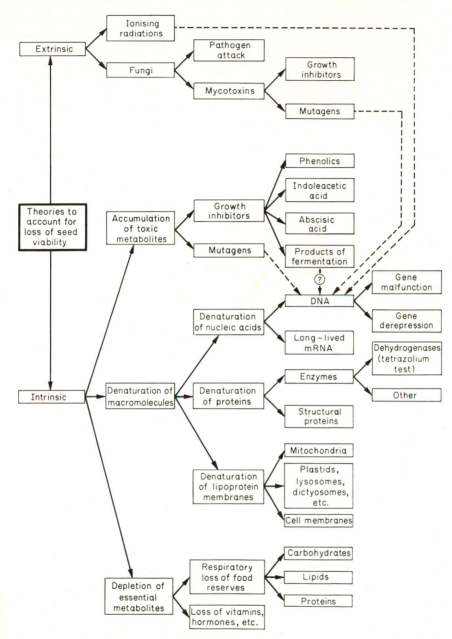

TABLE 9.5   *A classification of viability theories.*

with polyploid species, whereas Smith (1946) has pointed out that in the few studies which have been made on the longevity of polyploid as compared with diploid seeds there is no indication that polyploidy has any beneficial effect. Thus it would seem to follow that the loss of viability which occurs under normal circumstances is not due to radiation. Secondly it has been shown that the damaging effect of X-rays on barley seeds as determined by seedling growth (Caldecott, 1954) or induction of chromosome aberrations (Ehrenberg, 1955) *increases* with reduction in moisture content whereas it has been shown that nuclear damage (this Chapter) and loss of viability (Chapter 2) *decrease* with decrease in moisture content under ordinary storage conditions. Thirdly, extrapolation of experimental dose-response curves of ionising radiation to background level suggest that the background level could not account for the rapid loss of viability which occurs under many circumstances.

From these arguments it will be seen that although ionising radiation can have effects very similar to those observed under normal conditions, particularly with regard to the induction of nuclear damage, background radiation cannot be held responsible for loss of viability under ordinary storage conditions. It should be pointed out though that these arguments do not preclude alternative theories involving nuclear damage: what they do suggest is that, if nuclear-damage theories apply, then the damage involved is not of the same type as that produced by ionising radiations.

STORAGE FUNGI

There is no doubt that under many conditions storage fungi can accelerate the death of seeds (Chapter 3). This is one of the reasons why this is the only group of organisms associated with seeds to which a separate chapter has been allocated. The activity of fungi is known to increase with temperature, moisture content and oxygen pressure; consequently this explanation of loss of viability appears to be compatible with the major effects of storage environment outlined in Chapter 2. As Christensen points out, in circumstances where fungi can cause death, this may be the result of direct pathogenic attack or the result of the production of toxins.

In spite of this, the theory cannot be accepted for universal application for at least three reasons. First, although the evidence shows that sterile seeds survive longer than non-sterile seeds, the sterile seeds still lose viability. For example, Hummel, Cuendet, Christensen and Geddes (1954) obtained a wheat sample free from internal mycelium which they surface sterilised and declared as probably sterile. The

TABLE 9.6 *The viability of mouldy and mould-free[a] wheat after storage for 19 days at 35°C. (Data in first three columns from Hummel, Cuendet, Christensen, and Geddes, 1954.)*

Initial moisture content, %	Condition of wheat	Viability after storage for 19 days, %	Probable mean viability period[b]
14.9	mould-free	70	24
	mouldy	41	18.5
16	mould-free	30	16.5
	mouldy	10	14
18	mould-free	8	13.5
	mouldy	2	11
20.2, 24.2,	mould-free	0	—
27.7, and 30.8	mouldy	0	—

[a] Sample free from internal mycelium and surface-sterilised.
[b] Calculated from the percentage germination data in column 3 and scales c, d, and e of the nomograph for wheat in Appendix 3.

results for viability after 19 days' storage at 35°C are shown in Table 9.6. Although the mean viability period was not given it is possible to estimate the probable mean viability period and it will be seen that in the unsterile samples the presence of storage fungi probably reduced the mean viability period about 18–23 per cent. Secondly, it is known that storage fungi are ineffective at relative humidities below about 65 per cent (i.e. at an equilibrium moisture content of about 12.5 per cent in cereals and 14 per cent in peas and beans), yet loss of viability occurs at moisture contents below this. Furthermore, the same type of relationship between temperature, moisture content and loss of viability extends above and below the moisture content critical to fungal activity (Chapter 2). Thirdly, although oxygen pressure affects fungal activity in stored seeds (Peterson, Schlegel, Hummel, Cuendet, Geddes and Christensen, 1956) and thus could affect viability indirectly, it has been shown that oxygen continues to have a deleterious effect on viability in peas at 60°C and 12 per cent moisture content (Roberts and Abdalla, 1968) – conditions under which the seeds are almost certainly free from fungal activity.

It is concluded, then, that fungi can accelerate loss of viability within the range of temperatures and relative humidities in which they can be active. Even within this range viability is still lost in the absence of microorganisms though the period of viability is then extended. Outside the environmental range of microorganism activity obviously other theories must apply without possibility of interaction with a fungal theory.

The practical implications are that under ordinary commercial storage conditions fungi can be very important factors in accelerating deterioration of seed. At present there are no alternative methods for controlling the activity of storage fungi apart from controlling the environment. Christensen and López (1963) have reviewed the results which have been obtained with standard fungicides and conclude they are very ineffective. But even if it were possible to stop the activity of fungi by methods other than the manipulation of the environment – say by the use of improved antibiotics, should they become available – it still would not be possible to achieve long-term storage in typical ambient environments: it would still be necessary to reduce moisture content and temperature. And at these low temperatures and moisture contents seeds would still lose viability – albeit slowly. Thus for developing systems of long-term storage we have to be concerned with intrinsic theories of loss of viability.

ACCUMULATION OF GROWTH INHIBITORS

At various times the suggestion has been made that loss of viability is due to the accumulation of toxic substances of various sorts. I shall restrict this discussion to some of the main specific suggestions which have been made and tend to recur from time to time. Although not strongly advocating the theory, Crocker (1948) and Curtis and Clark (1950) suggested that loss of viability might be caused by the accumulation of the products of respiration and, more specifically, Wyttenbach (1955) suggested that loss of viability of seeds of *Medicago sativa*, *Trifolium pratense* and *Lotus corniculatus* is due to the accumulation of lactic acid. The idea, however, is difficult to reconcile with two facts. First there is a negative relationship between oxygen pressure and period of viability (Chapter 2); secondly measurements of the respiration of stored seeds tend to show extremely low RQ values – typically less than 0.7 and approaching 0.6, e.g. in barley (James and James, 1940; Merry and Goddard, 1941), in perennial ryegrass and red clover (Ching, 1961), in apple seeds (Stokes, 1965) and in peas (Roberts and Abdalla, 1968). According to Milner, Christensen and Geddes (1947a and b), fungal metabolism on wheat at moisture contents high enough to support the growth of moulds is characterised by a rather higher RQ of 0.8. The reason for the very low RQ values of stored seeds is not clear but, whatever the reason, the low values do not suggest that the accumulation of fermentation products is likely to create difficulties in stored seeds.

Another group of substances about which there has been specula-
tion is the fatty acids which tend to accumulate as seeds lose viability.
Barton (1961) who reviewed this subject concluded: '. . . even though
there is often an association with viability, the results are not consistent enough to use it as a reliable index of viability.' In rice seeds
claims have been made that loss of viability is the result of the accumulation of supra-optimal concentrations of indoleacetic acid and other
indole derivatives (Sircar and Biswas, 1960; Sircar and Dey, 1967).
Later it was suggested that an accumulation of phenolics, including
coumarin and ferulic acid may contribute to the inability of non-
viable rice seeds to germinate (Dey, Sircar and Sircar, 1967). The
most recent candidate in this catalogue of inhibitors which is said to
prevent non-viable rice seeds from germinating is abscisic acid (Sircar, 1967; Dey and Sircar, 1968). At this time we do not know if the
list of inhibitors is complete or if this work will be corroborated by
work on other species.

ACCUMULATION OF MUTAGENS

Growth inhibitors with mutagenic properties form a special category.
Many of the investigators of the chromosome aberrations and genetic
mutations which are the concommitants of loss of viability have been
concerned with the question of what causes the damage. It is probably
true to say that where opinions have been expressed the majority of
workers in this field have suggested that chromosome damage is
caused by the accumulation of automutagenic substances within the
seed. One of the first to suggest this was Stubbe (1935) and the evidence which has accumulated subsequently has been comprehensively reviewed by D'Amato and Hoffman-Ostenhof (1956) who
concluded that 'the most probable causes for the loss of germinability
are either lethal chromosome changes brought about by automutagenic substances accumulating during seed ageing, or a more general
poisoning of the embryo due to the accumulation of autotoxic substances.' James (1961) suggested that these toxic mutagenic substances
may be respiratory end-products. In cases where mutagens can be
shown it would be difficult to determine whether they are derived
from the seed's own metabolism (i.e. are truly automutagenic) or from
the associated microflora – unless, of course, the seeds were aged
under conditions which preclude the activity of microflora. In this
connexion it is worth pointing out that it is known that some of the
products of storage microflora are mutagenic, e.g. aflatoxin (a product
of the storage fungus *Aspergillus flavus*) has been shown to induce
chromosome breakage in *Vicia faba* (Lilly, 1965). Thus at present it

is difficult to categorise mutagen theories as either 'intrinsic' or 'extrinsic'. However, the majority of workers who favour the hypothesis assume that the mutagens are intrinsically derived.

Most of the research in this field is centred on attempts to extract and assay mutagenic substances. Much of the evidence cited in the main review of this subject (D'Amato and Hoffman-Ostenhof, 1956) is, however, not critical: in some cases control extracts from fresh seeds were not tested, while in a few cases comparable mutagenic activity was found in fresh seeds; in many cases extracts were only tested on other species or, if they were tested on the same species, they were found to have no mutagenic activity; and in other cases toxic and anti-mitotic activity were demonstrated, but no mutagenic activity. Nevertheless, if all the questionable evidence is ignored there still remains some convicing evidence furnished by Gisquet, Hitier, Izard and Mounat (1951) who found that oil obtained from old tobacco seeds had mutagenic activity on the same species (indicated both by genetical and cytological evidence) whereas extracts from young seeds were inactive. More recently Jackson (1959) reported an experiment in which leachates from aged or fresh onion seeds were applied to fresh onion seeds. Jackson cautioned that the experiment was preliminary, lacked statistical control and the results should be considered tentatively; nevertheless the indications were that chromosome breakage can be caused by leachates from aged seeds.

Recent investigations on peas, beans and barley (Abdalla and Roberts, 1968) have failed to find any mutagenic activity. Aqueous, ethanol and ether extracts were investigated, as were leachates prepared according to Jackson's method; and although aqueous extracts had growth-inhibiting properties, no differences were found between extracts from fresh seeds and seeds which had lost 50 per cent viability by either of the two different ageing treatments which were employed.

Reciprocal transplant experiments between young and old embryos and endosperms in *Triticum durum* carried out by Floris (1966) gave the following results: young embryo onto young endosperm – 94 per cent germination; young embryo onto old endosperm – 76 per cent; old embryo onto young endosperm – 13 per cent; old embryo on to old endosperm – 18 per cent. The last two figures are not significantly different but the first two are and suggest that inhibitory substance(s) accumulating in the endosperm can contribute to the inhibition of germination. Subsequent work (Floris, 1970) showed that endosperms from seeds which had been stored in the laboratory for up to four years had no significant effect on the percentage germination of the young embryo on to which they were transplanted, though some depression of seedling growth was observed. The viability of the four-year-old embryos (as indicated by data from transplants to young endosperms) was 31 per cent. Five-year-old endosperms of about the same percentage viability did, however, cause a depression of germination of young embryos to which they were attached, reducing the germination from 90 per cent in the young homotransplants to 53 per cent in the old embryo/young endosperm heterotransplants. These experiments indicate therefore that germination inhibitors produced in the endosperm can cause a significant depression of germination under conditions where considerable loss of viability has taken place.

Using similar transplant experiments Corsi and Avanzi (1969) concluded that chromosome damage in the embryo induced by ageing is not the result of senescence in the endosperm, although low levels of chromosomal damage can be induced by old endosperms. In this context it is interesting to note that the transmission of a mutagenic effect from an X-irradiated endosperm is much greater (Avanzi, Corsi, D'Amato, Floris and Meletti, 1967; Meletti, Floris and D'Amato, 1968) – thus again emphasising a difference between the effects of irradiation as opposed to 'natural' ageing phenomena mentioned earlier.

METABOLIC DEPLETION OF ESSENTIAL RESERVES

Oxley (1948) suggested that the continued life of the seed depends on the use of some labile organic matter present in the embryo; as this substance becomes exhausted, the seed loses viability. It has often been pointed out (e.g. Milner and Geddes, 1954) that any storage conditions which reduce respiratory activity tend to prolong seed viability and consequently the possibility has frequently been raised (e.g. Went and Muntz, 1949) that loss of viability may be the result of the depletion of respiratory substrate. This theory is not invalidated, though it is made more difficult to classify, by the knowledge that under most circumstances measurements of the respiration of stored seeds tend to reflect the activity of the associated microflora rather than the seeds themselves (Milner and Geddes, 1954).

There is no doubt that a number of changes take place in the carbohydrate, fat and protein reserves of seeds during storage. Most of the work in this field has arisen out of the practical problems concerning the bulk storage of grain for use as food. Although in such cases viability *per se* is unimportant, much attention has been paid to loss of viability in relation to chemical changes since viability has often been found to be a good index of these other changes. Typical changes

in the chemistry of cereal seeds with loss of viability have been reviewed by Zeleny (1954): they include decreases in protein and non-
reducing sugars, and increases in reducing sugars and free fatty acids.
In much of the work the changes have been associated with microorganism activity: for example there is evidence that treatments
which increase the mould count of seeds – high oxygen or low carbon
dioxide concentrations (Peterson *et al.*, 1956), or inoculation with
moulds (Golubchuck, Sorger-Domenigg, Cuendet, Christensen and
Geddes, 1956) – tend to increase the fatty-acid content of wheat;
likewise an increase in moisture content has been shown to increase
the fatty-acid content and decrease the amount of non-reducing
sugars of maize (Olafson, Christensen and Geddes, 1954). However,
in experiments on barley and wheat carried out at 12 per cent moisture
content and temperatures between 30° and 40°C, it has also been
shown that there is an association of loss of viability with decrease in
protein and total sugars, and an increase in free fatty acids and reducing sugars (F. H. Abdalla, private communication, 1970); and in view
of the low moisture contents used in these experiments it is unlikely
that the activity of microorganisms contributed to these changes (see
Fig. 2.6). Thus although microorganisms can accelerate these
changes, they are apparently not essential to them.

In spite of the fact that these catabolic changes lead to some loss of
the major food reserves of the seeds, it does not seem probable that
they are the cause of loss of viability. Barton (1961) has pointed out
that inspection of seeds which have lost viability shows that they still
contain large amounts of basic food reserves and James (1967) has
drawn attention to the fact that some seeds with large food reserves
deteriorate more rapidly than seeds with small reserves. Furthermore,
although the major factors which decrease viability (Chapter 2) –
increase in temperature, moisture content and oxygen pressure –
would be expected to increase respiration, it has been shown (Roberts
and Abdalla, 1968) that the deleterious effects of oxygen are also found
at a combination of temperature and moisture content (60°C and 12
per cent moisture content) at which respiration is likely to be inhibited. Consequently the effect of oxygen on loss of viability would
appear to be operating through some process other than respiration.

With regard to non-carbohydrate respiratory substrates, e.g. lipids
and proteins, the evidence reviewed by Owen (1956) suggests that
depletion of these reserves will also not account for loss of viability.
Finally, the possibility of the loss of substances only required in small
amounts should be considered. Theories of this type cannot be dis-
proved unless the substance is specified. All that can be said then at

the present time is that although age-dependent decreases have been found in some specific substances – thiamine and ascorbic acid for example (see the review by Owen, 1956) – decreases of this type have not yet been related to loss of viability.

DENATURATION OF PROTEINS

Ewart (1908) suggested 'Longevity depends not on the food material or seed coats, but upon how long the inert proteid molecules into which the living protoplasm disintegrates when drying, retain their molecular grouping which permits of their recombination to form the active protoplasmic molecule when the seed is moistened and supplied with oxygen.' It is hardly necessary to add that at the beginning of this century it would have been difficult for Ewart to find experimental support for his statement. Nevertheless, the basic idea that proteins may lose their integrity needs careful consideration since it is now more amenable to experimental investigation.

It has already been mentioned in Chapter 2 that Groves (1917) developed an equation for relating the viability period of wheat seeds to high temperatures (equation 2.2). This was based on an equation by Lepeschkin (1913) for studying the effect of temperature on the death of plant materials which, in turn, was derived from the work of Buglia (1909) on the coagulation of blood serum. Robbins and Petch (1932) also suggested from experiments carried out at high temperatures that the curve relating loss of viability of wheat and maize to moisture content of the seed was the same general shape as that relating coagulation of proteins to moisture content. Thus there is some indication that the data on the effects of both temperature and moisture content on seed viability would be compatible with the idea that loss of viability is related to the denaturation of proteins. However, although Crocker was partly responsible for the development of this idea (Crocker and Groves, 1915) he later favoured more a nuclear-degeneration theory (Crocker, 1938) and eventually developed the opinion (Crocker, 1948) that: 'This theory has the fault of being very general. There are so many different kinds of proteins in the embryo, and this work does not throw any light on the particular proteins which coagulate with time. Furthermore, it throws no light on the possibility of the degeneration of some particular mechanism in the cell.'

This may help to explain why those concerned with protein-denaturation theories have tended to concentrate on individual enzymes. Much of this work on loss of enzyme activity is now difficult to separate from theories based on the instability of DNA or long-lived mRNA since we do not yet have a catalogue of which enzymes are present in the resting seed and which have to be synthesised *de novo* on imbibition. The enzymes which have received greatest attention in viability theories have been those concerned with respiration and, as can be seen in Chapter 7, there is a reasonably good correlation between viability and the general activity of dehydrogenases as indicated by the tetrazolium test. However, it will be seen that, although this is probably the best chemical test for viability so far, the correlation is by no means perfect; the lack of correlation under certain circumstances is particularly well-known, for example, in barley (MacLeod, 1953). Attention has also been given to a number of specific dehydrogenases and oxidases but, if anything, the correlations with viability have been inferior to the general dehydrogenase test. Catalase has received a good deal of attention but other enzymes which have been studied are phenolase, malic dehydrogenase, alcohol dehydrogenase and cytochrome oxidase (see reviews by Barton, 1961 and James, 1967).

DENATURATION OF LIPOPROTEIN CELL MEMBRANES

Based on her work which has been reviewed here, Berjak (1968b) has suggested that loss of viability may be the result of the loss of integrity of cell membranes. The theory has the virtue of accounting for accelerated loss of viability with increased oxygen pressure since she points out that the membrane lipids have unsaturated fatty acid components and these have been suggested to be subject to peroxidative processes under aerobic conditions. This type of process involves initial removal of hydrogen (e.g. by reaction with endogenously-formed free radicals); as a result the lipid molecule becomes a free radical, which in turn can react with oxygen to form a peroxide, thus changing the nature of the membrane. This type of theory would also account for the accelerated loss of viability resulting from ionising radiation mentioned earlier. Berjak suggests that the damage may be most important in mitochondria; this would result in loss of respiratory efficiency which, in turn, would lead to loss of viability. In order to validate this theory it would be necessary first to examine ultra-structural damage under a wide range of storage conditions and extend the observations to other species.

DENATURATION OF NUCLEIC ACIDS

This type of theory has been separated from those involving the accumulation of mutagens since it is not essential to infer the accumulation of chemical mutagens in order to postulate that loss of viability may be the result of nuclear damage. Greer and Zamenhof (1962) have carried out experiments *in vitro* on the effects of temperature on dry DNA and DNA in solutions. They have shown that there is an increase in depurination (release of adenine and guanine) from the DNA under both conditions with increase in temperature. Furthermore, their data indicate that the relase is greater in solution. Thus one could expect both increase in temperature and moisture content to lead to increases in depuration of DNA. They conclude that 'depurination by heat may be considered a possible cause of spontaneous mutation and may take place in resting cells since the alteration can be registered in the absence of DNA synthesis.' Of all the theories mentioned so far it is those based on nuclear damage which have tended to find most favour in the previous major reviews of seed viability (e.g. Crocker, 1938; Crocker and Barton, 1953; Barton, 1961). The workd described at the beginning of the chapter showed that, with the exception of very severe storage conditions, there is a very good correlation between loss of viability and nuclear damage. Yet the theory is not without its difficulties if one asks the question: what aspect of nuclear damage leads to an inability of the seed to germinate?

For example, although chromosome breakage is a reliable index of loss of viability, the visible damage itself is unlikely to be a causal factor in seed deterioration. After all, cells containing large amounts of this sort of damage are still capable of at least one cell division (Figs. 9.1–9.3); and if a cell retains enough integrity to divide, presumably it is also capable of the cell elongation necessary for germination. But then it might be suggested that although chromosome breakage itself is unimportant, the gene mutations accompanying it are relevant. It could be said that, since most mutations are deleterious, loss of viability is the result of the accumulation of mutations. However, since most mutations are also recessive one would expect that replication of chromosomes, i.e. increased ploidy levels, would confer an improvement in longevity. But in connexion with theories based on background radiation it has already been pointed out that a high ploidy apparently confers no advantage under 'natural' storage conditions although it does when seeds are subjected to ionising radiation; thus although a simple mutational hypothesis could explain 'ageing' induced by ionising radiation one cannot apply such a simple hypothesis to ageing under more normal circumstances. Finally it is by no means certain yet whether integrity of chromosomal DNA is essential for germination. Although the concept of long-lived messenger RNA in seeds cannot yet be accepted unreservedly (Barker, Bray and Detlefsen, 1971), there is at least a theory which receives considerable support suggesting that the early stages of germination do not require transcription of the nuclear messages (see Chapter 12, p. 381). It may be, then, that DNA is not required during the early stages of germination.

If theories based on the stability of nucleic acids are to be postulated then the four major choices would seem to be as follows. (1) Ageing is caused by dominant lethal mutations. There is evidence for example that loss of viability caused by heat in eggs of the wasp *Habrobracon* is due to the induction of dominant lethal mutations (Whiting, 1946). (2) Ageing is the result of derepression of genes which should remain repressed, thus leading to a 'senseless' combination of active genes; this idea which was mentioned previously (Roberts, Abdalla and Owen, 1967) is based on a theory of ageing in mammals suggested by Medvedev (1964) in which it is postulated that a mutation may not completely inactivate a cistron but nevertheless lead to lack of complimentarity between cistron and repressor so that the gene becomes derepressed. (3) Ageing is the result of the denaturation of other forms of DNA. In this case it might be suggested that the nuclear damage observed is a reflection of a general deterioration of DNA but that it may be that damage to cytoplasmic DNA (e.g. in the mitochondria or plastids) is relevant to loss of viability. (4) Ageing is the result of the denaturation of long-lived messenger RNA on which the cell is dependent during the early stages of germination.

All these theories could involve the idea of events, like conventional mutations, occurring at random but with a mean frequency dependent on temperature, moisture content and oxygen. The process would not demand the production of stable mutagens, although the involvement of free-radicals which can be considered as short-lived mutagens could well be involved. Such theories, based on the idea of events occurring at random may be termed 'stochastic'.*

STOCHASTIC VERSUS ACCUMULATION AND DEPLETION
THEORIES OF AGEING

The survival curves which are observed in practice – a normal distribution of deaths in time (Chapter 2) – would fit in well with accumulation or depletion theories: it could be imagined that under a given set of environmental conditions some toxic substance tends to accu-

* '[Ageing] theories that depend on cell loss or on mutation, which is a random process, are called *stochastic* theories, from the Greek *stochadzein*, to aim at: they depend, in other words, on random hits – as if one of the Deaths were shooting not with buckshot but with microbullets that occasionally knocked out a cell, until enough were destroyed to kill the subject.' (Comfort, 1965.)

mulate, or some essential metabolite tends to be depleted, after a particular period but each individual seed's end-point in the process varies at random around this mean period. However, these are not the only possibilities and in many unicellular resting bodies there is evidence that such theories cannot apply: their survival curves are more compatible with the notion that each individual is victim to a constant probability of being the subject of a lethal accident. Before discussing whether seed survival curves would be compatible with a stochastic theory of ageing, it may be useful to spend a few moments discussing first the properties of the survival curves of typical unicellular organisms.

Consider the survival curves of spores and non-dividing vegetative cells of bacteria. Typically these are linear when log survivors are plotted against time. This is usually expressed as:

$$N = N_0 e^{-lt} \tag{9.1}$$

where N is the number of survivors at time t out of an initial population N_0, and l is a constant. Alternative ways to express the same relationship are to say that the death rate remains constant with the passage of time or, however long a bacterial cell has survived its chance of dying in the following unit interval of time remains the same. The shape of the population survival curve is identical, for example, with that of a batch of tumblers in a cafeteria (Brown and Flood, 1947). The tumblers do not age: they are simply subject to a constant probability each day of being broken. The analogy applies to the bacteria and, given this type of survival curve, it would be difficult to apply metabolic accumulation or depletion hypotheses to account for loss of viability in the bacterial population. In fact one might say that bacteria die but they do not age since one of the definitions we have of ageing is that the probability of death increases with age.

Next, consider a common variation which can occur to the typical bacterial survival curve. In some staphylococci, for example, the survival curve is sigmoid but very assymetrical so that there is a sharp shoulder and a long tail to the curve. Consequently, although it is still very unlike a typical seed survival curve it is much closer to it than the typical bacterial survival curve. Now it is well-established that the sigmoid survival curve of staphylococci is due to the fact that the cells tend to clump together and what is really being examined is the survival of clumps of cells rather than individual cells (for discussions of this see Wyss, 1951 and Lea, 1955). In the case of bacteria it is obviously only necessary for one cell in a clump to survive for the

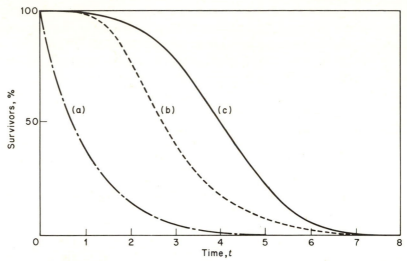

FIGURE 9.20 Examples of theoretical survival curves for bacteria and seeds. (a) Survival curve for a population of single cells in which the probability of survival $= 1 - e^{-t}$. (b) Survival curve for clumps of 10 cells in which each cell has the same probability of survival as in curve (a). (c) Survival curve typical of seeds, which is a negative cumulative normal distribution; in the example illustrated the mean viability period is 4 arbitrary time units and $K_\sigma = 0.33$. (From Roberts, Abdalla and Owen, 1967.)

clump to be capable of producing a new colony and be classed as living. If, as indicated by equation (9.1) the proportion of cells which survive a given period is e^{-t} where t is proportional to time, then the probability that an individual cell will die during this period is $1 - e^{-t}$ (Fig. 9.20; curve a). If the cells occur in clumps of n individuals, the probability of all n cells of a clump being killed will be $(1 - e^{-t})^n$. Hence the proportion of clumps which survive after time proportional to t will be $1 - (1 - e^{-t})^n$. It can be shown experimentally that the survival curves of staphylococci conform very closely to this expression when the appropriate value for n is inserted according to the degree of clumping (see Fig. 9.20, curve b for a theoretical survival curve when $n = 10$).

This behaviour of clumped bacterial cells suggested a possible model for the behaviour of seeds (Roberts, 1967). Could seeds be considered as a clump of cells in which the chance of an individual cell becoming non-functional remains the same, irrespective of how long the cell has survived? Obviously the 'clump' of cells in a seed would be large though it seems improbable that all the cells in the embryo are vital to its ability to germinate. But it seemed reasonable to suppose that there could be a group of 'key' cells most of which have to remain functional if the seed is to remain capable of germination. If the number of functional key cells drops below some critical value then the seed would not germinate and would be declared dead. The question to be answered was whether the model elaborated in this way would produce the negative cumulative normal distribution found in practice (Fig. 9.20; curve c). Initial crude mathematical analogies suggested it might (Roberts, 1967) but the development of an analytical solution was not straightforward, though eventually this was achieved by R. J. Owen (Roberts, Abdalla and Owen, 1967). However, the analytical solution is not easy to handle and for a further investigation of the hypothesis we have used a simulation approach* in order to construct theoretical survival curves based on this model.

In the model it was assumed that a seed contains a number of key cells; lethal accidents occur at random to these cells and each cell has the same probability of being the subject of a lethal accident during the passage of each unit of time; when a critical number of cells have died the seed is declared dead. To simulate the life-span of an individual seed, the following conditions were arbitrarily fixed and fed into the programme: the number of key cells, the probability per unit of time of a cell accident, and the critical number of cell deaths which causes the seed to lose viability. The programme was then run and the lifespan of the model seed recorded in arbitrary time units. For each set of conditions the programme was repeated 100 times. From the periods of survival of the individual model seeds the survival curve for a population of 100 seeds under those conditions could then be constructed.

From an inspection of the simulated survival curves constructed for a number of different sets of conditions it has been possible to define the limits within which the hypothesis would fit experimental observations on real seeds. The simulated survival curves had to agree with the following experimental observations (Chapter 2): (1) the curves have to describe negative cumulative normal distributions; (2) irrespective of the rate of 'ageing' (which in the model is a function of the probability of cell death per unit time) the Coefficient of Variation of the distribution of seed deaths in time for a given species (i.e. for a number of key cells and critical number of cell deaths in the model), has to remain the same, i.e. equation (2.10) has to apply; (3) the Coefficient of Variation has to have a value between 20 and 40 per

---

*I am indebted to Miss Sarah Clay and Mr David Brooke of the Department of Agriculture at Reading for writing the simulation programme in Focal and running it on a Digital PDP8/L computer.

cent (to comply with the observed values in the five species which have
been investigated in detail); in other words, the value of K_σ in
equation (2.10) has to be between 0.2 and 0.4.

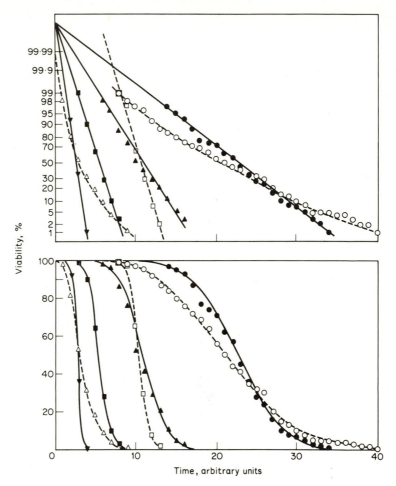

FIGURE 9.21   Simulated survival curves of model seed populations in which
percentage viability is plotted on a probability scale in the upper graph and the
identical untransformed data are shown in the lower graph. In all cases the model
assumes that cells become non-functional at random and that when a certain
proportion of 'key' cells becomes non-functional the seed is unable to germinate.
The solid curves were generated by assuming that each seed contains 100 key
cells and a seed loses viability when 20 of these cells become non-functional.
Four different rates of ageing were simulated by programming four different
probabilities of cell deaths/unit time: ● 0.01; ▲ 0.02; ■ 0.04; and ▼ 0.08.
The broken curves illustrate the three main ways in which the model will no
longer show the properties of the survival curves of real seed populations, viz:

Broadly speaking, it has been found that these conditions are met if the number of key cells is somewhere between 100 and 1,000 and if the critical number of cell deaths is between 10 and 20. Figure 9.21 illustrates a family of curves which meet these conditions (100 key cells per seed in which the critical number of cell deaths is 20). The curves have been plotted on both a probability scale and on a linear scale. It will be seen that in this family of curves the distribution is normal (as indicated by their straight-line nature when plotted on a probability scale); as the probabilty of cell death is altered the co-efficient of variation remains the same as indicated by their common intercept (compare Fig. 2.1) and calculation shows that the Coefficient of Variation has a value of 21 per cent. Thus all the stated conditions are met.

In addition to these simulated survival curves Fig. 9.21 also shows for comparison some other curves which did not meet the required conditions; however this model shows that the biological conditions that have to be met by a stochastic theory such as this are not un-reasonable. It also indicates that it is not essential to postulate that *individual* cells of the embryo show typical ageing behaviour – they might behave like bacterial cells; ageing then becomes a property of non-ageing cells associated in groups.

There have been two reasons for presenting this model. The first is to emphasise that we should keep an open mind as to the types of hypothesis which need further investigation; it would seem that those involving accumulation or depletion of metabolites on the one hand, or stochastic theories on the other are both equally valid approaches at this time. The second reason is that the model shows a possible

---

FIGURE 9.21 *continued*

No. of key cells	Critical no. of non-functional cells	Probability of cell death/unit time	Fault of model
□ 1000	100	0.01	Too many key cells; distribution is normal but Coefficient of Variation too small (intercept on probability scale too high).
○ 50	10	0.01	Too few key cells; survival curve asymetrical (extended tail to distribution).
△ 100	3	0.01	Critical number of non-functional cells is too small: survival curve markedly asymetrical (extended tail to distribution).

connexion between the behaviour of seeds and the resting cells of unicellular organisms which carry out a similar function; it may be, then, that there is more to be learned about the viability of seeds from studies on more simply organised structures.

ACKNOWLEDGEMENT

I am deeply grateful to Dr Patricia Berjak and Professor T. A. Villiers for generously providing me with photographs of their work (Figs. 9.6–9.19) and access to their results before publication elsewhere.

References to Chapter 9

ABDALLA, F. H., and ROBERTS, E. H., 1968. Effects of temperature, moisture, and oxygen on the induction of chromosome damage in seeds of barley, broad beans, and peas during storage. *Ann. Bot.*, **32**, 119–36.

ABDALLA, F. H., and ROBERTS, E. H., 1969. The effects of temperature and moisture on the induction of genetic changes in seeds of barley, broad beans, and peas during storage. *Ann. Bot.*, **33**, 153–67.

ABDUL-BAKI, A. A., 1969. Relationship of glucose metabolism to germinability and vigor in barley and wheat seeds. *Crop Sci.*, **9**, 732–37.

ABDUL-BAKI, A. A., and ANDERSON, J. D., 1970. Viability and leaching of sugars from germinating barley. *Crop Sci.*, **10**, 31–34.

ABU-SHAKRA, S. S., and CHING, T. M., 1967. Mitochondrial activity in germinating new and old soybean seeds. *Crop Sci.*, **7**, 115–18.

ANDERSON, J. D., 1970. Physiological and biochemical differences in deteriorating barley seed. *Crop Sci.*, **10**, 36–39.

ANDERSON, J. D., BAKER, J. E., and WORTHINGTON, E. K., 1970. Ultrastructural changes of embryos in wheat infected with storage fungi. *Pl. Physiol., Lancaster,* **46**, 857–59.

ASHTON, T., 1956. Genetical aspects of seed storage. In *The Storage of Seeds for Maintenance of Viability*, by E. B. Owen, 34–38. Commonwealth Agricultural Bureaux, Farnham Royal, England.

AVANZI, S., CORSI, G., D'AMATO, F. D., FLORIS, C., and MELETTI, P., 1967. The chromosome breaking effect of the irradiated endosperm in water soaked seeds of durum wheat. *Mutation Res.*, **4**, 704–7.

AVERY, A. G., and BLAKESLEE, A. F., 1936. Visible mutations from aged seeds. *Amer. Nat.*, **70**, 36–37.

BARKER, G. R., BRAY, C. M., and DETLEFSEN, M. A., 1971. An examination of the evidence for stable messenger RNA in seeds. *Biochem. J.*, **124**, in the press.

BARTON, L. V., 1961. *Seed preservation and longevity*. Leonard Hill, London.

BERJAK, P., 1968a. A lysosome-like organelle in the root cap of *Zea Mays*. *J. Ultrastruct. Res.*, **23**, 233–42.

BERJAK, P., 1968b. *A study of some aspects of senescence in embryos of Zea mays* L. Ph.D. Thesis, University of Natal.

BERJAK, P., and VILLIERS, T. A., 1970. Ageing in plant embryos. I. The establishment of the sequence of development and senescence in the root cap during germination. *New Phytol.*, **69**, 929–38.

BERJAK, P., and VILLIERS, T. A., 1971. Ageing in plant embryos. II. Repair and compensatory mechanisms. In the press.

BERJAK, P., and VILLIERS, T. A. Ageing in plant embryos. III. Precocious senescence in the root cap. In preparation.

BERJAK, P., and VILLIERS, T. A. Ageing in plant embryos. IV. Cellular failure. In preparation.

BERJAK, P., and VILLIERS, T. A. Ageing in plant embryos. V. Aspects of damage at the control level. In preparation.

BLAKESLEE, A. F., and AVERY, A. G., 1934. Visible genes from aged seeds. *Amer. Nat.*, **68**, 466.

BROWN, G. W., and FLOOD, M. M., 1947. Tumbler mortality. *J. Amer. Statist. Assoc.*, **42**, 562.

BUGLIA, G., 1909. Über die Hitzegerinnung von flussingen und festen organischen Kolloiden. *Z. Chem. Ind. Kolloide*, **5**, 291–93.

CALDECOTT, R. S., 1954. Inverse relationship between the water content of seeds and their sensitivity to X-rays. *Science*, **120**, 809–10.

CALDECOTT, R. S., 1961. Seedling height, oxygen availability, storage and temperature: their relation to radiation-induced genetic and seedling injury in barley. In *Effect of Ionising Radiations in Seeds*, 3–24. International Atomic Energy Agency, Austria.

CARTLEDGE, J. L., and BLAKESLEE, A. F., 1933. Mutation rate increased by ageing seeds as shown by pollen abortion. *Science*, **78**, 523.

CARTLEDGE, J. L., and BLAKESLEE, A. F., 1934. Mutation rate increased by ageing seeds as shown by pollen abortion. *Proc. natn. Acad. Sci. USA*, **20**, 103–10.

CARTLEDGE, J. L., BARTON, L. V., and BLAKESLEE, A. F., 1936. Heat and moisture as factors in the increased mutation rate from *Datura* seeds. *Proc. Am. phil. Soc.*, **76**, 663–85.

CHING, T. M., 1961. Respiration of forage seed in hermetically sealed cans. *Agron. J.*, **53**, 6–8.

CHING, T. M., and SCHOOLCRAFT, I., 1968. Physiological and chemical differences in aged seeds. *Crop Sci.*, **8**, 407–9.

CHRISTENSEN, C. M., and LÓPEZ, F. L. C., 1963. Pathology of stored seeds. *Proc. int. Seed Test. Ass.*, **28**, 701–11.

CLOWES, F. A. L., 1969. Anatomical aspects of structure and development. In *Root Growth*, ed. W. J. Whittington. *Proc. 15th Easter School Agric. Sci., Nottingham.* Butterworth, London.

COMFORT, A., 1964. *Ageing, the Biology of Senescence*, 2nd edit. Routledge and Kegan Paul.

COMFORT, A., 1965. *The process of Ageing.* Weidenfeld and Nicolson, London.

CORSI, G., and AVANZI, S., 1969. Embryo and endosperm response to ageing in *Triticum durum* seeds as revealed by chromosomal damage in the root meristem. *Mutation Res.*, **7**, 349–55.

CROCKER, W., 1938. Life span of seeds. *Bot. Rev.*, **4**, 235–74.

CROCKER, W., 1948. *Growth of Plants.* Rheinhold, New York.

CROCKER, W., and BARTON, L. V., 1953. *Physiology of seeds.* Chronica Botanica Co., Waltham, Mass.

CROCKER, W., and GROVES, J. F., 1915. A method for prophesying the life duration of seeds. *Proc. natn. Acad. Sci.*, **1**, 152–55.

CURTIS, H. J., 1967. Radiation and ageing. *Symp. Soc. exp. Biol.*, **21**, 51–63.

CURTIS, O. F., and CLARK, D. G., 1950. *An introduction to plant physiology*, 569. McGraw-Hill, New York.

D'AMATO, F., 1951. Spontaneous chromosome aberrations in seedlings of *Pisum sativum*. *Caryologia*, **3**, 285–93.

D'AMATO, F., and HOFFMAN-OSTENHOF, O., 1956. Metabolism and spontaneous mutations in plants. *Adv. Genet.*, **8**, 1–28.

DEY, B., and SIRCAR, S. M., 1968. The presence of an abcisic acid like factor in nonviable rice seeds. *Physiol. Pl.*, **21**, 1054–59.

DEY, B., SIRCAR, P. K., and SIRCAR, S. M., 1967. Phenolics in relation to non-viability of rice seeds. *Proc. int. Symp. Pl. Growth Substances, Calcutta*.

DE VRIES, H., 1901. *Die Mutationstheorie*, Band I. Veit and Co., Leipzig.

DODSON, E. O., and YU, C. K., 1962. Depression of germination treated with severe centrifugal force. *Can. J. Bot.*, **40**, 1714–17.

EHRENBERG, L., 1955. Factors influencing radiation induced lethality, sterility and mutation in barley. *Hereditas*, **41**, 123–46.

EWART, E. J., 1908. On the longevity of seeds. *Proc. Roy. Soc., Victoria*, **21**, 1–210.

FLORIS, C., 1966. The possible role of the endosperm in the ageing of the embryo in the wheat seed. *Giorn. Bot., Ital.*, **73**, 349–50.

FLORIS, C., 1970. Ageing in *Triticum durum* seeds: behaviour of embryos and endosperms from aged seeds as revealed by the embryo-transplantation technique. *J. exp. Bot.*, **21**, 462–68.

FRÖIER, K., GELIN, O., and GUSTAFSSON, Å., 1941. The cytological response of polyploidy to X-ray dosage. *Bot. Notiser*, **2**, 199–216.

GABER, S. D., and ROBERTS, E. H., 1969. Water sensitivity in barley seeds. II. Association with micro-organism activity. *J. Inst. Brew.*, **75**, 303–14.

GENKEL, P. A., and SHIH-HSÚ, C., 1958. The role of plasmodesmata in the loss of germination ability of coffee (*Coffee robusta* Linn) seeds. *Pl. Physiol. (Fiziologia Rasterii. transln)*, **5**, 303–7.

GISQUET, P., HITIER, H., IZARD, C., and MOUNAT, A., 1951. Mutations naturelles observées chez *N. tabacum* L. et mutations experimentales provoquées par l'extrait a froid de graines vieilles prematurement. *Ann. Inst. expt. Tab. Bergerac*, **1**, 1–31.

GOLUBCHUK, M., SORGER-DOMENIGG, H., CUENDET, L. S., CHRISTENSEN, C. M., and GEDDES, W. F., 1956. Grain storage studies. XIX. Influence of mold infestation and temperature on the deterioration of wheat during storage at approximately 12 per cent moisture. *Cereal Chem.*, **33**, 45–52.

GREER, S., and ZAMENHOF, F., 1962. Studies on depurination of DNA by heat. *J. molec. Biol.*, **4**, 123–41.

GROSCH, D. S., 1965. *Biological effects of radiations*. Blaisdell, New York.

GROVES, J. F., 1917. Temperature and life duration of seeds. *Bot. Gaz.*, **63**, 169–89.

GUNTHARDT, H., SMITH, L., HAFERKAMP, M. E., and NILAN, R. A., 1953. Studies on aged seeds. II. Relation of age of seeds to cytogenic effects. *Agron. J.*, **45**, 438–41.

HARRISON, B. J., 1966. Seed deterioration in relation to storage conditions and its influence upon germination, chromosomal damage and plant performance. *J. natn. Inst. agric. Bot.*, **10**, 644–63.

HARRISON, B. J., and MCLEISH, J., 1954. Abnormalities of stored seeds. *Nature, Lond.*, **173**, 593–94.

HUMMEL, B. C. W., CUENDET, L. S., CHRISTENSEN, C. M., and GEDDES, W. F.,

1954. Grain storage studies. XIII. Comparative changes in respiration, viability, and chemical composition of mold-free and mold-contaminated wheat upon storage. *Cereal Chem.*, **31**, 143–50.

HUNTER, R. E., and PRESLEY, J. T., 1963. Morphology and histology of pinched root tips of *Gossypium hirsutum* L. seedlings grown from deteriorated seeds. *Can. J. Pl. Sci.*, **43**, 146–50.

JACKSON, W. D., 1959. The life-span of mutagens produced in cells by irradiation. *Proc. 2nd Australian Conf. Radiation Biol., Melbourne, 1958*, 190–208. Butterworth, London.

JAMES, E., 1961. Perpetuation and protection of germ plasm as seed. In *Germ Plasm Resources*, ed. R. E. Hodgson, 317–26. Publication No. 66, Amer. Assoc. Adv. Sci., Washington DC.

JAMES, E., 1967. Preservation of seed stocks. *Adv. Agron.*, **19**, 87–106.

JAMES, W. O., and JAMES, A. L., 1940. The respiration of barley germinating in the dark. *New Phytol.*, **39**, 145–76.

KATO, Y., 1951. Spontaneous chromosome aberrations in mitosis of *Allium fistulosum* L. (a preliminary note). *Bot. Mag., Tokyo*, **64**, 152–56.

KOSTOFF, D., 1935. Mutations and the ageing of seeds. *Nature, Lond.*, **135**, 107.

KRISHNASWAMI, R., 1968. The relationship between response to radiations and nature of polyploidy in some crop plants. *Caryologia*, **21**, 303–10.

LEA, D. E., 1955. *Actions of radiations on living cells*, 2nd edit. Cambridge University Press.

LEPESCHKIN, W. W., 1913. Zur Kenntnis der Einwirkung supramaximaler Temperaturen auf die Pflanze. *Ber. Dtsch. bot. Ges.*, **30**, 703–14.

LILLY, L. J., 1965. Induction of chromosome aberrations by aflatoxin. *Nature, Lond.*, **207**, 433–34.

LINDOP, P. J., and SACHER, G. A. (eds), 1966. *Radiation and ageing*. Taylor and Francis, London.

MACLEOD, A. M., 1953. The quality of cereals and their industrial uses. *Chem. and Ind.*, 289–91.

MATSUMURA, S., and NEZU, M., 1961. Relation between polyploidy and effects of neutron-radiation on wheat. In *Effects of Ionizing Radiations on Seeds*, 543–52. International Atomic Energy Agency, Vienna.

MEDVEDEV, ZH. A., 1964. The nucleic acids in development and ageing. *Adv. Geront. Res.*, **1**, 181–206.

MELETTI, P., FLORIS, C., D'AMATO, F., 1968. The mutagenic effect of the irradiated endosperm in water-soaked seeds of *durum* wheat. *Mutation Res.*, **6**, 169–72.

MERRY, J., and GODDARD, D. R., 1941. A respiratory study of barley grain and seedlings. *Proc. Rochester Acad. Sci.*, **8**, 28–44.

MILNER, M., CHRISTENSEN, C. M., and GEDDES, W. F., 1947a. Grain storage studies. VI. Wheat respiration in relation to moisture content, mold growth, chemical deterioration, and heating. *Cereal Chem.*, **24**, 182–99.

MILNER, M., CHRISTENSEN, C. M., and GEDDES, W. F., 1947b. Grain storage studies. VII. Influences of certain mold inhibitors on respiration of moist wheat. *Cereal Chem.*, **24**, 507–17.

MILNER, M., and GEDDES, W. F., 1954. Respiration and heating. In *Storage of Cereal Grains and their Products*, eds J. A. Anderson and A. W. Alcock, 152–213. Amer. Assoc. Cereal Chemists, St Paul.

MOUTSCHEN-DAHMEN, M., MOUTSCHEN, J., and EHRENBERG, L., 1959. Chromosome disturbances and mutation produced in plant seeds by oxygen at high pressures. *Hereditas*, **45**, 230–44.

NAVASHIN, M. S., 1933a. Origin of spontaneous mutations. *Nature, Lond.*, **131**, 436.

NAVASHIN, M. S., 1933b. Ageing of seeds is a cause of chromosome mutations. *Planta*, **20**, 233–43.

NAVASHIN, M. S., and GERASSIMOVA, H. N., 1935. [Nature and causes of mutations. I. On the nature and importance of chromosomal mutations taking place in resting plant embryos due to their ageing.] *Biol. Zh.*, **4**, 593–634. Cited by Ashton, 1956.

NAVASHIN, M. S., and GERASSIMOVA, H. N., 1936a. Natur und Ursachen der Mutationen. I. Das Verhalten und die Zutologie der Pflanzen, die aus infolge Alterns mutierten Keimen stammen. *Cytologia*, **7**, 324–62.

NAVASHIN, M. S., and GERASSIMOVA, H. N., 1936b. Natur und Ursachen der Mutationen. III. Uber die Chromosomenmutationen, die in den Zellen von ruhenden Pflanzenkeimen bei deren Altern auftreten. *Cytologia*, **7**, 437–65.

NAVASHIN, M. S., and CHKVARNIKOF, P., 1933a. Process of mutation in resting seed accelerated by increased temperature. *Nature, Lond.*, **132**, 482–83.

NAVASHIN, M. S., and SHKVARNIKOF, P., 1933b. [The acceleration of the process of mutation in seeds increased by temperature.] *Priroda, Leningrad*, No. 10, 54–55. Cited by Ashton, 1956.

NICHOLS, C., 1941. Spontaneous chromosome aberration in *Allium. Genetics*, **26**, 89–100.

NILAN, R. A., and GUNTHARDT, H. M., 1956. Studies on aged seeds. III. Sensitivity of aged wheat seeds to X-irradiation. *Caryologia*, **8**, 316–21.

NILSSON, N. H., 1931. Sind die induzierten Mutanten nur selektive Erscheingun? *Hereditas*, **15**, 320–28.

OLAFSON, J. H., CHRISTENSEN, C. M., and GEDDES, W. F., 1954. Grain storage studies. XI. Influence of moisture content, commercial grade, and maturity on the respiration and chemical deterioration of corn. *Cereal Chem.*, **31**, 333–40.

OWEN, E. B., 1956. *The storage of seeds for maintenace of viability*. Commonwealth Agricultural Bureaux, Farnham Royal, England.

OXLEY, T. A., 1948. *The Scientific Principles of Grain Storage*. Northern Publishing Co., Liverpool.

PALENZONA, D. L., 1961. Effects of high doses of X-rays on seedling growth in wheats of different ploidy. In *Effects of Ionizing Radiation on Seeds*, 533–42. International Atomic Energy Agency, Austria.

PETERSON, A., SCHLEGEL, V., HUMMEL, B., CUENDET, L. S., GEDDES, W. F., and CHRISTENSEN, C. M., 1956. Grain storage studies. XXII. Influence of oxygen and carbon dioxide concentrations on mold growth and grain deterioration. *Cereal Chem.*, **33**, 53–66.

PETO, F. H., 1933. The effect of ageing and heat on the chromosomal mutation rate in maize and barley. *Can. J. Res.*, **9**, 261–64.

READ, J., 1959. *Radiation biology of* Vicia faba *in relation to the general problem*. Blackwell, Oxford.

ROBBINS, W. J., and PETCH, K. F., 1932. Moisture content and high temperature in relation to the germination of corn and wheat grains. *Bot. Gaz.*, **93**, 85–92.

ROBERTS, E. H., 1967. The control of viability in cereal seed. *Proc. int. Symp.*

Physiol. Ecol. Biochem. Germination, Greifswald, ed. H. Borriss, 975–81. Ernst-Moritz-Arndt-Universität, Greifswald.

ROBERTS, E. H., and ABDALLA, F. H., 1968. The influence of temperature, moisture, and oxygen on period of seed viability in barley, broad beans, and peas. *Ann. Bot.*, **32**, 97–117.

ROBERTS, E. H., ABDALLA, F. H., and OWEN, R. J., 1967. Nuclear damage and the ageing of seeds with a model for seed survival curves. *Symp. Soc. exp. Biol.*, **21**, 65–100.

SAX, K., and SAX, H. J., 1962. Effect of X-rays on the ageing of seeds. *Nature, Lond.*, **194**, 459–60.

SAX, K., and SAX, H. J., 1964. The effect of chronological and physiological ageing of onion seeds on the frequency of spontaneous and X-ray induced chromosome aberrations. *Radiat. Bot.*, **4**, 37–41.

SHKVARNIKOV, P., 1935. [Influence of temperature and moisture on the process of mutation in resting seeds.] *Semenovodstvo*, No. 1, 46–52. Cited by Ashton, 1956.

SHKVARNIKOV, P., 1936. Einfluss hoher Temperatur ant die Mutationsrate bei Weizen. *Planta*, **25**, 471–80.

SHKVARNIKOV, P., 1939. [Mutation in seeds and its significance in seed production and plant breeding.] *Bull. Acad. Sci. URSS Ser. Biol.*, 1009–54. Cited by Ashton, 1956.

SHKVARNIKOV, P., and NAVASHIN, M. S., 1934. Über die Beschleunigung des Mutationsvorganges in ruhenden Samen unter dem Einfluss von Temperatur erhöhung. *Planta*, **2**, 720–36.

SCHKVARNIKOV, P., and NAVASHIN, M. S., 1935. [Acceleration of the mutation process in resting seeds under the influence of increased temperatures.] *Biol. Zh.*, **4**, 25–38. Cited by Ashton, 1956.

SIRCAR, S. M., 1967. Biochemical changes of rice seed germination and its control mechanism. *Trans. Bose Res. Inst.*, **30**, 189–98.

SIRCAR, S. M., and BISWAS, M., 1960. Viability and germination inhibitor in rice seed. *Nature, Lond.*, **187**, 620–21.

SIRCAR, S. M., and DEY, B., 1967. Dormancy and viability of rice (*Oryza sativa* L.). *Proc. int. Symp. Physiol. Ecol. Biochem. Germination, 1963*, ed. H. Borriss, 969–73. Ernst-Moritz-Arndt-Universität, Greifswald.

SMITH, L., 1943. Relation of polyploidy to heat and X-ray effects in cereals. *J. Hered.*, **34**, 130–34.

SMITH, L., 1946. A comparison of the effect of heat and X-rays on dormant seeds of cereals, with special reference to polyploidy. *J. agric. Res.*, **73**, 137–58.

STADLER, L. J., 1929. Chromosome number and the mutation rate in *Avena* and *Triticum*. *Proc. natn. Acad. Sci.*, **15**, 876–81.

STOKES, P., 1965. Temperature and seed dormancy. *Handb. PflPhysiol.*, **15**, 746–803.

STUBBE, H., 1935. Samenalter und Genmutabilitat bei *Antirrhinum majus* L. *Biol. Zbl.*, **55**, 209–15.

TAKAYANAGI, K., and MURAKAMI, K., 1968a. New method of seed viability tests with exudates from seed. *15th Int. Seed Test. Congr., New Zealand*. Preprint No. 25.

TAKAYANAGI, K., and MURAKAMI, K., 1968b. Rapid germinability test with exudates from seed. *Nature, Lond.*, **218**, 493–94.

THRONEBERRY, G. O., and SMITH, F. G., 1955. Relation of respiratory and enzymic activity to corn seed viability. *Pl. Physiol., Lancaster*, **30**, 337–43.

WENT, F. W., and MUNTZ, P. A., 1949. A long term test of seed longevity. *El Aliso*, **2**, 63–75.

WHITING, A. R., 1945. Dominant lethality and correlated chromosome effects in *Habrobracon* eggs X-rayed in diplotene and in late metaphase I. *Biol. Bull., Woods Hole*, **89**, 61.

WYSS, O., 1951. Chemical factors affecting growth and death. In *Bacterial Physiology*, eds C. H. Werkman and P. W. Wilson, 178–213. Academic Press, New York.

WYTTENBACH, E., 1955. Der Einfluss verschieder Lagerungsfaktoren auf die Haltbarkeit von Feldsämereien (Luzerne, Rotklee, und gemeinem Schotenklee) bei länger dauernder Aufbewahrung. *Landw. Jb. Schweiz*, **4**, 161–96.

YU, C. K., and DODSON, E. O., 1961. Depression of germination of seed treated with severe centrifugal force. *Can. J. Bot.*, **40**, 1714–17.

ZELENY, L., 1954. Chemical, physical, and nutritive changes during storage. In *Storage of cereal grains and their products*, eds J. A. Anderson and A. W. Alcock, 46–76. Amer. Assoc. Cereal Chemists, St Paul.

CHAPTER 10

Loss of Viability and Crop Yields

E. H. Roberts

The deterioration leading to loss of viability in seeds can affect the yield of a crop in two ways: first the decreased germination can lead to a sub-optimal population of plants per unit area; secondly the deterioration – of which seed viability may be an index – may result in a poorer performance by the surviving plants. Providing one is aware of the first problem, theoretically it could be overcome by increasing seed rates. But the necessary adjustment may be greater than that indicated by a simple germination test, since the decreased vigour of the surviving seedlings (various aspects of which have been discussed in Chapters 4, 6, 7 and 8) may well result in a lower field emergence than can be obtained from fresh seed.

The problem of reduced plant population could obviously be important, particularly in those species which are unable to compensate for a reduced density (for example by tillering) and thus have a population-yield curve which shows a relatively sharp yield peak at a particular population density: it will be less important in those crops which have a flat-topped peak or show an asymptotic relationship between population and yield. The relationships between plant population and yield are now reasonably well-understood and the subject has been comprehensively reviewed by Willey and Heath (1969). There would, therefore, be little point in detailing here the way in which crop yield would be decreased simply as a result of a decrease in the population of a crop through the failure of some seeds to emerge. Accordingly in this chapter I shall concentrate on the performance of crops after they have been established and, where population is important, at a density which is approximately optimal.

There have been a number of investigations in which reduced yields have been reported from old seeds; in other reports no effect of age has been found and exceptionally – e.g. in mung beans (*Phaseolus aureus*) (Rodrigo, 1939) – increased yields have been recorded. Much of the early literature has been reviewed by Barton (1961) and will not be re-examined here. The reason for this is that much of it is difficult to interpret because yield has been related to chronological age and, as we have seen (Chapters 2 and 9), the chronological age of a sample does not necessarily give any indication of the degree to which a population of seeds has deteriorated.

One of the first more important investigations was carried out by Barton and Garman (1946) on seeds, up to 13 years old, of China aster (*Callistephus chinensis*), verbena (*Verbena teucriodes*), pepper (*Capsicum frutescens*), tomato (*Lycopersicon esculentum*) and lettuce (*Lactuca sativa*). This work emphasised that the chronological age of the seeds is unimportant if the seeds are stored under conditions under which

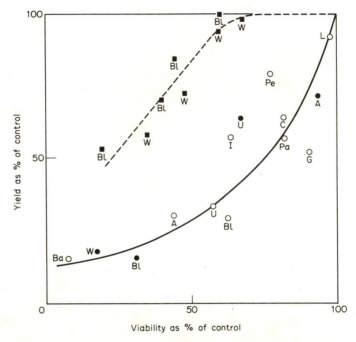

FIGURE 10.1   The relationship between percentage viability and the fresh-weight yield of lettuce plants produced from the surviving seeds.

- ■ Seedling yield after storage in air at 10 per cent moisture content and 35°C for 8, 9, 10, and 11 days, expressed as a percentage of yield obtained from untreated seeds of 80–90 per cent germination.
- ● Yield of mature plants obtained from seeds kept in open storage at 18°C for 5 years expressed as a percentage of the yield obtained from similar seed stored in CO_2 (73–92 per cent germination).
- ○ Yield of mature plants obtained from seeds stored in CO_2 for 10 years at 18°C expressed as a percentage of that obtained from one-year-old seed (75–96 per cent germination).

Varieties: A, All-the-year-round; Ba, Balloon; Bl, Black-seeded Bath; C, Continuity; G, Giant White; I, Imperial; L, Little Gem; Pa, Paris White; Pe, Peerless; L, Little Gem; W, White Heart.

(Calculated from the original data of Harrison, 1966.)

little or no loss of viability occurs. But in tomato and lettuce they showed that if there had been loss of viability then decreased yields from the plants produced from the surviving seeds could be expected.

In her review Barton (1961) concluded that 'the actual age of the seed is much less important than the environment in which it is held.' In spite of this conclusion, until recently the literature has contained little in the way of a critical analysis of the effect of storage conditions on the growth and yield of surviving plants. However, since Barton's review Harrison (1966) has produced some very interesting data on the effects of different 'ageing' treatments on the performance of lettuce and onion plants.

Harrison carried out two types of experiment on lettuce: in the first he used 'slow-ageing' treatments in which the seeds were stored for 5–10 years in open storage or sealed in air or carbon dioxide at 18°C; in the second type of experiment he employed 'rapid-ageing' treatments in which the seeds were stored at 10 per cent moisture content at 35°C for 8–11 days. The main results have been transformed from Harrison's original data and are shown in Fig. 10.1.

In the rapid-ageing treatments it will be seen that both cultivars which were investigated showed roughly the same relationship between percentage viability and the yield derived from the surviving seeds. Significant decreases in growth were only obtained once viability had dropped below about 50 per cent. However, it is important to note that these results are based on seedling growth only: the mean fresh weights of the seedlings from the different treatments at harvest varied between about 0.2 and 0.4 gm. In the slow-ageing treatments the results again indicate a consistent relationship between loss of viability and yield irrespective of the cultivar or ageing treatment, but it will be seen that this relationship is distinctly different from that shown by the rapid ageing treatments. In the slow-ageing treatments even a small loss of viability results in a severe loss of yield in plants derived from the surviving seeds (see Figs. 10.1 and 10.2).

The reason for the two distinct types of relationship between loss of viability and yield is not known but it would seem that there are two main possibilities: first it is possible that conditions leading to a rapid loss of viability produce a different relationship from those resulting in a slow loss of viability; or alternatively the relationship between loss of viability and seedling yield may be different from that between loss of viability and final yield. However, from the practical point of view it is the relationship shown by the slow-ageing experiments which is important because the storage conditions employed are similar to those found in normal practice and the yield figures reflect the economic yield of the crop.

FIGURE 10.2   White Heart lettuces grown from 5-year-old seed stored at about 18°C.

Storage	14A Unsealed in air	14B Sealed in air, about 6 per cent moist. content	14C Sealed in carbon dioxide, about 6 per cent moist. content
Viability, %	18	73	99
Yield/plant, gm fresh weight	113	485	513

(From Harrison, 1966.)

Figure 10.3 summarises some results obtained by Harrison (1966) on onion seed which was 5 years old and had been stored at 18°C for 4 years sealed in various gases. Harrison points out that nitrogen appears to be superior to either argon or carbon dioxide in its effect on final yield, though it shows no significant difference on percentage viability as compared with the other two gases. It is difficult to imagine the reason for this superiority of nitrogen – particularly when compared with an inert gas – and the phenomenon obviously needs further investigation. However, in spite of this apparently anomolous behaviour, I have taken what may be an unwarranted liberty of indicating in Fig. 10.3 the major trend between viability and yield of the

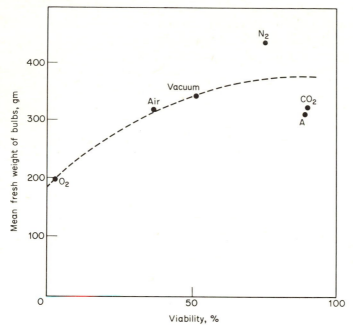

FIGURE 10.3   The relationship between percentage viability and yield in onion. The seeds were 5 years old and had been stored in sealed containers in the gases indicated at 18°C and 8 per cent moisture content for 4 years. (Drawn from the original data of Harrison, 1966.)

surviving plants. Although this has been done with some trepidation it is at least partially justified by the similarity of the apparent trend with that revealed more clearly in the work to be described next.

In view of the fact that in three species, peas (*Pisum sativum*), broad beans (*Vicia faba*) and barley (*Hordeum distichon*), percentage viability appeared to be a reliable index of the deterioration of the surviving seed as indicated by the accumulation of chromosome aberrations (Abdalla and Roberts, 1969a) (see Chapter 9), we were interested to find out whether percentage viability would also indicate the growth potential of the surviving seeds under field conditions. In preliminary field trials three storage treatments were applied to the three species so that they lost approximately 50 per cent viability and the final economic yields (weight of grain per unit area) were compared with control treatments in which there had been no loss of viability (Abdalla and Roberts, 1969b). For any species all field plots had identical populations which were similar to those used in commercial practice. The ageing treatments were as follows: peas—25°C, 18 per cent moisture content (m.c.) for 100 days (54 per cent viability);

35°C, 18 per cent m.c. for 24 days (53 per cent viability); 45°C, 12.3 per cent m.c. for 17 days (53 per cent viability); beans—25°C, 18.5 per cent m.c. for 45 days (48 per cent viability); 35°C, 18.5 per cent m.c. for 17 days (48 per cent viability); 45°C, 11.5 per cent m.c. for 37 days (53 per cent viability); barley—25°C, 18 per cent m.c. for 54 days (50 per cent viability); 35°C, 18 per cent m.c. for 48 days (48 per cent viability); 45°C, 12.1 per cent m.c. for 17 days (47 per cent viability). Thus in all species the treatments included different rates of ageing and treatments in which a high temperature was combined with a low moisture content, or a low temperature was combined with a high moisture content. Although the final yields of the bean trial were lost because of bird damage, the details of the treatments have been mentioned here because they will be referred to in another connexion later. In the trials on peas and barley final yields were obtained satisfactorily but no significant differences were shown between the treatments and controls for either weight of seeds or weight of straw.

A second series of field trials was then carried out on these three species with the following objectives: (1) to confirm that deterioration associated with a reduction of viability to 50 per cent has no significant effect on final yield, (2) to find out whether seed deterioration associated with a reduction of viability below 50 per cent affects final yield and, if it does, (3) to find out whether the particular combination of environmental factors during storage or the rate of loss of viability is important in determining yield, or whether percentage viability alone is sufficient as an index of yield potential.

TABLE 10.1　　*The storage treatments used to obtain the seeds for the field trials illustrated in Fig. 10.4. (From Abdalla and Roberts, 1969b.)*

Species	Storage treatment		Period of storage (days) and viability (%) shown in brackets	
	Temp., °C	Moisture content, %	Control	Treatments
Barley	25	17.8	0(100)	49(79)　56(65)　63(40)　84(15)
	45	12.6		14(81)　18(60)　21(37)　25(21)
Peas	25	18.2	0(100)	56(85)　98(67)　126(40)　147(24)
	45	12.4		35(80)　42(60)　47(45)　56(20)
Broad bean	25	18.3	0(90)	35(75)　49(55)　70(32)　77(15)
	45	11.7		26(70)　35(54)　49(37)　63(13)

For each species two contrasting storage environments were used – a relatively low temperature combined with a high moisture content, or a high temperature combined with a relatively low moisture content; samples were withdrawn from the two storage environments to obtain a range of viability levels as shown in Table 10.1. This table also shows that the rate of loss of viability in the first storage treatment is about half that in the second. The results of the final grain yield are shown in Fig. 10.4.

These trials confirmed the first series in that seed deterioration associated with a reduction of viability down to 50 per cent had no significant effect on final yield. However, there is evidently a trend of decreasing yield with decreasing viability, but the slope of the curve is gradual so that significant reductions in yield only become evident when viability has dropped to values below about 50 per cent. The most striking feature of these results is the close similarity between the contrasting seed-storage environments in their effect on yield. The fact that the curves showing yield against percentage viability are virtually identical in any species strongly suggests that percentage viability is an excellent index of the loss of yield potential of the surviving seeds under agricultural conditions. It would appear to be unimportant which factor – temperature or moisture content – is mainly responsible for the deterioration, or how rapidly the deterioration has occurred.

Although these results suggest that a simple germination test can act as an index of the potential yield of the surviving seeds for a given species, the relationship between viability and yield may be different in different species. On the one hand there is the type of relationship shown by peas, beans, barley and onion (Figs. 10.3, 10.4) in which a significant loss of final yield only occurs when a considerable loss of viability has taken place. On the other hand there is the type of relationship shown so far only in a single species – lettuce – in which there is a considerable loss of yield when only a small loss of viability has taken place.* From the practical point of view it is very important to

---

*It should be confessed that the interpretation of the lettuce data given here is not identical with Harrison's original interpretation. Harrison (1956) concluded that 'Seed deterioration can express itself in reduced vigour of seedlings and maturer plants and this *may or may not* [my italics] be accompanied by a decline in the germination index.' However, I suggest that the impression that there may be a loss of yield potential without a concomitant drop in viability may be the result of the characteristic shape of the curve relating viability to final yield (Fig. 10.1): this indicates that very little deterioration as indicated by viability is necessary to cause a considerable fall in yield in this species.

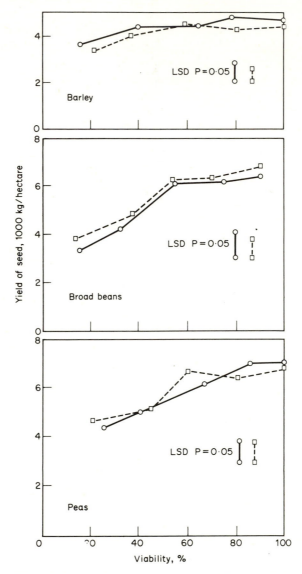

FIGURE 10.4 The relationship between percentage viability and the yield of crops grown from the surviving seeds of barley, broad beans and peas. The crops were grown from seeds stored at 45°C and approximately 12 per cent moisture content (□) and from seeds stored at 25°C and approximately 18 per cent moisture content (○) for various periods (details given in table 10.1) so that viability had deteriorated to the levels indicated. The points at the extreme right of each curve represent the values obtained from the control seeds which received no adverse storage treatments. LSD = least significant difference. (From Abdalla and Roberts, 1969b.)

know to which category a species belongs. In the storage of seeds for crop production some loss of viability can be tolerated in the first type, but not in the second.

Evidence for a diminution in the initial deleterious effects of poor storage conditions during the subsequent development of the crop

So far the emphasis has been on the relationship between seed deterioration as indicated by loss of viability and the final economic yield of plants grown under agricultural conditions. The final yield may be considered as a measure of the total cumulative growth of the crop. The relationship between seed deterioration and the initial stages of growth has been considered in Chapter 8 which dealt with vigour, since vigour can often be considered as a reflection of seedling growth. I should now like to consider the intermediate period of growth between the seedling stage and the development of the mature plant.

Detailed examination of root growth in broad beans and peas has shown that the rate of growth is affected if the storage conditions have led to some loss of viability (Abdalla and Roberts, 1969b): the decreased growth rates were particularly noticeable if the viability had fallen to about 60 per cent or less (Fig. 10.5). Further decreases in viability to about 30 per cent had little further effect on the root growth of the survivors.

The effects on root growth lasted at least until the roots were about 30 cm or more long. In broad beans, however, there was evidence that when the roots were approaching this length the roots from the storage treatments showing poor viability began to conform with the growth rates of the control roots from seeds showing high viability. In other words the initially low growth rates did not persist. But in peas the evidence was less clear.

An examination of the increase in dry weight of pea shoots over the first six weeks in the glasshouse showed that two alternative seed-storage treatments giving roughly the same loss of viability (45°C and 12.8 per cent m.c. for 28 days giving 69 per cent viability; or 35°C and 18.4 per cent m.c. for 21 days giving 63 per cent viability) led to roughly the same decrease in mean growth rate of the shoots from the surviving seeds: in both cases the growth rate was about 25 per cent less than that of the control plants from seeds which showed 100 per cent viability (Abdalla and Roberts, 1969b).

Further investigations of the growth of peas, beans and barley were made at later stages of growth in the field. Three deleterious seed-

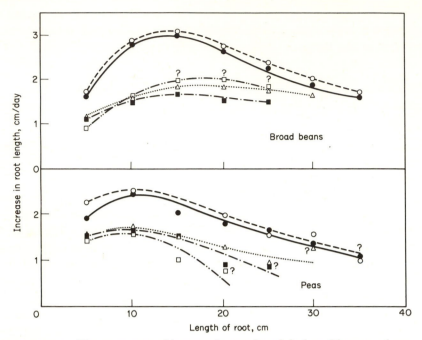

FIGURE 10.5   The mean rate of increase in root length in broad beans and peas which had been stored at 25°C and 18.2 per cent moisture content until viability had dropped to the following percentage germination values:

		Broad beans	Peas
o – – – – o	(control)	90	97
• ———— •		76	80
△ · · · · △		61	60
■ —·—· ■		46	45
□ —··—··□		33	25

Question marks indicate points for mean values derived from less than 5 roots. Seedlings were grown in aerated culture solutions at 19°C. (From Abdalla and Roberts, 1969b.)

storage treatments were used (viz. those mentioned earlier in connexion with the yield-trial investigations) in which viability was reduced to about 50 per cent. These treatments were compared with the plants produced from the control seeds in which viability had been maintained at approximately 100 per cent. Detailed measurements were made after about 5–6 weeks and again after about 8½ weeks. After 5–6 weeks in all three species it was found that in general the treatments which led to a decrease in mean plant height also led to an increase in the plant-to-plant variation in plant height. This was particularly pronounced in peas where the normal distribution of plant heights in the control plants was changed by all the seed-storage treatments to a markedly skew distribution with the mode amongst the smaller plants. In most cases these effects of seed-storage treatments were paralleled by corresponding decreases in tiller or leaf number and increases in the plant-to-plant variation in leaf number.

From the measurements of the individual plant heights, the mean relative linear growth rates (cm/cm/week) for increase in plant height were calculated for the period between 5–6 and $8\frac{1}{2}$ weeks. None of the growth rates were significantly different from the controls, though there was no exception to the rule that the growth rate recorded for all 9 storage treatments had *greater* values than the controls. Accordingly it may be concluded that the adverse storage treatments cause no diminution in the rate of increase in the linear growth of the shoots during the later stage of growth – in fact there was some indication of a slight compensatory increase in growth rate at this stage. The differences in plant height must therefore be attributed to the effect of seed-storage treatments on the early stages of growth.

In summary, it has already been shown that in these three species the storage treatments which reduced viability to about 50 per cent have no significant effect on final yield of grain or straw. Nevertheless such treatments do affect the early growth of the roots and shoots of the plants; some individuals are affected more than others so that the variability of the plants is increased. Eventually these early effects on the rate of growth tend to disappear and there is even some possibility of compensatory growth during the later stages of development; thus the early slower rates of growth may be of little consequence when it comes to final yield – unless the deterioration was so severe during storage that it leads to a drop in viability to below about 50 per cent. These generalisations apply at least for peas, beans and barley but, judging by the different relationship between viability and final yield, would not hold for any species which behaves like lettuce.

Possible explanations for the decreased growth rates in plants from seed populations which have lost viability

In Chapter 9 it was shown that loss of viability is associated with an accumulation of chromosome damage in the surviving seeds. Cytological examinations have shown that the visible aberrations produced as a result of poor storage conditions rapidly disappear during the growth of the seedling: all visible aberrations in pea roots were no longer present in the meristem when the roots were about 10 cm long, i.e. after a few cell divisions (Abdalla and Roberts, 1969b). This diplontic selection, which is a well-known phenomenon – particularly

from work on the induction of aberrations by ionising radiation –
comes about because of loss of viability of the aberrant cells after
division, presumably because the daughter cells will tend to contain
gross deletions and other forms of genetic imbalance.

On the other hand minor chromosome damage, as was indicated by
the induction of recessive chlorophyll-deficiency mutations, are obviously not subject to such selection and not only persist to the
maturity of the plant but through to successive generations. It seems
probable that some chromosome damage – less severe than the visible
aberrations but more severe than the damage which behaves as single
recessive point mutations – may also occur. This postulated intermediate damage may also be selected out during cell division, but it
may take longer to disappear than the more obvious aberrations. It
could be that part of the reduction in growth rate is due to these
chromosome disturbances of intermediate severity, and these may
tend to disappear with time so that growth rate eventually returns to
normal. The hypothesis that reduced growth rate is the result of
nuclear damage accumulated during storage would also explain
satisfactorily the increased plant-to-plant variation which occurs as a
result of seed deterioration during storage.

In relation to seed deterioration rather more work has been carried
out on chromosome damage than on the damage occurring to the
cytoplasmic organelles. This disparity should not mislead us into
ignoring the possibility that the lower growth rates found in plants
derived from aged seeds could be due to the malfunction of some of
the cytoplasmic organelles. The mitochondria and plastids, for
example, are self-replicating bodies which contain their own DNA.
Damage to these organelles might persist during the development of
the plant though, as with chromosome damage, there would probably
be a tendency for the damaged organelles or the cells containing them
to be selected out during development. Furthermore, as mentioned
in Chapter 9, there is evidence that repair mechanisms can operate
providing the initial damage was not too great. Consequently the
observed gradual return to normal growth rates in plants derived
from aged seeds would also be compatible with the hypothesis that
the loss of early vigour is due to damage to cytoplasmic organelles.
In Chapter 9 the work of Berjak and Villiers on the ultra-structure
of maize seed embryos was discussed in some detail. These workers,
together with Abu-Shakra and Ching whose investigations on maize
seed mitochondria were also described, suggested that it may be
damage to mitochondria which is critical in causing loss of viability.
Whether or not this is so one could envisage that such damage in the

surviving seeds might seriously affect initial growth. This hypothesis at least has the strong attraction that it lends itself to experimental investigation.

Other explanations could of course be postulated based on the various theories mentioned in Chapter 9 which have been suggested to account for loss of viability. However, at the present time it would not be very profitable to discuss these in detail since most of them would not easily account for the relatively persistent effects on growth following germination.

Whatever the explanation of the decreased growth rates, the practical implications for crop production are reasonably clear. Small losses of viability in crops like peas, broad beans, barley and onion are not critical but, because of the lower rates of early growth, the crop will probably be more susceptible to adverse conditions during emergence and early establishment. For example such a crop might be expected to be more susceptible to soil-capping, pests, diseases and weed competition. However, under favourable conditions the decreased early growth rates will have little effect on final yield. In contrast, in crops like lettuce even a small loss of viability apparently indicates a degree of deterioration which will seriously affect the final yield of the crop. In all cases any crop which has been derived from a seed stock which had lost a significant amount of viability should not be used for producing further seed since the evidence indicates (Chapter 9) that such seed would contain a large amount of genetic mutation.

It would be misleading to end this chapter without mentioning that the views expressed here are not entirely orthodox. In Chapter 8 a great deal of evidence is cited which indicates that the results of an ordinary germination test do not give reliable information on seedling vigour, i.e. on the early growth of seedlings; and deterioration in the vigour of a seed lot can occur before there is a significant decline in germination. This is true and maintenance of maximum vigour is certainly important in order to retain maximum probability of good establishment of the crop. The point brought out here, though, is that in some crops at least, if seedlings are not actually lost during early growth, the early effects of low vigour tend to disappear during growth and thus the final result may not be quite so catastrophic as indicated by a vigour test. Thus it cannot be assumed automatically that a decrease in seedling vigour will lead to a significant loss of final yield.

The fact that it is possible to hold different views in this subject indicates that there is a need for more quantitative work on the effect of seed deterioration on final yield – particularly using designs in which effects on population density (i.e. establishment) are separated from other effects on plant growth.

References to Chapter 10

ABDALLA, F. H., and ROBERTS, E. H., 1968. The effects of temperature, moisture, and oxygen on the induction of chromosome damage in seeds of barley, broad beans, and peas during storage. *Ann. Bot.*, **32**, 119–36.

ABDALLA, F. H., and ROBERTS, E. H., 1969a. The effects of temperature and moisture on the induction of genetic changes in seeds of barley, broad beans, and peas during storage. *Ann. Bot.*, **33**, 153–67.

ABDALLA, F. H., and ROBERTS, E. H., 1969b. The effect of seed storage conditions on the growth and yield of barley, broad beans, and peas. *Ann. Bot.*, **33**, 169–84.

BARTON, L. V., 1961. *Seed Preservation and Longevity*. Leonard Hill, London.

BARTON, L. V., and GARMAN, H. R., 1946. Effect of age and storage condition of seeds on the yield of certain plants. *Contr. Boyce Thompson Inst.*, **14**, 243–55.

HARRISON, B. J., 1966. Seed deterioration in relation to seed storage conditions and its influence upon seed germination, chromosomal damage and plant performance. *J. nat. Inst. agric. Bot.*, **10**, 644–63.

RODRIGO, P. A., 1939. Study on the vitality of old and new seeds of Mungo (*Phaseolus aureus* Roxb.). Philipp. J. Agric., **10**, 285–91.

WILLEY, R. W., and HEATH, S. B., 1969. The quantitative relationships between plant population and crop yield. *Adv. Agron.*, **21**, 281–321.

CHAPTER II

Dormancy: a Factor Affecting Seed Survival in the Soil

E. H. Roberts

Definitions of dormancy

It is unfortunate that the term dormancy, when applied to seeds, has been used in at least two distinct ways. Sometimes any seed which is not in the process of germinating – e.g. a dry seed in storage – is said to be dormant. However, more commonly – and more usefully – the term is used in a more restricted sense to indicate that the seed, although viable, will not germinate under conditions normally considered to be adequate for germination, viz. when provided with a suitable temperature, adequate water and oxygen. In this chapter the term dormancy will be used in the more restricted sense. Before discussing its relationship with viability, however, it is necessary to distinguish between various types of dormancy.

Three types of dormancy are now recognised. 'Some seeds are born dormant, some achieve dormancy and some have dormancy thrust upon them.' This is the paraphrase which Harper (1957) has aptly used to describe the three types of dormancy which he terms 'innate', 'induced' and 'enforced'. Alternatively, innate dormancy has been described as primary (Crocker, 1916), natural (Brenchley and Warrington, 1930), inherent (Bibbey, 1948), and endogenous (Schafer and Chilcote, 1969); induced dormancy is often known as secondary dormancy (Crocker, 1916); and enforced dormancy has been termed environmental dormancy (Bibbey, 1948) but it has also been known as induced dormancy (Brenchley and Warrington, 1930). In spite of the proliferation of terminology there is little possibility of confusion except in the case of the term induced dormancy which has been used in the two senses mentioned above. However, in this chapter I shall use the terms as defined by Harper (1957) since his was the first comprehensive and clearly defined terminology (Thurston, 1960).

The adjective innate describes the dormancy which is present immediately the new embryo ceases to grow when it is still attached to the parent plant: such dormancy prevents the seed from germinating viviparously and also usually for some time after the ripe seed is shed or harvested. The seeds of the vast majority of species have a period of innate dormancy. There is usually a large variation amongst the dormancy periods of the individual seeds from a single plant. Very often the distribution of dormancy periods amongst a seed lot subjected to constant environmental conditions is normal. Although the mean dormancy period and the standard deviation of the distribution varies with environment, in the only case where these relationships have been examined in detail, i.e. in rice (Roberts, 1965), it has been shown that the Coefficient of Variation has a constant value. In other words the standard deviation of the dormancy periods of the individual seeds, expressed as a percentage of the mean dormancy period, remains the same. Even pure-line cultivars of rice show a relatively high Coefficient of Variation of about 50 per cent (Popay and Roberts, 1970b).

In some cases there is evidence that a single plant produces seeds with distinct differences in degree of innate dormancy or, in other words, a discontinuous distribution of dormancy periods of the individual seeds. The classical example of this is in *Xanthium pensylvanicum* in which two seeds are produced in each capsule: the lower seed shows very little innate dormancy while the dormancy of the upper seed is very pronounced (for a recent discussion of the physiology see Esashi and Leopold, 1968). Another two examples where two distinct types of seed are produced are *Salsolla volkensii* and *Aellenia antrani* in which chlorophyllous and achlorophyllous seeds are produced: in both species the chlorophyllous seeds are practically non-dormant whereas the achlorophyllous seeds are dormant for a considerable period (Negbi and Tamari, 1963).

In some cases the different types of seeds produced by plants of the same genetic constitution can be linked with the environmental conditions to which the parent plant was subjected during the period of seed formation and maturation. For example in *Halogeton glomeratus* very dormant brown seeds are produced in long days, and much less dormant black seeds are produced in shorter days (Williams, 1960). Sometimes there may be more than two types of seed which show morphological differences and the different morphological types may be associated with different degrees of innate dormancy. The relationship of different dormancy responses with distinct morphological differences has been examined in detail, for example, in *Chenopodium album* by Williams and Harper (1965); they showed that the polymorphism (sometimes called by others polyspermy or heterospermy), involving at least three kinds of seeds, was paralleled by different dormancy responses. It has since been shown that in this species the day length to which the parent plant was exposed has a marked effect on seed morphology and dormancy; as with *Halogeton glomeratus* mentioned above, long days result in the production of more dormant seeds (Karssen, 1970). Cavers and Harper (1966) later used the term polymorphism, by analogy, to describe variation in dormancy which is not paralleled by morphological differences. Many other examples of such polymorphism have been reviewed by Koller (1964) – though he uses the term physiological heterogeneity.

In no case does a plant produce seeds all with identical dormancy periods: there is always some variation. If the term polymorphism as applied to seed germination is to retain its usefulness, its application should be restricted to cases where there is evidence of a *discontinuous* distribution of dormancy periods or degree of dormancy (Popay and Roberts, 1970b). In order to demonstrate polymorphic characteristics it is not sufficient to show that seeds can be divided into two categories by applying a given treatment: after all, any germination test which does not give either 0 or 100 per cent germination will separate seeds into two distinct groups – those which germinate and those which do not.

Sometimes after a seed has lost its innate dormancy a similar type of dormancy may be induced. When such induced or secondary dormancy occurs, it is usually the result of seeds being supplied with water but in an environment where some other particular factor is unfavourable for germination. Vegis (1963) has suggested that high temperatures and a limited oxygen supply are often the most important factors for inducing secondary dormancy. For example, dormancy may be induced in *Avena fatua* under conditions where the oxygen supply to the embryo is restricted (Hay and Cumming, 1959); this induced dormancy has very similar characteristics to primary dormancy (Hay, 1962). Dormancy can be induced in *Avena ludoviciana* by subjecting the imbibed seed to a temperature of 76°F (24.5°C) or above (Thurston, 1960). Kidd (1914) showed that dormancy could be induced in *Brassica nigra* by high concentrations of carbon dioxide in the atmosphere. In all these cases the dormancy which has been induced persists for a considerable time: it is this persistence of dormancy after the inhibitory factor has been removed which distinguishes induced dormancy from enforced dormancy.

Enforced dormancy describes the condition when viable seeds do not germinate because of some limitation in the environment. It is mostly used to describe the dormancy of seeds buried beneath the soil surface, which is removed immediately when the seeds are exposed. In order that the term remains useful, it is assumed that an adequate supply of water is not the limiting factor – otherwise the term enforced dormancy would have to be extended to describe the condition of dry seeds in storage. Enforced dormancy has been variously attributed to high carbon dioxide levels, darkness and lack of fluctuating temperatures – factors which become more pronounced beneath the soil surface (Crocker, 1948; Harper, 1957). The relative importance of these factors will be discussed later.

Is there a functional relationship between innate dormancy and period of viability?

It has often been suggested (e.g. Ramiah, 1937; Sahadevan, 1953; Toole and Toole, 1953; Caldwell, 1959; Mouton, 1960; Kernick, 1961; Berrie, 1964) that dormancy may be connected with viability – the more pronounced the dormancy, the greater the period of viability. In her review on seed viability Owen (1956) suggested that the relationship between these factors was one of the more important problems requiring further investigation. In spite of the fact that an association between dormancy and viability has often been implied, the evidence to support the idea is meagre, except for some work on lettuce seed by Toole and Toole (1953). These workers found that seeds of some lettuce varieties become dormant and do not germinate when imbibed and held at 30°C. Such seeds were compared with 'dry' seeds: both types were placed in an atmosphere of 85–90 per cent RH and 30°C for periods up to 105 days. The 'dry' seeds quickly absorbed water from the atmosphere and lost viability, whereas the imbibed seeds of much higher moisture content germinated well when removed to suitable conditions. Toole and Toole suggested that the dormancy of the imbibed seeds suppressed life processes that lead to seed deterioration and that a similar explanation could be given for survival of seeds in moist soil. One other example has been brought to my attention by T. A. Villiers (private communication, 1970) who has observed that if the cold treatment on moist seeds (stratification treatment) necessary to break dormancy in *Fraxinus* is withheld the seeds last for at least 6 years whereas if the seeds are stratified they only remain viable for about 3 years. Suggestive though this evidence is, however, it does not establish a causal relationship between dormancy and viability.

The possibility of a relationship between innate dormancy and viability has been investigated in some detail in rice (Roberts, 1963a). Six cultivars of rice were chosen for study, differing widely not only

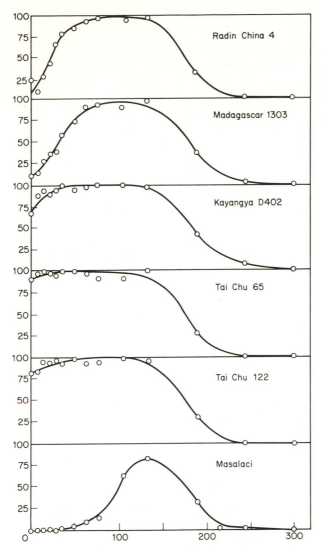

FIGURE 11.1   Percentage germination of six cultivars of rice stored under the same conditions (see text) plotted against time. (From Roberts, 1963a.)

in genotype but also in colour and morphology of the seeds (caryopses with attached lemmas and paleas) so that they could easily be recognised. One cultivar of *Oryza glaberrima* was included and five of *O. sativa*. Of the *O. sativa* cultivars, three belonged to the *indica* sub-species and two to the *japonica* sub-species. All seeds were sun-dried to 10–11 per cent moisture content immediately after harvest and refrigerated until the experiment started. At the beginning of the experiment seeds of all six cultivars were thoroughly mixed and moistened to 13.5 per cent moisture content. The seeds were then sealed in glass ampoules and stored at 27°C (except for short periods of time when the temperature control broke down and temperature rose to about 33°C). At regular intervals, an ampoule was broken open, the seed separated into cultivars and germination tests carried out (Fig. 11.1). The rising parts of the curves show the increasing percentage germination resulting from loss of dormancy whereas the falling parts of the curves show the pattern of decrease in germination due to the loss of viability. It will be seen that although there are marked differences in dormancy amongst the cultivars – ranging from Tai Chu 65 which showed practically no dormancy to Masalaci which took 100 days for half the seeds to break dormancy – the pattern of loss of viability was identical in all cases.

Furthermore it was argued that whereas in rice it is possible to alter the period of viability by altering temperature and moisture content, as in equation (2.11), the period of dormancy can only be altered appreciably by altering temperature; alterations in moisture content have very little effect on dormancy in rice (Roberts, 1962). The relationship between temperature and mean dormancy period in rice, at least over the range 27–47°C (Roberts, 1962, 1965), is given by the equation:

$$\log d = K_d - C_d t \qquad (11.1)$$

where d is the mean dormancy period, t is temperature and K_d and C_d are constants. The slope constant, C_d, is apparently the same for all cultivars but the value of K_d, the intercept constant, is peculiar to each cultivar. The Q_{10} for the rate of loss of dormancy (derived from this equation) is just over 3 and, of course, since C_d is a constant for all cultivars – the Q_{10} is the same for all cultivars (Roberts, 1965). On the other hand, the Q_{10} calculated for the rate of loss of viability in rice over the same temperature range was much higher viz. 4.9 (Roberts, 1961).

Thus on two grounds it must be concluded that in rice, at least, there is no association between innate dormancy and period of viability. First there are marked genotypic differences in dormancy period but not in viability period. Secondly the two properties are affected differently by environmental factors: temperature affects both the rates of loss of dormancy and loss of viability but the Q_{10} values are different and, furthermore, whereas moisture content is an important factor affecting the loss of viability, it has little effect on the loss of dormancy. The conclusion that there is no functional relationship between dormancy and viability agrees with the view put forward by Barton (1961): 'Embryo dormancy is also important in germination and hence in tests for viability, but it is not directly related to longevity.'

Although there is little evidence of a functional relationship between innate dormancy and period of viability, it is perhaps worth mentioning that, speaking very generally, cultivated species very often not only show less innate dormancy than wild species but also a much reduced period of viability under comparable environmental conditions. In this sense it could be held that there is a very rough correlation between the degree of dormancy and the ability to remain viable for long periods. The classical long-term seed-burial experiment started by Duvel in 1902 is often cited as evidence of the longer periods of viability of wild versus cultivated species. In this experiment 107 species of wild and cultivated seeds were buried in sterilised soil in flower pots with porous clay covers; most of the seeds of the cultivated plants (with the exception of *Nicotiana tabacum* and *Trifolium pratense* which both survived for 39 years) died after one year, whereas 34 of the wild species were still showing some ability to germinate after 39 years (Toole and Brown, 1946). Unfortunately this evidence is not entirely satisfactory because it is not clear whether the cultivated species failed to survive because of loss of viability or because they germinated (lack of enforced dormancy). Clearer evidence was provided by Lewis (1958) who showed that, in general, cultivated species of cereals, grasses and legumes did not last as well as some weed species when stored under similar conditions. More recently Berrie (1964) showed that under the same storage conditions seeds of the wild oat *Avena ludoviciana* remain viable for much longer periods than the cultivated oat *Avena sativa*.

It should be pointed out, however, that the reason for an apparent correlation between dormancy and viability in wild versus cultivated species may not be the result of a functional relationship, but the result of selection pressure operating on both properties in wild species and the removal of this pressure in cultivated species. There is in fact positive selection against prolonged periods of dormancy in most cultivated plants.

The relationship between dormancy and the ability of seeds to survive in the soil

Although there is little evidence of a functional relationship between the degree of innate dormancy and period of viability, when seeds are buried in the soil it is essential that they are subjected to some form of dormancy if they are to survive. This is because in most soils seeds are subjected, from time to time at least, to moisture conditions adequate for germination. Non-dormant seeds near the surface can grow into new plants after germination. However, if they are buried at some depth, they will germinate but run out of food reserves before the shoots reach the surface and start photosynthesis. The majority of seeds found in the soil are very small and have to be very near the surface for germination to lead to the successful establishment of a new plant.

It is not surprising then that most wild species have evolved mechanisms for preventing germination when the seeds are not near the soil surface.* It is a common experience of agriculturalists, gardeners and naturalists that when soil is disturbed, the disturbance is immediately followed by a flush of germination of a number of species. The rapidity of the response indicates that, before the soil disturbance, the seeds were being prevented from germination by an enforced dormancy.

In addition to the enforced dormancy response, other dormancy mechanisms have evolved which result in an adjustment between species and environment. Very often for example temperate species show an innate dormancy which is removed by a period of cold (preferably near 5°C) applied to the imbibed seed. Treating imbibed seeds with low temperatures to break innate dormancy is often known as 'stratification'; this was originally a horticultural term which refers to the practice of over-wintering seeds between layers of peat or sand and watering them from time to time; essentially it means the treatment of imbibed seeds with a low temperature. Under natural conditions a stratification requirement ensures that the seeds will not germinate until the beginning of the favourable growing season after winter.

In other environments other innate dormancy mechanisms have evolved which ensure that seeds will only germinate when environmental conditions are such that there is a reasonable probability that the seedling will be able to reach maturity and start a new cycle.

---

*Exceptions to this are usually associated with special ecological environments. Koller (1964) has pointed out, for example, that *Citrullus colocynthis* and *Calligonum comosum* typically inhabit gravelly or coarse sandy soils in the desert where surface conditions (moisture and temperature) are precarious. Germination of both species is strongly inhibited by light at all temperatures and thus the seeds only germinate when they are beneath the surface.

Went (1957) has cited some very interesting examples of dormancy-breaking responses of species growing in the Colorado desert. When soils from the desert were moistened in the laboratory the seedlings which emerged depended on the temperature at which the experiment was carried out: at a low temperature (10°C) mainly winter annuals germinated, at higher temperatures (26–30°C) only summer annuals germinated, while at intermediate temperatures a third group of plants could be recognised. Many other interesting examples have been cited by Harper (1957), Koller (1962, 1969) and Mayer and Poljakoff-Mayber (1963).

In some cases innate dormancy does not occur at one specific time in the year but the mechanisms are such that loss of dormancy occurs in different members of the population at different times of the year. In other words some plants not only have mechanisms which result in the efficient dispersal of seeds in space, but also a mechanism which ensures the dispersal of the onset of germination in time. It is easy to imagine the survival value of such mechanisms to the species. Salisbury (1961) has suggested a classification of germination behaviour into 'quasi-simultaneous', 'continuous' and 'intermittent' to describe the distribution of onset of germination in time.

In our terms 'quasi-simultaneous' refers to a unimodal distribution of germination in time with a small Coefficient of Variation; 'continuous' refers to a unimodal distribution with a very high Coefficient of Variation; whereas 'intermittent' refers to a multimodal distribution. Salisbury uses these terms to describe the germination patterns in natural environments, i.e. under variable environmental conditions. In such cases the intermittent pattern could be the result of changes in environment on what would have been a unimodal distribution (either quasi-simultaneous or continuous) under constant conditions, or it could be the result of seed polymorphism. (If one is dealing with more than a single population of plants is could also be due to discontinuous genetic variation.) Some examples of seed polymorphism have already been quoted; but there are also a number of cases of multi-modal patterns of germination in annual weed species in circumstances where there is no evidence of polymorphism.

In some experiments carried out by H. A. Roberts (1964) seeds of a number of weed species were sown in Warwickshire and the soil was disturbed on five to six occasions each year. Figure 11.2 shows the combined results of the observations made over a period of six years. It will be seen that there was a tendency towards two modes of emergence mor or less coincident in five of the eleven species (*Capsella bursa-pastoris, Senecio vulgaris, Veronica persica, Chenopodium album*

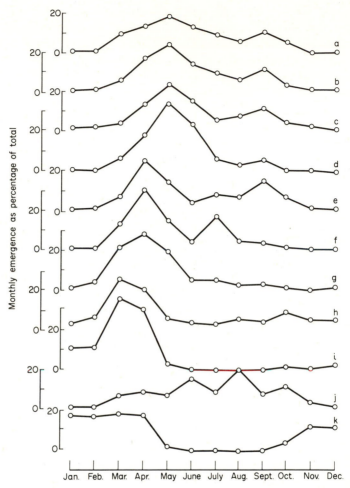

FIGURE 11.2   Monthly distribution of emergence: (a) *Capsella bursa-pastoris*;
(b) *Senecio vulgaris*; (c) *Veronica persica*; (d) *Chenopodium album*; (e) *Stellaria
media*; (f) *Urtica urens*; (g) *Thlaspi arvense*; (h) *Tripleurospermum maritimum* ssp.
indorum; (i) *Fumaria officinalis*; (j) *Poa annua*; (k) *Veronica hederifolia*.
Mean of data from four experiments and six years' observations. (From H. A.
Roberts, 1964.)

and *Stellaria media*). In some experiments of Popay and Roberts
(1970b) which were confined to *Capsella bursa-pastoris* and *Senecio
vulgaris*, seed was sown at two sites twenty miles apart in the north of
England just beneath the soil surface which was then left undisturbed
(Fig. 11.3); again there were coincident peaks of germination which
can be seen more clearly since the data for the two years of observation
were not combined: there were three major coincident peaks in 1966

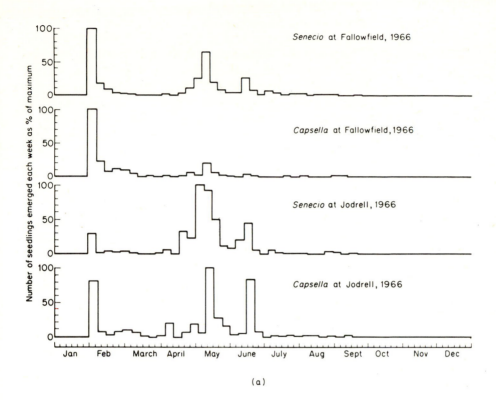

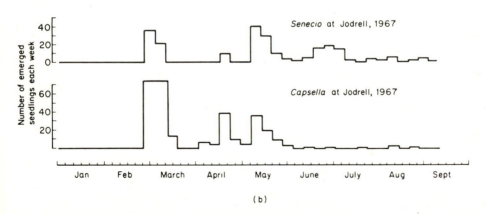

FIGURE 11.3 (a) Weekly emergence of seedlings at Fallowfield, Manchester in 1966, and at Jodrell Bank, Cheshire in 1966; values for emergence are given as the percentage of the maximum number of seedlings which emerged in any one week. (b) Weekly emergence of seedlings at Jodrell Bank in 1967; actual values. (Redrawn from Popay and Roberts, 1970b.)

and three in 1967, although in 1967 *Senecio* showed an additional peak not matched in *Capsella*. In the case of *Chenopodium album* in which Williams and Harper (1965) have demonstrated seed polymorphism, it could be argued that the different modes represent different polymorphic types. However, the peaks coincide with other species in which there is no evidence of seed polymorphism. In addition it is quite evident from Fig. 11.3 and from some further results of H. A. Roberts (1964) shown in Fig. 11.4 that the number and timing of the peaks can vary from year to year. From these results it is difficult to avoid two conclusions: first, that a number of weed species have very similar dormancy-breaking responses and secondly, that some environmental factors causing the loss of dormancy operate strongly on more than one occasion each year (there is no need to suppose, of course, that it will be the same combination of factors that stimulates germination on each occasion).

In these cases of multi-modal patterns of loss of dormancy, it is not always clear whether the dormancy which is being lost is innate, secondary or enforced. In some cases it is possible that the behaviour is to be explained in terms of all three types of dormancy. For example Courtney (1968) has shown that the innate dormancy of fresh seeds of *Polygonum aviculare* is overcome by the low temperatures of autumn and winter, and seedlings emerge from late February onwards. In some of the seeds dormancy is enforced by burial and where

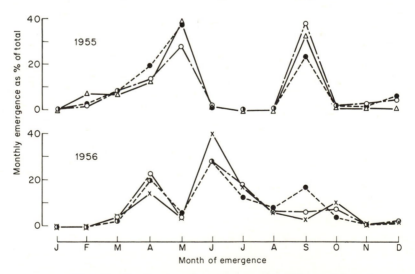

FIGURE 11.4   Monthly distribution of emergence of seedlings of *Capsella bursa-pastoris* in 1955 and 1956 at Wellesbourne, Warwickshire from seed sown in 1952 (△), 1953 (●), 1954 (○) and 1955 (×). (From H. A. Roberts, 1964.)

enforced dormancy occurs, induced dormancy is superimposed by the high (23–25°C) temperatures in late May. The induced dormancy is removed again by low temperatures the following winter. Although all forms of dormancy are involved in the survival of buried seed populations, it is becoming increasingly apparent, as we shall see in the next section, that a very large proportion of buried seeds are in a condition of enforced dormancy. Thus, assuming an adequate period of viability, it is generally this property of enforced dormancy which is vital to the survival of buried seeds.

The environmental factors which enforce dormancy in buried seeds

Although it has long been known that many seeds need to be brought to the soil surface before they will germinate, this phenomenon has not received much serious attention until recently. Crocker (1948) discussed the major factors which are likely to stimulate germination at or near the soil surface: these included reduced carbon dioxide concentration, fluctuating temperatures and light. Bibbey (1948) suggested that the raised carbon dioxide and lowered oxygen concentrations in the soil atmosphere are largely responsible for maintaining the enforced dormancy of buried seeds. This view has continued to be popular until recently. Harper (1957) pointed out that if a soil sample is taken from the field and spread thinly under favourable conditions of temperature and moisture supply, most of the buried viable weed seeds germinate quickly. He went on to suggest: 'In the case of a few weed species dormancy may now be broken because a light stimulus is present – but the greater number of weeds which will germinate under these conditions are insensitive to a light stimulus. It can easily be shown that the conditions of temperature and moisture supply existing at the depth of buried viable weed seeds are suitable for rapid germination and it is therefore argued by elimination that enforcement of dormancy results from changed conditions in the soil atmosphere. Either reduced oxygen tension or increased carbon dioxide tension have been suggested as the effective factor in enforcing dormancy.'

These arguments were entirely logical at the time, nevertheless it has since become evident that the most generally effective factor in removing the enforced dormancy of buried seed is light. Sauer and Struik (1964) reported some experiments on the effect of disturbing soil either in the light or the dark on the subsequent germination of buried seed populations. This work indicated the importance of light

TABLE II.I   *Mean values for emergence after 4 weeks*
of seedlings from soil in light and dark. (From Wesson
and Wareing, 1969a.)

	Grasses	Dicots
Light	14.1	8.3
Dark	0.5	0.7
Dark as % of light	4.2%	7.8%

but, as Sauer and Struik pointed out, the work was preliminary and
further investigation was needed. The main evidence for the importance of light comes now from two basically simple experiments
carried out by Wesson and Wareing (1967, 1969a).

In the first, 3 kg soil samples were taken from a field which had
been down to pasture for the previous six years. These samples, which
were derived from cores in which the upper 2 cm of soil was discarded,
were taken in complete darkness and sieved in a dark room. Each
sample was then divided into two halves and placed in germination
trays in a glasshouse; half of the trays were kept in the light and half
in the dark. Three identical experiments were carried out over twelve

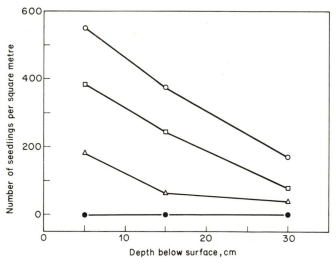

FIGURE II.5   The number of weed seedlings to emerge from field plots at three
different depths.

- □   Plots uncovered: recorded after 5 weeks.
- ○   Plots covered with glass: recorded after 5 weeks.
- ●   Plots covered with asbestos (dark); recorded after 5 weeks.
- △   Asbestos covers removed from Treatment 3 and replaced by glass. Emergence
      recorded for a further 3 weeks. (From Wesson and Wareing, 1969a.)

months and Table 11.1 shows the mean values for seedling emergence after 4 weeks. The only dicotyledonous species which germinated in the dark were *Sinapsis arvensis* and *Spergula arvensis*; but even these two species were always represented in much larger numbers in the illuminated trays. A similar experiment carried out by Feltner and Vesecky (1968) in Kansas has confirmed the importance of light, but in this case detailed identifications of the seedlings were not made.

The second experiment of Wesson and Wareing which will be mentioned here was even more striking. Holes 75 cm square were dug to depths of 5, 17.5 and 30 cm in a $2\frac{1}{2}$-year old pasture. Plots of each depth were either left uncovered, or covered with a sheet of either glass or light-proof asbestos. The results (Fig. 11.5) indicate a complete dependence of emergence upon light at all depths. The fact that the dark treatments gave some germination in the previous experiment could not be explained. Nevertheless, both experiments emphasise the importance of light.

These experiments exposed a very important factor which explains why the inhibitory effect of darkness had previously been under-estimated. Wesson and Wareing (1969a) discovered that many of the species found to germinate after disturbance of the soil are unaffected or inhibited by light when fresh seed is examined (Table 11.2); they were therefore forced to conclude that, either a wide range of species possess some previously unobserved light-dependent seeds, or that burial induces a light dependence in previously non-light-sensitive seeds. Wareing and Wesson (1969b) then went on to investigate these possibilities in experiments in which the light-sensitivity of the same seed populations were examined before and after burial. In these

TABLE 11.2 *List of dicotyledonous species which Wesson and Wareing (1969a) found to be stimulated by light after burial.*

Aphanes arvensis	*Polygonum persicaria
*Atriplex hastata	*Rumex crispus*
Cerastium vulgatum	*Senecio jacobea*
*Chenopodium rubrum	*Senecio vulgaris*
Chrysanthemum segetum	*Sinapsis arvensis
*Hypochoeris radicata	*Sonchus asper*
Leontodon autumnalis	*Spergula arvensis
Myosotis arvensis	*Stellaria media
*Papaver dubium	*Trifolium repens
*Plantago lanceolata	*Tripleurospermum maritima*
*Plantago media	*Veronica persica

*Species in which the germination of *fresh* seed was not found to be promoted by light.

experiments it was shown that a number of species (viz. *Papaver dubium, Plantago lanceolata, Spergula arvensis* and *Stellaria media*) which after harvest were either insensitive to light or, in the case of *Stellaria media*, partly inhibited by light, were inhibited from germinating by burial and, during burial, developed the condition of enforced dormancy which could immediately be removed on exposure to light. Similar results have since been obtained in *Poa trivialis* by E.G. Budd (private communication, 1970). Further experiments on *Spergula arvensis* (Wesson and Wareing, 1969b) in which buried seeds were aerated intermittently (30 minutes every 24 hours) with nitrogen or a mixture of nitrogen and 12.5 per cent carbon dioxide, indicated that the enforcement of dormancy in the soil depended on the effect of a gaseous inhibitor (other than carbon dioxide) produced by the seeds themselves. Accordingly Wesson and Wareing concluded that weed seeds may fail to germinate immediately after being shed because of innate dormancy such as an after-ripening or chilling requirement. If these seeds become buried, even after the loss of this innate dormancy, they may still fail to germinate because of the presence of inhibitors in the soil air. Over a period these seeds may become 'light requiring'. Consequently the germination of these buried seeds, following disturbance of the soil, is stimulated by exposure to light.

In most species of light-sensitive seeds which have been sufficiently investigated there is evidence that light-sensitivity depends on the phytochrome pigment. In Chapter 12 (p. 372) there will be a detailed discussion of the possible physiological mode of action of phytochrome. Here it will be sufficient to describe those aspects which are essential to an understanding of the ecology of seed germination. The existence of this reversible pigment has been known for some time. Its history starts with the observation of Flint and McAlister (1935, 1937) that red light promotes whereas far-red (or infra-red) light inhibits the germination of a light-requiring variety of lettuce seed. Later it was shown (Borthwick, Hendricks, Parker, Toole and Toole, 1952; Borthwick, Hendricks, Toole and Toole, 1954) that the stimulatory effect of red radiation and the inhibitory effect of far-red is reversible: if seeds are exposed to alternating periods of red and far-red light, whether or not they germinate depends on the quality of light to which they were last subjected. On the basis of such reversibility experiments Borthwick and his colleagues postulated a photoreceptor capable of existing in two states represented thus: $P_r \rightleftharpoons P_{fr}$. This hypothetical pigment, phytochrome, was eventually isolated and purified from oat seedlings (Siegelman and Firer, 1964): it consists of a protein coupled to a chromophore; the chromophore – which undergoes the reversible changes – is a phycocyanin (a tetrapyrrole compound) and it has been suggested that the change in the chromophore is brought about by the shift of two hydrogen atoms (Hendricks, 1968).

It is now known then that phytochrome does in fact exist in two inter-convertible forms. P_{fr} is the physiologically active form and has a peak absorption in the far-red spectrum at about 725 nm. The P_r form does not absorb at this wavelength and irradiation with light of this quality converts the P_{fr} form to P_r. The P_r form on the other hand shows peak absorption at 665 nm (i.e. in red light) and irradiation at this wavelength shifts the equilibrium towards a predominance of the P_{fr} form. The shift from P_r to P_{fr} at this wavelength is not complete since there is a spectral overlap and the P_{fr} form also absorbs to some extent in this region: thus the equilibrium at 665 nm is approximately 81 per cent P_{fr} and 19 per cent P_r (for further discussion see Siegelman, 1969, and Black, 1970).

The reversible promotion and inhibition of the germination of light-sensitive seeds by alternate exposures to red and far-red light is an interesting physiological trick, but the full ecological significance of this extraordinary response has only recently emerged. Cumming (1963) carried out laboratory experiments on the seeds of several species of *Chenopodium* in which he examined the germination responses to light sources of differing qualities. The results of these experiments, together with observations that had been made on the spectral energy distribution of sunlight in comparison with the light transmitted through green leaves, led Cumming to postulate that the phytochrome system equips the seed with a neat environment-sensing device. He pointed out that if one considers the red/far-red ratios based on the energy within 35 nm either side of 640 and 740 nm, there was more germination with red/far-red ratios similar to sunlight (1.3) than with ratios similar to that provided by sunlight transmitted through green vegetation (0.70–0.12). The restrictive effect of low as compared with high red/far-red ratios was obtained over a wide range of energies and photoperiods: in other words, the germination response is relatively independent of light intensity or duration, but it is very sensitive to light quality – particularly as affected by filtration through a leaf canopy.

Experiments substantiating the ecological significance of Cumming's ideas have been carried out by C. D. Piggott (private communication, 1967). Field experiments, supported by experiments in which the spectral transmission of leaf canopies was imitated by

artificial filters, convincingly showed that the major factor suppress-
ing the germination of *Urtica dioica* under a woodland canopy is the
low red/far-red ratio of the woodland light.

More detailed laboratory investigations (Taylorson and Borthwick,
1969) have confirmed that in six weed species (*Chenopodium album*,
Amaranthus retroflexus, *Potentilla norvegica*, *Rumex obtusifolius*, *Barbarea vulgaris* and *Lepidium virginicum*) percentage germination is
controlled by the red/far-red ratio of light. Sunlight was reported to
have a red/far-red ratio of 1.2 (based on measurements at 650 and
730 nm), whereas sunlight filtered through tobacco leaves had a ratio
of 0.18. Leaves of maize and soya beans were also shown to have very
similar transmission properties. Experiments using light of these two
qualities showed that light filtered through leaves is more inhibitory
than sunlight in all six weed species. The levels of germination
achieved varied from species to species: for example *Amaranthus* and
Barbarea gave fairly high germination percentages even in filtered
light as compared with the dark controls (this is postulated to be due
to the low levels of P_{fr} required to trigger germination in these species).
But on the other hand seeds of *Chenopodium* and *Rumex* were inhibited
by leaf-filtered light more or less to the values of the dark controls.

These experiments strongly suggest that, not only does the phytochrome system provide an environment-detecting device which indicates when the seed is at the soil surface,[*] but the mechanism also
senses, in some cases at least, whether there is a leaf canopy above the
seed. The discovery of this response may be important in explaining
the commonly observed fact that the presence of grass often inhibits
the germination of seeds at the soil surface, e.g. in *Plantago* spp.
(Sagar and Harper, 1960), *Capsella bursa-pastoris* and *Senecio vulgaris*
(Popay and Roberts, 1970b). This response to light quality represents
a subtle adaptation to the environment since it prevents a seed from
germinating until conditions are sufficiently favourable for there to
be a reasonable chance of it reaching maturity. Piggott was initially
drawn to his study on germination mentioned above because of his

[*] The term 'at the soil surface' would normally be taken to mean on the surface of
the soil where the seed is fully exposed to light. But Wells (1959) has shown that in
sandy soil light may be transmitted to a depth of about 10 cm. As would be expected
he showed that the longer wavelengths penetrate more than the shorter wavelengths.
His data show that 10 cm below wet sand the red/far-red ratio (655 nm/735 nm) is
0.63. Wells assumed that because red light penetrates reasonably well, light-sensitive
seeds would be stimulated to germinate; this is possible, but the suggestion was made
before the significance of the red/far-red ratio was realised, and it is now evident
that the ratio found at 10 nm below sand is less satisfactory for stimulating germination than the unfiltered light.

interest in the fact that many weed species immediately began to grow in woodland as soon as trees were cut down. If a number of these species were forced to germinate artificially and then placed in the woodland, they were capable of making some growth but there was not sufficient light beneath the woodland canopy to enable them to continue growth to maturity. The important environmental factor so far as the plant growth is concerned is the light intensity (radiant energy/unit time). It would be difficult to think of a mechanism by which a seed could react satisfactorily to intensity since photochemical reactions normally show reciprocity, i.e. like photographic emulsions, the extent to which a photochemical reaction takes place depends on the total energy received, so that a long period at low light intensity is equivalent to a short period at high light intensity. Thus if germination depended directly on a simple photochemical reaction which, say, required one minute of sunlight, it would be found that beneath a leaf canopy – where the light intensity may be 1/50 to 1/100 of that in the open – germination would still be initiated, although it would now take 50 to 100 minutes of exposure. Such a mechanism would not prevent germination in unsuitable light intensities for growth but, by employing a mechanism which relates to light quality, rather than quantity, the seed receives reliable information about the suitability of the light intensity of the environment.

It may be concluded that the phytochrome system enables the seed to survive in an enforced dormant condition when it is in an environment which would be unsuitable for growth. Not only are seeds prevented from germinating beneath the soil surface but, in some cases, they are also prevented from germinating when at the soil surface but beneath a leaf canopy. Further aspects of light-sensitivity in seeds have been discussed in Chapter 6 (p. 164).

Although the discovery of these light responses has shifted the emphasis away from considering the possible enforcement of dormancy by the gaseous composition of the soil atmosphere, the effects of raised carbon dioxide and lowered oxygen partial pressures cannot be completely ignored. Figure 11.6 shows the germination of seeds of *Capsella bursa-pastoris* (after chilling) and *Senecio vulgaris* in different combinations of oxygen and carbon dioxide in the light and in the dark (Popay and Roberts, 1970a). These results show that although darkness is the most influential factor in enforcing dormancy, raised carbon dioxide partial pressures and decreased oxygen partial pressures can also have a contributory effect. Under most conditions, however, it would seem that the effects of carbon dioxide and oxygen are not likely to be very large, particularly when it is recalled that in

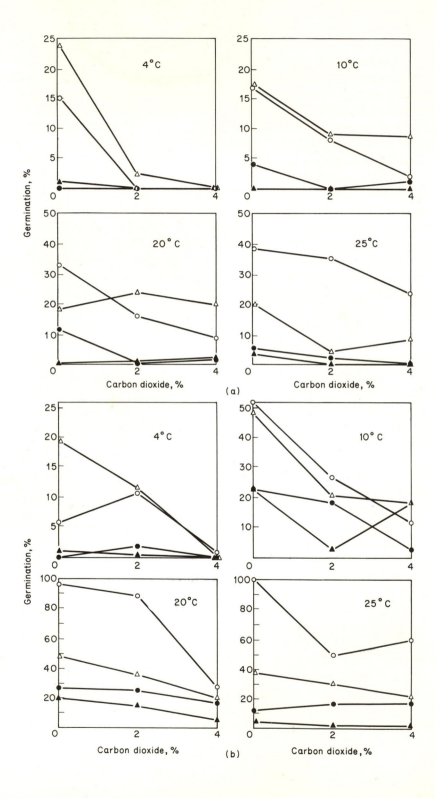

most soils the level of carbon dioxide in the soil atmosphere seldom exceeds 1 per cent and the level of oxygen seldom falls below about 19 per cent (Russell, 1961). Nevertheless one should not be too dogmatic since little is known about variations in gas composition within the soil on a microscale: the composition in the immediate vicinity of a seed and its associated microflora may in fact be significantly different from the average figure for the soil atmosphere as a whole.

Before leaving a discussion on the possible inhibitory effects of carbon dioxide on germination, one further complication has to be mentioned: in certain cases low concentrations of carbon dioxide, usually around 2.5–5 per cent, are capable of removing innate dormancy from a proportion of the seeds in some species, e.g. in some legumes (Lipp and Ballard, 1959), in peanut (*Arachis hypogea*) (Toole, Bailey and Toole, 1964), in rice (Tseng, 1964), in *Avena fatua* (Hart and Berrie, 1966), and in barley (Major and Roberts, 1968); in addition Thornton (1936) found that treatment with carbon dioxide could bring about the germination of light-requiring lettuce seeds in the dark and could prevent the development of thermodormancy in the dark. A raised level of carbon dioxide then can not only contribute to enforced dormancy but, under some circumstances, it can remove innate and enforced dormancy and prevent the development of induced dormancy.

So far we have considered two of the factors – light and decreased carbon dioxide concentrations – which Crocker (1948) suggested may be important in removing the dormancy of buried seeds; we now need to consider the third – temperature fluctuations. In addition to this, however, the level of nitrate ions in the soil could also be important since nitrate has long been recognised as one of the major dormancy-breaking agents for a wide range of species (Toole, Hendricks, Borthwick and Toole, 1956). Steinbauer and Grigsby (1957) studied the germination of 85 species of weed plants in 15 families and half showed more germination in the presence of nitrate. In some cases reduced forms of nitrogen have an effect but in many species only nitrate and nitrite are effective (Roberts, 1969). Thus the nitrification

FIGURE 11.6   Germination of (a) *Senecio vulgaris* and (b) *Capsella bursa-pastoris* after 28 days, under different gas mixtures at different temperatures, in light (diffuse laboratory light) and in darkness. The *Capsella* seeds had previously been kept imbibed for two weeks at 4°C to meet their chilling requirement for removing innate dormancy. The lowest concentration of CO_2 indicated is 0.03 per cent. Note the different scales for percentage germination in each graph. 10 per cent oxygen: light (△), dark (▲); 21 per cent oxygen: light (○), dark (●). (From Popay and Roberts, 1970a.)

of ammonium ions in the soil in response to increased activity of soil microflora could be a significant factor in the removal of dormancy in buried seeds.

Evidence is now accumulating that one should not consider the effect of nitrate ions on their own: because nitrate, fluctuating temperatures and light can show strong positive interactions in overcoming dormancy. For example in *Capsella bursa-pastoris* it was shown (Popay and Roberts, 1970a) that, in the presence of light, nitrate or nitrite ions could overcome innate dormancy in a small proportion of seeds at constant temperatures; but when the seeds were subjected to alternating temperatures, the stimulatory effect of nitrate or nitrite was dramatically increased, although the alternating temperatures in the absence of nitrate or nitrite had only a small effect (Fig. 11.7). Unfortunately the possibility that this interaction could overcome enforced dormancy (i.e. stimulate the germination of seeds in the dark) was not investigated in this species. The interaction of nitrate with alternating temperatures can be disconcertingly complex: it promotes germination in *Digitaria sanguinalis* at lower ranges of temperature alternations (15–25°C or 20–30°C), but retards germination at higher temperature alternations (20–35°C or 20–40°C); with-

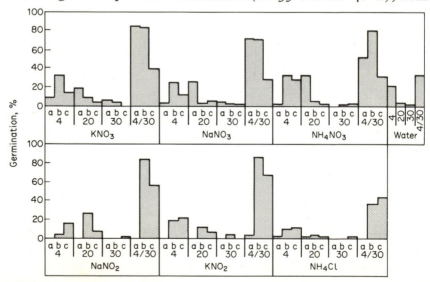

FIGURE 11.7 Germination of *Capsella bursa-pastoris* in the presence of inorganic nitrogen sources at different concentrations and temperatures. 4/30°C refers to diurnal fluctuations in the ratio of 19/5 hr between these two temperatures. The tests were continued for 28 days, during which the seeds were exposed to low intensity daylight for a few minutes each day. (a) 10^{-1}M solution; (b) 10^{-2}M solution; (c) 10^{-3}M solution. (From Popay and Roberts, 1970a.)

out nitrate the higher temperature alternations are preferable (Toole and Toole, 1941).

In some cases it is known that nitrate enhances a light stimulation but has no effect in the dark, e.g. in seeds of *Lepidium virginicum* (Toole, Toole, Borthwick and Hendricks, 1955a), *Jussiaea suffruticosa* (Wulff and Medina, 1969) and *Potentilla norvegica* (Taylorson, 1969). In other cases, which have been reviewed by Toole, Hendricks, Borthwick and Toole (1956), nitrate can substitute for a light requirement. In a number of ecotypes of Johnson grass (*Sorghum halepense*) Taylorson and McWhorter (1969) showed that nitrate, light or fluctuating temperatures had only small effects on their own; however, if either nitrate or light were combined with fluctuating temperatures a much greater response was obtained; the greatest response was obtained when all three factors – nitrate, light and fluctuating temperatures – were combined. It is also known that the response can change with the age of the seed: Henson (1970) showed that in *Chenopodium album* either light or nitrate can stimulate germination to roughly the same extent. The effect of either is very marked on old seed; in young seed, however, the stimulation caused by either factor alone is usually small but, when both factors are combined, there is a very marked synergism (Table 11.3).

TABLE 11.3 *The influence of light ('daylight' fluorescent tubes: 16 lm/ft^2) and nitrate* ($10^{-1}M$) *on the percentage germination in 10 days of old (32–35 months) and young (8–11) months) seed of* Chenopodium album *under constant and alternating temperature regimes (results of four experiments). (From Henson, 1970.)*

Temperature	Light	Old seed		Young seed	
		Water	KNO$_3$	Water	KNO$_3$
Constant	Light: 14 h/day	50	79	13	75
23°C	Dark	29	66	2	21
LSD (P = 0.05)		11.1		4.2	
Alternating	Light: 14 h/day	73	87	36	97
15–25°C	Dark	30	75	6	33
LSD (P = 0.05)		10.5		5.8	
Constant	Light: single 16 min	62	87	8	76
23°C	Dark	43	72	1	9
LSD (P = 0.05)		8.8		4.9	
Alternating	Light: single 16 min	44	79	3	65
15–25°C	Dark	24	53	2	3
LSD (P = 0.05)		9.2		4.1	

There are a number of examples of light-sensitive seeds in which dormancy of some of the population can be overcome in darkness by alternating temperatures alone, e.g. in *Brassica juncea* and *Nicotiana tabaccum* (Toole, Toole, Borthwick and Hendricks, 1955b). In *Digitaria sanguinalis* light is necessary for maximum germination at 20–30°C alternations but, if the range is increased to 20–35°C, almost maximum germination is possible in the dark (Toole and Toole, 1941). There are also many more examples in which a fluctuation in temperature has little effect on its own but increases the response to light, e.g. in *Lepidium virginicum*, *Lepidium campestre*, *Sisymbrium officinale* and *Verbascum thapsus* (Toole *et al.*, 1955a and b; Toole, Toole, Hendricks and Borthwick, 1957), and in *Potentilla norvegica*, *Potentilla recta*, *Chenopodium album*, *Barbarea vulgaris* and *Rumex obtusifolius* (Taylorson, 1969). Finally, an extreme case may be cited, *Lycopus europaeus*, where there is an absolute requirement for both light and fluctuating temperatures (Thompson, 1969).

From these examples it can be seen that in certain cases the light requirement normally necessary to overcome enforced dormancy may be substituted, or at least modified, by another factor or combination of factors – particularly nitrate and fluctuating temperatures.

Practically all the work on the environmental factors which break dormancy has been carried out on temperate species, but there is some indication at least that weed species in the tropics show very similar characteristics with respect to light and nitrate and thus it would seem that the typical characteristics and responses described here may well be universal. In an investigation on six common East African weeds, Huxley and Turk (1966) showed that four species were light-sensitive (*Bideus pilosa*, *Galinsoga parviflora*, *Tagetes minuta* and *Euphorbia hirta*); in addition 0.2 per cent KNO_3, which was investigated in the presence of light, was shown to increase germination in two of these species (*Tagetes* and *Euphorbia*). However, there is a great need in both tropical and temperate species for more work in this area. If we are to understand the ecological factors which affect loss of dormancy, and thus the patterns of seed survival in the field, it is obviously necessary to study loss of dormancy on a multifactorial basis because of the important interactions which take place.

From this discussion it would be expected that agricultural practices will have a profound effect on the removal of dormancy from weed seed populations. The most important practice will be tillage operations which expose seeds to light or bring them near the surface of the soil where they are subject to large diurnal temperature fluctuations. But, in addition, applications of nitrogen fertiliser may have

TABLE 11.4 *Effect of nitrogen level on emergence of* Avena ludoviciana. (*From Watkins, 1966.*)

Nitrogen applied (lb/ac), January, 1965	Available soil nitrogen (ppm), May, 1965	Plants/yard², August, 1965
0	8	74
23	–	160
46	–	239
92	142 (approx.)	286

LSD (P = 5%) = 41.

some effect. For example, Taylorson (1969) reported that photo-sensitisation of seeds of *Potentilla norvegica* is caused by KNO_3 at an optimum concentration of 2,000–4,000 ppm although detectable responses occur at less than 100 ppm; and, he pointed out, 30 lb/ac (33 kg/ha) of KNO_3 distributed uniformly in the surface inch would approximate to 100 ppm in soils. Watkins (1966) has provided practical evidence that nitrogen-fertiliser application can influence the germination of weed seeds. He investigated the effects of application of three forms of fertiliser – urea, ammonium sulphate and calcium ammonium nitrate – on the emergence of wild oats (*Avena ludoviciana*) in Queensland. No differences were found amongst the fertilisers (presumably because the urea and ammonium forms would be nitrified to nitrate in the soil),* but all stimulated seedling emergence. The biggest effect was obtained from applications of fertiliser in January, the results of which are shown in Table 11.4. H. A. Roberts (1962) also found some indication that increased applications of nitrogen fertiliser lead to a greater loss of soil weed seed populations, presumably the result of increased germination.

---

*A point which also needs to be considered is that, in a number of circumstances, nitrite can be an even more effective dormancy-breaking agent than nitrate – e.g. in rice (Roberts, 1963b), in barley (Major, 1966) and in *Capsella bursa-pastoris* (Popay and Roberts, 1970a); the greater efficacy of nitrite has been discussed previously in relation to a dormancy hypothesis which would account for the difference (Roberts, 1969). Although nitrite levels in the soil are usually very low (Russell, 1961), it has been shown that nitrite can accumulate to significant amounts in urea-treated soil: in some experiments on loam in 5-inch pots, Court, Stephen and Waid (1962) showed that the nitrite content of urea-treated soil can increase transitorily to 24 per cent of the applied nitrogen, reaching a peak during the fourth week after application.

Patterns of seed survival in the soil

The factors affecting pattern of survival of seeds in the soil are evidently not only of ecological interest but are of potential practical importance to agriculturalists concerned with methods of weed control. It has now been made clear that a pre-requisite for survival depends on the ability of the seeds to remain viable in the sense discussed in Chapter 2, but the pattern of survival will often depend more immediately on additional factors. These additional factors will include loss due to the attack by other soil organisms but probably even more important, particularly where the soil is disturbed, will be the losses from the population resulting from germination caused by the removal of enforced dormancy.

In undisturbed soil some species can maintain viable seed populations for very long periods. Evidence of longevity comes mainly from two types of observation. First there are the long term burial experiments in which an attempt is made to bury seeds under conditions which approach those found in the field; secondly there have been observations on the appearance of weed species on land which was previously undisturbed for many years. Typically the second type of observation has been made on the appearance of arable weeds following the cultivation of what was previously permanent grassland, or from taking soil samples from beneath permanent grassland. An experiment enabling observations to be made on the germination ability of deliberately buried seeds has already been mentioned – that which Duvel laid down in 1902. An even earlier experiment was laid down by Beal in 1879 under slightly more artificial conditions (Darlington, 1951). In this experiment seeds of 20 different wild species were mixed with sand and buried about 18 inches beneath the surface of the soil in uncorked pint bottles, the mouths of which were tilted downwards to prevent the bottles from filling with water. Germination tests were then carried out at intervals. Nearly half the species had lost all viability by the fifth year, but eleven species still showed some ability to germinate after 20 years. Two species, *Oenothera biennis* and *Rumex crispus*, were capable of germination after 70 years.

As an example of observations made on soil taken from permanent pastures, the following may be cited. Brenchley (1918) produced a number of weed seedlings from soil which had been taken from beneath old pastures that had been under grass for 32 years; these included *Polygonum aviculare*, *Atriplex patula*, *Papaver rhoeas* and *Anagallis arvensis*. It is obvious then, from both types of evidence, that seeds of a number of wild species can remain viable for very long periods when buried in the soil.

One question may be raised in passing: does the soil environment confer any particular advantage in preserving viability? Kjaer (1948) reported that 9 out of 17 species showed a greater percentage viability after ten years in the soil as compared with ten years in dry storage. Crocker (1948) cited similar evidence. However, this is a subject which needs further investigation: for although the literature quoted is suggestive, the evidence is by no means critical since the comparisons were not made at identical temperatures and seed moisture contents.

Although the ability to maintain viability is one of the prerequisites, it is certainly not the only property required for survival in the soil. The overall problem of survival has been clearly defined by Schafer and Chilcote (1969) who have proposed the following model to describe the components of a buried seed population:

$$S = P_{ex} + P_{end} + D_g + D_n \qquad (11.2)$$

in which S represents the total buried seed population of a species at a point in time ($=100$ per cent), P_{ex} represents the percentage of seeds in a state of exogenous dormancy (enforced dormancy), P_{end} is the percentage of seeds in a state of endogenous dormancy (innate dormancy and induced dormancy), D_g represents the percentage of seeds which are undergoing germination *in situ* (they may be destroying themselves if below a critical depth, or they may be producing new plants if near to the surface), and D_n represents the percentage of seeds which have lost viability.

Schafer and Chilcote point out that in practice the quantitative determination of D_g is limited to relatively short periods following burial. This is due to the fact that emerging organs of the germinating seed deteriorate rapidly (unless the seed is near enough to the surface to develop into a functional plant) and hence the evidence of *in situ* germination is transitory. In certain circumstances it may be useful to analyse the term D_g into its two components: seeds which germinate in 'deep' positions and fail to emerge and those which are near the surface and emerge into seedlings. For the present discussion we shall consider that:

$$D_g = D_{gd} + D_{ge}$$

where D_{gd} is that percentage of seeds which germinates deeply and so fails to emerge and D_{ge} is that percentage which germinates near the surface and emerges.

The term D_n includes not only loss of viability due to physiological ageing but also loss due to damage by predatory organisms; it also includes those seeds which were originally non-viable as a result of events that transpired during seed development and maturation (see Chapter 5). In certain circumstances, then, it may also be useful to analyse this term into its components and for the present discussion we shall consider that:

$$D_n = D_{ni} + D_{na} + D_{np}$$

where D_{ni} is that percentage of seed which was initially non-viable, D_{na} is that percentage which lost viability because of physiological ageing (the loss of viability discussed in Chapter 2) and D_{np} is that percentage which has lost viability due to the action of predators.

At the beginning of this chapter it was pointed out that many workers now categorise dormancy into three types: innate, induced and enforced. Schafer and Chilcote include both innate and induced in their term P_{end}. However, in many circumstances it is useful to distinguish between all three forms and in the present discussion we shall represent the three forms of dormancy as P_{inn} for innate, P_{ind} for induced, and P_{enf} for enforced. If we adopt all these modifications Chilcote and Schafer's equation becomes:

$$S = P_{inn} + P_{ind} + P_{enf} + D_{gd} + D_{ge} + D_{ni} + D_{na} + D_{np} \qquad (11.3)$$

So far there have been few attempts to estimate the relative importance of the various factors which lead to the depletion of seed populations in the soil. Lewis (1960) mentioned briefly that in some work carried out at the Welsh Plant Breeding Station it was found that rapid depletion of seed numbers in various types of soil to a depth of 8 inches (25 cm) was found to be attributable to germination *in situ* (D_{gd}); but no details were given. In a study of a wide range of weed seeds in the soil Taylorson (1970) concluded that the major factor causing loss of seeds from the population was germination (probably D_{gd}) – though pathogen and insect attack (D_{np}) could not be ruled out. The difficulties of categorising the losses of seeds are emphasised by Rampton and Ching (1970) who made a specific attempt to apply Schafer and Chilcote's equation to a study of various grass and clover seeds in the soil; they found it particularly difficult to distinguish between the terms D_g and D_n. Schafer and Chilcote (1970) have carried out a detailed analysis of seeds of *Lolium multiflorum* and *Lolium perenne* buried 10 cm deep in silty loam in Corvallis, Oregon. The seeds were buried in saran mesh packets and the surface of the ground was sown to grass after the packets had been buried. Both species on burial showed negligible innate dormancy (1.3 per cent or less) and were about 96 per cent viable. Samples were removed for analysis after various periods of time. *In situ* germination (D_{gd}) was determined by counting the number of seeds with emerging radicles and coleoptiles at the time of removal from the soil. The remaining seeds were placed in 14-day germination tests in alternating temperatures (25/15°C) in the light (8 hr/day). Those which germinated in the test were termed 'non-dormant', but in our terminology would be considered to be in a state of enforced dormancy (D_{enf}) at the time of removal from the soil. Those which did not germinate in this test were tested for viability using two methods of estimation: in one sample an estimate was made using a tetrazolium test, and in a parallel sample by a further germination test preceded by 5 days' stratification in 0.1 per cent KNO_3; the results of both viability estimates were in good agreement. Those seeds which germinated in the second viability test were termed 'dormant' but, since these seeds were not dormant at the time of burial, in our terminology they would be considered to be in a state of induced dormancy (D_{ind}) when they were removed from the soil.

Some of the mean values calculated from Schafer and Chilcote's results are shown in Table 11.5. These results show that the major loss in both species was due to *in situ* germination (D_{gd}). The *in situ* germination in the case of *Lolium multiflorum*, however, was not as great and this behaviour was associated with a greater degree of both enforced and induced dormancy in this species. These results are also compatible, as Schafer and Chilcote point out, with the observation that in western Oregon, where both species are grown for seed, *Lolium multiflorum* is a notorious weed whereas volunteer plants of *Lolium perenne* are seldom observed.

TABLE 11.5 *Estimate of the proportion of Ryegrass seed in various categories after 60 days' burial at 10 cm in silty loam. (Data from Schafer and Chilcote, 1970.)*

Terms as in equation (11.3)*	Category	*Lolium multiflorum*	*Lolium perenne*
P_{enf}	Enforced dormancy, %	30	0
P_{ind}	Induced dormancy, %	7	1
D_{na}	Non-viable, %	12	15
D_{gd}	*In situ* germination, %	49	85

*The other terms are not included in this table for the following reasons: innate dormancy (P_{inn}) was negligible; the seed was buried too deep for any seeds which germinated *in situ* to emerge (D_{ge}); the proportion of seed initially non-viable (D_{ni}) was negligible; and there was apparently no evidence of loss due to predators (D_{np}).

Another interesting analysis has been carried out by Courtney (1968) on seeds of *Polygonum aviculare*. In terms of equation (11.3) the analysis was not complete, but it throws an interesting light on the effects of soil disturbance. Seeds were buried 2 inches (5 cm) below the surface of the soil. At the time the seeds were buried they were found to be 86 per cent viable (i.e. $D_{ni} = 14$ per cent). After a year in undisturbed soil 65 per cent were still viable, i.e. there had been a 24 per cent loss from the viable portion of the population; about one-fifth of this loss was accounted for by seedlings which had emerged (D_{ge}). After the second year 55 per cent of the original number were still viable, i.e. there had been a further loss of 15 per cent of those viable at the beginning of the second year. Courtney concluded: 'It appears far more seeds were lost because they were decayed [D_{na} or D_{np}?] or germinated but failed to emerge [D_{gd}].' In a parallel experiment but in which the soil was disturbed (mixing the soil every two months from March to October) the viable seeds decreased during the first year from 86 to 38 per cent, i.e. a 56 per cent loss rather less than half of which, it was reported, could be attributed to emergence (D_{ge}). After two years only 10 per cent of the seeds originally added were still viable. In this case about half the loss was due to emergence (D_{ge}); the remainder was lost by decay or failure to emerge $(D_{na} + D_{np}? + D_{gd})$.

The detailed observations on all these species suggest that the most important factor in the depletion of weed seed populations in the soil is loss of dormancy resulting in germination which either leads to emergence (D_{ge}) or to the death of the seedling (D_{gd}). From the earlier discussion on enforced dormancy, it would be expected that there would be an increase in germination following soil disturbance – as indeed Courtney found in *Polygonum aviculare*. And although there appear to be some exceptions in which increased soil disturbance does not increase germination – e.g. in seeds of *Sonchus asper*, *Taraxacum officinale*, *Sagina procumbens*, *Juncus butonius* and *Graphalium uliginosum* (Chancellor, 1965) – nevertheless it would be expected, particularly from the work of Wesson and Wareing already described, that in general the more the soil is disturbed the more rapid would be the depletion of the seed population in the soil. This expectation has been shown to be fulfilled in practice by Dawkins and Roberts (1967) who, summarising the results of several years' work at the National Vegetable Research Station in Warwickshire, concluded that the decrease in the number of viable weed seeds (all species combined) in undisturbed soil is about 22 per cent per year; in soil dug twice a year it is 30 per cent; in soil dug four times a year it is 36 per cent;

and in soil in which frequently cultivated vegetable crops are grown it is 45 per cent per year.

The model originated by Schafer and Chilcote (1969) which has been elaborated here has emphasised that – except for the work described above – there are very few quantitative data on the relative importance of the factors which lead to the depletion of seed populations in the soil. Nevertheless it would seem from the discussion so far that loss of viability as described in Chapter 2 (D_{na}) is relatively unimportant as a cause of mortality in populations of weed seeds in the soil: loss is much more likely to be the result of germination ($D_{gd} + D_{ge}$) and possibly to some extent – though there is little evidence on this point – the result of attack by predators (D_{np}). But if, for the sake of argument, that loss due to germination and predators were unimportant then the survival curve would depend on D_{na} and we might expect a survival curve for seeds in the soil of the type described in Chapter 2, i.e. a negative cumulative normal frequency distribution. The evidence of the classical buried-seed experiments mentioned earlier would suggest that the mean viability period for this distribution would be large – many years in fact. However, the present evidence suggests that the potential period of survival, the limits of which would be set by this viability curve, is not normally reached: loss of seeds is controlled almost entirely by the secondary factors and these secondary factors ($D_{gd} + G_{ge} + D_{np}$) will control the shape of the survival curve. If we assume that the probability of a seed entering any of these categories remains more or less constant from year to year then, in any one year, the probability of a surviving seed being lost from the viable population would remain the same, irrespective of how long that seed had been in the soil. Under these conditions each year one would lose a constant proportion of those seeds which were present at the beginning of that year. The survival curve would in fact be of an identical form to that of typical bacterial survival curves or the survival of tumblers in a cafeteria as already described in equation (9.1). In these cases it does not matter how long a bacterium or a tumbler has survived, the probability of it dying, or being accidentally broken, in the next unit of time remains the same. Similarly, the relationship between time and the number of surviving seeds in the soil could be expressed as:

$$S = S_0 e^{-gt} \qquad (11.4)$$

in which S is the number of survivors at time t out of an initial population S_0, and g is a constant. The symbol g has been chosen here on the assumption that the probablity of loss due to predators is small and the most important factor in the loss of a seed is the probability of it germinating $(D_{gd} + D_{ge})$. On this basis, g may be regarded as the germination constant: the greater the probability of a seed germinating, the greater the value of g. The relationship would be seen most clearly if $\log_e$ number of survivors is plotted against time: the result would be a straight line with a negative slope, and the gradient of the slope would be proportional to g.

Now the estimation of viable seed populations in the soil is not easy, but a substantial amount of information has been obtained by H. A. Roberts and his colleagues at the National Vegetable Research Station in Warwickshire. H. A. Roberts (1962, 1963) showed that on land under cleaning crops of vegetables, the population of weed seeds declined by an approximately constant proportion of about 45 per cent per year. In other words, the survival curves are described by equation (11.4). This work confirmed and extended the work of Brenchley and Warrington (1930) who had examined the change in populations of seeds under one year of vegetable cropping on fields with different initial seed populations. Again (except in the case of extremely weedy fields which showed a greater reduction) there was a loss of approximately half the initial population after one year's cropping. In other words, as Roberts found, weed seeds have a half-life of about a year under conditions of frequent cultivation in England.* H. A. Roberts (1963) argued that at this rate of loss of weed seeds, about 7 years of cleaning crops would be required for the seed population to fall to 1 per cent of that originally present. Assuming

---

*Investigations carried out by Chepil (1946) on five common weeds in the prairie region of Western Canada suggested that the loss of viable seeds from soil populations occurs more rapidly in these circumstances – 75 per cent or more seeds were lost after one year in undisturbed soil. However, this work is not strictly comparable because the experiments involved the incorporation of known numbers of seeds into sterilised soil rather than investigations on the behaviour of natural populations of seeds in the soil. Consequently it is not clear whether the different rates of depletion found in England and Canada are due to different techniques or the result of the different environments. Chepil's work is particularly interesting in other respects: it showed that when seeds were left at the surface, practically all were lost after one year, presumably because of lack of enforced dormancy. Practically all the seeds of one species, *Salsolla pestifer*, were also lost even when incorporated in the soil to a depth of 6 inches (15 cm) because of an incapability of developing enforced dormancy. The others (*Thlaspi arvense, Sinapsis arvensis, Sisymbrium altissimum* and *Amaranthus retroflexus*) all developed enforced dormancy so that, on average, 25.2, 16.5, 27.3, and 24.8 per cent of the initial population respectively survived one year when the soil was undisturbed; simulated ploughing, followed by several surface cultivations, resulted in corresponding values of 15.8, 8.0, 22.3 and 10.4 per cent.

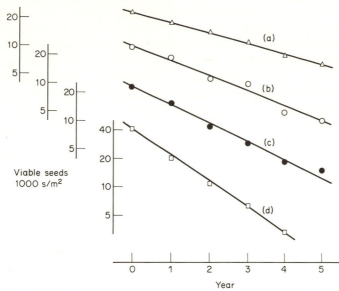

FIGURE 11.8   Survival curves showing the number of viable seeds in the soil
(to the depths indicated) in the absence of further seeding: (a) undisturbed soil,
23 cm; (b) cultivated twice a year, 23 cm; cultivated four times a year, 23 cm;
(d) cultivations involved in rotation of vegetable crops, 15 cm. (From H. A.
Roberts, 1970.)

an initial population of 50,000,000 seeds/acre to a depth of 6 inches
(15 cm) (i.e. a not exceptionally weedy field), after 7 years there would
be 100 seeds/yard². Of these, only a small proportion would be
expected to germinate and emerge at any one time (because of enforced dormancy), and thus he suggests that a weed population at this
level would result in a weed stand of perhaps 2–3 plants/yard². At
this density the weeds might well have no competitive effect on the
crop. However, unless the degree of crop competition were sufficient
to ensure that few seeds were produced, there might be a rapid increase
in the seed population. Accordingly he pointed out that his analysis
gives some experimental and theoretical support to the old adage,
'One year's seeding: seven years' weeding!'.

As mentioned previously, Dawkins and Roberts (1967) concluded
that the rate of depletion of seeds depends on the frequency of soil
disturbance (Fig. 11.8). Thus using their figures for rates of depletion
per year of the soil population – 22 per cent for uncultivated soil,
30 per cent for soil dug twice a year, 36 per cent for soil dug four
times a year, and 45 per cent for frequently cultivated soil – one can
calculate the value of g in equation (11.4) to be 0.248, 0.357, 0.446
and 0.598, respectively, under these four conditions of soil distur-

bance. It must be remembered, of course, that in these cases g is a composite constant for a typical combination of common British weed species; each species itself will probably have its own individual constant for a particular cultivation regime. There is evidence (H. A. Roberts, 1962) that under frequent cultivation quite big differences may be found between weed species in the rates at which their buried seed population decline, and there is reasonable agreement for the depletion of any one species at three sites in England – Rothamsted, Woburn and Wellesbourne (Roberts, 1958). Nevertheless quite a large number of species apparently have a similar value of g to the composite value encompassing all species. Figure 11.9 shows the survival of seeds of a number of common British weed species under two different regimes of soil disturbance. In these experiments, on undisturbed plots the rates of decrease of viable seeds of *Chenopodium album*, *Capsella bursa-pastoris*, *Poa annua* and *Solanum nigrum* were similar, from 21 per cent per year ($g = 0.236$) to 26 per cent per year ($g = 0.301$); for *Stellaria media* a rather higher value of 30 per cent a year ($g = 0.357$) was recorded (H. A. Roberts, 1967). On the plots dug twice a year (Fig. 11.9a) and on those dug four times (Fig. 11.9b), the rate of decrease of seeds of *Stellaria media* was also greater than that of other species and on the plots dug four times a year was equivalent to 56 per cent a year ($g = 0.821$). There were only small differences in rates of the other four species although, as would be expected, the decrease was rather more rapid on plots dug four times a year than

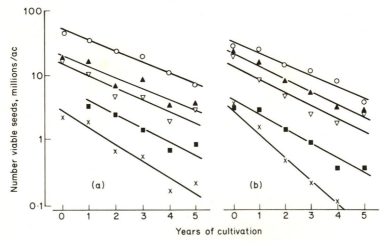

FIGURE 11.9 Numbers of viable seeds of individual species in the top 9 in. (23 cm) of soil on (a) plots dug twice a year and on (b) plots dug four times a year. ○, *Chenopodium album*; ▲, *Capsella bursa-pastoris*; ▽, *Poa annua*; ■, *Solanum nigrum*; ×, *Stellaria media*. (From H. A. Roberts, 1967.)

on those dug only twice. However, since the seed numbers of the individual species in this experiment were often small, H. A. Roberts (1967) suggests that no great significance can be attached to differences between the individual species, with the exception of the difference between *Stellaria* and the rest.

The main practical objective in attempting to understand the factors controlling seed survival in the soil is to develop efficient methods of weed control. From the arguments developed in this chapter it would seem that two complementary approaches will be particularly useful in this area in the future. First we need further information on the quantitative importance of the various components in equation (11.3) in controlling the value *g* in equation (11.4) under different conditions. Secondly, since the present evidence suggests that the most important factor in the depletion of weed seed populations in the soil is germination – D_{gd} and D_{ge} in equation (11.3) – we need to understand dormancy mechanisms in order to devise methods of breaking dormancy of weed seeds in the soil. The evidence from the few cases which have been studied suggests that seeds which germinate beneath the ground do so because of a loss of enforced dormancy (seeds enter category D_{gd} or D_{ge} from P_{enf}), and it is therefore the study of the mechanisms of enforced dormancy which is likely to yield the most useful results.

References to Chapter 11

BARTON, L. V., 1961. *Seed preservation and longevity*, 39. Leonard Hill (Books) Ltd, London.

BERRIE, A. M. M., 1964. From seed to seedling. *Times Sci. Rev. Winter, 1964*, 12–13.

BIBBEY, R. O., 1948. Physiological studies of weed seed germination. *Pl. Physiol., Lancaster*, **23**, 467–84.

BLACK, M., 1970. Seed germination and dormancy. *Sci. Prog., Oxf.*, **58**, 379–93.

BORTHWICK, H. A., HENDRICKS, S. B., PARKER, M. W., TOOLE, E. H., and TOOLE, V. K., 1952. A reversible photoreaction controlling seed germination. *Proc. Nat. Acad. Sci.*, **38**, 662–66.

BORTHWICK, H. A., HENDRICKS, S. B., TOOLE, E. H., and TOOLE, V. K., 1954. Action of light on lettuce-seed germination. *Bot. Gaz.*, **115**, 205–25.

BRENCHLEY, W. E., 1918. Buried weed seeds. *J. agric. Sci.*, **9**, 1–31.

BRENCHLEY, W. E., and WARRINGTON, K., 1930. The weed seed population of arable soil. I. Numerical estimation of viable seeds and observations on their natural dormancy. *J. Ecol.*, **18**, 235–72.

CALDWELL, F., 1959. Some notes on dormancy in cereal grains. *Agric. Hort. Engng Abstr.*, **9**, 189–92.

CAVERS, P. B., and HARPER, J. L., 1966. Germination polymorphism in *Rumex crispus* and *R. obtusifolius*. *J. Ecol.*, **54**, 367–82.

CHANCELLOR, R. J., 1965. Emergence of weed seedlings in the field and the effects of different frequencies of cultivation. *Rep. Seventh Brit. Weed Control Conf.*, 599–606.

CHEPIL, W. S., 1946. Germination of weed seeds. II. The influence of tillage treatments on germination. *Sci. Agric.*, **26**, 347–57.

COURT, M. N., STEPHEN, R. C., and WAID, J. S., 1962. Nitrite toxicity arising from the use of urea as a fertiliser. *Nature, Lond.*, **194**, 1264–65.

COURTNEY, A. D., 1968. Seed dormancy and field emergence in *Polygonum aviculare*, *J. appl. Ecol.*, **5**, 675–84.

CROCKER, W., 1916. Mechanics of dormancy in seeds. *Amer. J. Bot.*, **3**, 99–120.

CROCKER, W., 1948. *Growth of plants; twenty years research at Boyce Thompson Institute.* Reinhold, New York.

CUMMING, B. G., 1963. The dependence of germination on photoperiod, light quality, and temperature, in *Chenopodium* spp. *Can. J. Bot.*, **41**, 1211–33.

DARLINGTON, H. T., 1951. The seventy-year period for Dr Beal's seed viability experiment. *Amer. J. Bot.*, **38**, 379–81.

DAWKINS, P. A., and ROBERTS, H. A., 1967. Weed seed populations. *Rep. Nat. Vegetable Res. Station, Wellesbourne, 1966*, 75–76.

ESASHI, Y., and LEOPOLD, C., 1968. Physical forces in dormancy and germination of *Xanthium* seeds. *Pl. Physiol., Lancaster*, **43**, 871–76.

FELTNER, K. C., and VESECKY, J. F., 1968. Light quality and temperature effects on weed seed germination in two Kansas soils. *Trans. Kansas Acad. Sci.*, **71**, 7–12.

FLINT, L. H., and MCALISTER, E. D., 1935. Wave length of radiation in the visible spectrum inhibiting the germination of light sensitive lettuce seed. *Smithsonian Inst. Misc. Collections*, **94**, 1–11.

FLINT, L. H., and MCALISTER, E. D., 1937. Wave length of radiation in the visible spectrum promoting the germination of light sensitive lettuce seed. *Smithsonian Inst. Misc. Collections*, **96**, 1–8.

HARPER, J. L., 1957. The ecological significance of dormancy and its importance in weed control. *Proc. 4th Int. Congr. Crop Protection, Hamburg 1957, Vol. 1*, 415–20. Braunschweig, Hamburg.

HART, J. W., and BERRIE, A. M. M., 1966. The germination of *Avena fatua* under different gaseous environments. *Physiol. Pl.*, **19**, 1020–25.

HAY, J. R., 1962. Experiments on the mechanism of induced dormancy in wild oats, *Avena fatua* L. *Can. J. Bot.*, **40**, 191–202.

HAY, J. R., and CUMMING, B. G., 1959. A method for inducing dormancy in wild oats, *Avena fatua* L. *Weeds*, **7**, 34–40.

HENDRICKS, S. B., 1968. How light interacts with living matter. *Scientific American*, **219** (3), 174–86.

HENSON, I. E., 1970. The effects of light, potassium nitrate and temperature on the germination of *Chenopodium album* L. *Weed Res.*, **10**, 27–39.

HUXLEY, P. A., and TURK, A., 1966. Factors which affect the germination of six common East African weeds. *Expl. Agric.*, **2**, 17–25.

KARSSEN, C. M., 1970. The light promoted germination of the seeds of *Chenopodium album* L. III. Effect of the photoperiod during growth and development of the plants on the dormancy of the produced seeds. *Acta Bot. Neerl.*, **19**, 81–94.

KERNICK, M. D., 1961. *Agricultural and Horticultural Seeds*, 89, F.A.O., Rome.

KIDD, F., 1914. The controlling influences of carbon dioxide in the maturation, dormancy and germination of seeds. *Proc. R. Soc.*, B, **87**, 408–21, 609–25.

KJAER, A., 1948. Germination of buried and dry stored seed. II. 1934–44. *Proc. int. Seed Test. Ass.*, **14**, 19–26.

KOLLER, D., 1964. The survival value of germination-regulating mechanisms in the field. *Herb. Abstr.*, **34**, 1–7.

KOLLER, D., 1969. The physiology of dormancy and survival of plants in desert environments. *Symp. Soc. exp. Biol.*, **23**, 449–69.

LEWIS, J., 1958. Longevity of crop and weed seeds. I. First interim report. *Proc. int. Seed Test. Ass.*, **23**, 340–54.

LEWIS, J., 1960. The influence of water table, soil depth and soil type on the survival of certain crop and weed seeds. *Rep. Welsh Plant Breeding Station, 1959*, 60.

LIPP, A. E. G., and BALLARD, L. A. T., 1959. The breaking of seed dormancy of some legumes by carbon dioxide. *Aust. J. agric. Res.*, **10**, 495.

MAJOR, W., 1966. *Investigations into the physiology of dormancy in cereal seeds.* Ph.D. thesis, University of Manchester.

MAJOR, W., and ROBERTS, E. H., 1968. Dormancy in cereal seeds. II. The nature of the gaseous exchange in imbibed barley and rice seeds. *J. exp. Bot.*, **19**, 90–101.

MAYER, A. M., and POLJAKOFF-MAYBER, A., 1963. *The germination of seeds.* Pergamon Press, Oxford.

MOUTON, J. A., 1960. La dormance chez *Oryza sativa* L. Revision bibliographique. *J. Agric. Trop. Bot. Appliq.*, **7**, 588–90.

NEGBI, M., and TAMARI, B., 1963. Germination of chlorophyllous and achlorophyllous seeds of *Salsolla volkensii* and *Aellenia autrani. Israel J. Bot.*, **12**, 124–35.

OWEN, E. B., 1956. *The Storage of Seeds for the Maintenance of Viability.* Commonwealth Agric. Bureaux, Farnham Royal, England.

POPAY, A. I., and ROBETS, E. H., 1970a. Factors involved in the dormancy and germination of *Capsella bursa-pastoris* (L.) Medik. and *Senecio vulgaris* L. *J. Ecol.*, **58**, 103–22.

POPAY, A. I., and ROBERTS, E. H., 1970b. Ecology of *Capsella bursa-pastoris* (L.) Medik. and *Senecio vulgaris* L. in relation to germination behaviour. *J. Ecol.*, **58**, 123–39.

RAMIAH, K., 1937. *Rice in Madras.* Government Press, Madras.

RAMPTON, H. H., and CHING, T. M., 1970. Persistence of crop seeds in soil. *Agron J.*, **62**, 272–77.

ROBERTS, E. H., 1961. The viability of rice seed in relation to temperature, moisture content, and gaseous environment. *Ann. Bot.*, **25**, 381–90.

ROBERTS, E. H., 1962. Dormancy in rice seed. III. The influence of temperature, moisture, and gaseous environment. *J. exp. Bot.*, **13**, 75–94.

ROBERTS, E. H., 1963a. An investigation of inter-varietal differences in dormancy and viability of rice seeds. *Ann. Bot.*, **27**, 365–69.

ROBERTS, E. H., 1963b. The effect of inorganic ions on dormancy in rice seeds. *Physiol. Pl.*, **16**, 732–44.

ROBERTS, E. H., 1965. Dormancy in rice seed. IV. Varietal responses to storage and germination temperatures. *J. exp. Bot.*, **16**, 341–49.

ROBERTS, E. H., 1969. Seed dormancy and oxidation processes. *Symp. Soc. exp. Biol.*, **23**, 161–92.

ROBERTS, H. A., 1958. Studies on the weeds of vegetable crops. I. Initial effects of cropping on the weed seeds in the soil. *J. Ecol.*, **46**, 759–68.

ROBERTS, H. A., 1962. Studies on the weeds of vegetable crops. II. Effect of six years of cropping on the weed seeds in the soil. *J. Ecol.*, **50**, 803–13.

ROBERTS, H. A., 1963. The problem of weed seeds in the soil. In *Crop production in a weed-free environment*, ed. E. K. Woodford, 73–82. Blackwells, Oxford.

ROBERTS, H. A., 1964. Emergence and longevity in cultivated soil of seeds of some annual weeds. *Weed Res.*, **4**, 296–307.

ROBERTS, H. A., 1967. Effect of cultivation on the numbers of viable weed seeds in soil. *Weed Res.*, **7**, 290–301.

ROBERTS, H. A., 1970. Viable weed seeds in cultivated soils. *Rep. natn. Veg. Res. Stn, Wellesbourne, 1969*, 25–38.

RUSSELL. E. W., 1961. *Soil conditions and plant growth*, 9th ed., Longmans, London.

SAGAR, G. R., and HARPER, J. L., 1960. Factors affecting the germination and early establishment of plantains (*Plantago lanceolata, P. media* and *P. major*). In *The Biology of Weeds*, ed. J. L. Harper, 236–45. Blackwell, Oxford.

SAHADEVAN, P. C., 1953. Studies on the loss of viability of rice seeds in storage. *Madras agric. J.*, **40**, 133–43.

SALISBURY, E., 1961. *Weeds and aliens*. Collins, London.

SAUER, J., and STRUIK, G., 1964. A possible ecological relation between soil disturbance, light flash, and seed germination. *Ecology*, **45**, 884–86.

SCHAFER, D. E., and CHILCOTE, D. O., 1969. Factors influencing persistence and depletion in buried seed populations. I. A model for analysis of parameters of buried seed persistence and depletion. *Crop Sci.*, **9**, 417–19.

SCHAFER, D. E., and CHILCOTE, D. O., 1970. Factors influencing persistence and depletion in buried seed populations. II. The effects of soil temperature and moisture. *Crop Sci.* **10**, 342–45.

SIEGELMAN, H. V., 1969. Phytochrome. In *Physiology of Plant Growth and Development*, ed. M. B. Wilkins, 489–506. McGraw-Hill, London.

SIEGELMAN, H. W., and FIRER, E. M., 1964. Purification of phytochrome from oat seedlings. *Biochem.*, **3**, 418–23.

STEINBAUER, G. P., and GRIGSBY, B., 1957. Interaction of temperature, light and moistening agent in the germination of weed seeds. *Weeds*, **5**, 175–82.

TAYLORSON, R. B., 1969. Photocontrol of rough cinquefoil seed germination and its enhancement by temperature manipulation and KNO_3. *Weed Sci.*, **17**, 144–47.

TAYLORSON, R. B., 1970. Changes in dormancy and viability of weed seeds in soils. *Weed Sci.*, **18**, 265–69.

TAYLORSON, R. B., and BORTHWICK, H. A., 1969. Light filtration by foliar canopies; significance for light-controlled weed seed germination. *Weed Sci.*, **17**, 48–51.

TAYLORSON, R. B., and MCWHORTER, C. G., 1969. Seed dormancy and germination of ecotypes of Johnsongrass. *Weed Sci.*, **17**, 359–61.

THOMPSON, P. A., 1969. Germination of *Lycopus europaeus* L. in response to fluctuating temperatures and light. *J. exp. Bot.*, **20**, 1–11.

THORNTON, N. C., 1936. Carbon dioxide storage. IX. Germination of lettuce seeds at high temperatures in both light and darkness. *Contrib. Boyce Thompson Inst.*, **8**, 25.

THURSTON, J. M., 1960. Dormancy in weed seeds. In *The biology of weeds*, ed. J. L. Harper, 69–82. Blackwell, Oxford.

TOOLE, E. H., and BROWN, E., 1946. Final results of the Duvel buried seed experiment. *J. agric. Res.*, **72**, 201–10.

TOOLE, E. H., and TOOLE, V. K., 1941. Progress of germination of seed of *Digitaria* as influenced by germination temperature and other factors. *J. agric. Res.*, **63**, 65–90.

TOOLE, E. H., HENDRICKS, S. B., BORTHWICK, H. A., and TOOLE, V. K., 1956. Physiology of seed germination. *Ann. Rev. Pl. Physiol.*, **7**, 299–324.

TOOLE, E. H., TOOLE, V. K., BORTHWICK, H. A., and HENDRICKS, S. B., 1955a. Photocontrol of *Lepidium* seed germination. *Pl. Physiol.*, *Lancaster*, **30**, 15–21.

TOOLE, E. H., TOOLE, V. K., BORTHWICK, H. A., and HENDRICKS, S. B., 1955b. Interaction of temperature and light in germination of seeds. *Pl. Physiol.*, *Lancaster*, **30**, 473–78.

TOOLE, E. H., TOOLE, V. K., HENDRICKS, S. B., and BORTHWICK, H. A., 1957. Effect of temperature on germination of light-sensitive seeds. *Proc. int. Seed Test. Ass.*, **22**, 1–9.

TOOLE, V. K., and TOOLE, E. H., 1953. Seed dormancy in relation to seed longevity. *Proc. int. Seed Test. Ass.*, **18**, 325–28.

TOOLE, V. K., BAILEY, W. K., and TOOLE, E. H., 1964. Factors influencing dormancy of peanut seeds. *Pl. Physiol.*, *Lancaster*, **39**, 822–32.

TSENG, S. T., 1964. Breaking dormancy in rice seed with carbon dioxide. *Proc. int. Seed Test. Ass.*, **29**, 45.

VEGIS, A., 1963. Climatic control of germination, bud break, and dormancy. In *Environmental Control of Plant Growth*, ed. L. T. Evans, 265–87. Academic Press, New York.

WATKINS, F. B., 1966. Effect of nitrogen fertilizer on the emergence of wild oat (*Avena ludoviciana*). *Queensland J. agric. animal Sci.*, **23**, 87–89.

WELLS, P. V., 1959. Ecological significance of red light sensitivity in tobacco. *Science*, **129**, 41–42.

WENT, F. W., 1957. *Experimental control of plant growth*. Chronica Botanica Co., Waltham, Mass.

WESSON, G., and WAREING, P. F., 1967. Light requirements of buried seeds. *Nature, Lond.*, **213**, 600–1.

WESSON, G., and WAREING, P. F., 1969a. The role of light in the germination of naturally occurring populations of buried weed seeds. *J. exp. Bot.*, **20**, 402–13.

WESSON, G., and WAREING, P. F., 1969b. The induction of light sensitivity in weed seeds by burial. *J. exp. Bot.*, **20**, 414–25.

WILLIAMS, J. T., and HARPER, J. L., 1965. Seed polymorphism and germination. I. The influence of nitrates and low temperatures on the germination of *Chenopodium album*. *Weed Res.*, **5**, 141–50.

WILLIAMS, M. C., 1960. Biochemical analyses, germination, and production of black and brown seed of *Halogeton glomeratus*. *Weeds*, **8**, 452–61.

WULFF, R., and MEDINA, E., 1969. Germination of seeds in *Jussiaea suffruticosa*. *Pl. Cell Physiol.*, **10**, 503–11.

CHAPTER 12

Control mechanisms in the resting seed

H. Thomas

This review deals with the biochemical control mechanisms of seeds in the period of development between late maturation, when the metabolism of the seed is progressively inactivated, and early germination when seed metabolism becomes active again. The use of the term 'resting', when applied to plants and their organs, has been anything but consistent; in the present discussion the term will be applied in a general sense to seeds in which germination is prevented either because they have an inadequate water content or because there is some metabolic block.

Seed development

The mature condition in seeds is characterised by a low metabolic activity, a high level of resistance to adverse conditions such as desiccation, and a store of reserves sufficient to meet the needs of germination. The control of seed development can thus be conveniently considered from three distinct aspects: the control of the entry of the embryo into the quiescent (or dormant) condition so that it is resistant to unfavourable conditions; the control of the accumulation of reserve materials; and finally the biochemical organisation of the mature dry seed.

CHANGES IN THE METABOLIC ACTIVITY OF MATURING SEEDS AND THE DEVELOPMENT OF RESISTANCE TO ADVERSE CONDITIONS

The completion of seed maturation is marked by a period of rapid dehydration, during which the water content of the seed falls to a very low level. This process has profound effects on the ultrastructure of seed tissue. Accompanying loss of water from the seed is a decrease in respiration and a fall in protein synthesis (Kollöffel, 1970; Marré, 1967). Undoubtedly, many of the biochemical and cytological changes that accompany removal of water from the seed are a direct result of the reduced availability of the aqueous medium in which many enzymes and cell structures ordinarily function. On the other hand it will be shown that some of the changes occurring in the period leading to desiccation may have a significance beyond their immediate involvement in the development of the viable, mature, biochemically inactive seed.

Marked changes in cell organisation occur as the seed matures. From studies of the ultrastructure of developing seeds of species such as pea (Bain and Mercer, 1966a), *Phaseolus vulgaris* (Öpik, 1968), *Vicia faba* (Briarty, Coult and Bouter, 1969), lima bean (Klein and Pollock, 1968) and cotton (Yatsu, 1965), it is apparent that cell nuclei, mitochondria, plastids, endoplasmic reticulum (ER) and dictyosomes change from a relatively undeveloped state in early embryogenesis to a highly active state during reserve accumulation and finally decrease in activity to become indistinct in the desiccated seed. These changes are direct expressions of changes in the state of cell membranes (Bain and Mercer, 1966a) and appear to be active and developmentally significant processes related in some way to lipid accumulation (Bain and Mercer, 1966a; Öpik, 1968). In order to assess the developmental significance of desiccation it is appropriate to enquire whether these cytological changes can be related to biochemical changes in dehydrating seeds.

In castor bean and maize O_2 uptake increases as the dry weight of the seed increases early in development and then falls gradually during the remainder of the maturation period (Marré, 1967; Dalby and Davies, 1967). A similar pattern of change is shown by the RQ which rises from 1 to 1.5 early in embryogenesis and subsequently falls to 0.7 to 0.8 (Marré, 1967). These changes cannot be explained simply on the basis of changes in respiratory substrate and mitochondrial development: although most of the soluble and mitochondrial enzyme activities increase and decrease as O_2 consumption and RQ increase and decrease, there occur changes in the levels of certain glycolytic intermediates which suggest that there is a differential inactivation of glycolytic enzymes during maturation (Marré, 1967).

The observed gradual fall in RQ and O_2 consumption occurs during the constant fresh weight phase of seed development. Early in this phase the metabolic activity of the seed begins to fall, a process that seems to be genetically programmed. The nature of the stimulus that operates the genetic switch is unknown. It has been observed, however, that in the constant fresh weight phase there is often a significant fall in seed water content (Klein and Pollock, 1968; Marré, 1967) and it is possible that the consequent gradual increase in the

degree of water stress to which the seed is subject is partially responsible for the respiratory changes observed in seeds at this stage.
Changes in mitochondrial activity that occur during dehydration may
well be directly related to the degree of swelling of the mitochondria.
It has been demonstrated that there is a close relationship between
mitochondrial activity and the extent to which mitochondria are
swollen or contracted (Stoner and Hanson, 1966). Furthermore, not
all respiratory enzymes are affected equally by dehydration, so that
the differential changes in the activity of respiratory enzymes observed
during the constant fresh weight phase of seed development may be a
reflection of this differential response to water stress. For example,
Kollöffel (1970a) has shown that the activities of a number of mitochondrial enzymes in pea seeds actually increase when the relative
water content of the cotyledon decreases from 65 to 55 per cent while
further reduction in water content causes a sharp reduction in the
activities of succinate and malate oxidases but only a slight decrease
in succinate and malate dehydrogenase activity. The mitochondria of
roots are known to be sensitive to water stress (Nir, Poljakoff-Mayber
and Klein, 1970a). Furthermore, an initial rise and subsequent fall in
respiration and enzymic activity has been observed in water-stressed
organs such as wilting leaves (Vaadia, Raney and Hagan, 1961; Crafts,
1968). In other words, it is suggested that the maturing seed is subjected to a water stress severe enough to contribute to marked changes
in respiratory activity during maturation.

Marré (1967) lists a number of enzymes that decline in activity
during the constant fresh weight phase of castor bean development.
These include enzymes of the glycolytic (EMP) pathway, pentose
phosphate (PP) pathway, Krebs (TCA) cycle and cytochrome system.
In addition, there is a pronounced decrease in the activity of both the
amino acid activating enzymes and the soluble enzymes of protein
synthesis during the final stage of maturation; changes in these enzymes may, however, be peculiar to the seeds of some species and not
of others, since the protein synthesising systems of dry peanut cotyledons and wheat embryos seem to lack only translatable messenger
RNA (mRNA) and are able to translate synthetic messenger such as
polyuridylic acid (polyU) with much greater activity than the system
from castor bean (Marcus and Feeley, 1964, 1965; Sturani, 1968).
In the dehydration phase of castor bean maturation not only is there
a breakdown of polysomes such as has been observed in a number of
seeds such as pea (Bain and Mercer, 1966a), *Vicia faba* (Payne and
Boulter, 1969a), *Phaseolus vulgaris* (Walbot, 1970) and lima bean
(Klein and Pollock, 1968), but there appears to be damage to the ribo-

somes themselves. Abnormal ribosomes have been noted in the cotyledons of germinating *Pisum arvense* (Barker and Hollinshead, 1965) in which protein synthesis proceeds very slowly and in the cotyledons of germinating *Pisum sativum*, where levels of polysomes are so low as to be virtually undetectable (Thomas, unpublished).

There is much evidence that the decrease in enzyme activities during maturation is due to decreased synthesis coupled with normal turnover, and that decreased synthesis is, in turn, due to changes in mRNA metabolism and, in some seeds, ribosome function. These changes have been suggested by Öpik (1968) to be related to an increase in ribonuclease (RNase) activity (Ingle, Beitz and Hageman, 1965; Dalby and Davies, 1967). Dove (1967) showed that water-stressed tomato leaflets had a higher RNase activity than unstressed leaflets. Changes in ER and a loss of membrane-bound polysomes during dehydration (Payne and Boulter, 1969a) may account for some of the qualitative alterations in protein synthesis occuring in late maturation. Loss of polysomes has been observed in a number of water-stressed tissues such as maize coleoptiles (Hsiao, 1970), and maize roots (Nir *et al.*, 1970b) but on the other hand dehydration often has the effect not of disorganising ER but of organising normally vesicular ER into parallel lamellae (Schnepf, 1961). Thus, in general, it is possible to explain in terms of water stress effects the proposal (Marré, 1967) that decreased enzyme activities during maturation are due very largely to decreased enzyme synthesis.

An alternative view is that there is conversion of many enzymes to an inactive form during maturation. For example, Mayer and Shain (1968) have demonstrated that zymogen-like granules can be extracted from peas and that they represent an inactive form of a glucosidase that can be activated by mild proteolysis. Although it appears probable that enzymes can react to water stress by becoming converted to such inactive forms, evidence for this in other tissues is scanty. The true situation existing in seeds is possibly one in which decreased protein synthesis, increased protein breakdown, reversible conversion of enzymes to inactive forms and inactivation due simply to a scarcity of aqueous medium all contribute to the run-down in activity during desiccation.

The changes that occur in respiratory activity during the constant fresh weight period suggest that desiccation-resistant respiratory systems may be 'selected' in preparation for total inactivation in the dry seed. Marked changes in seed ribosomal systems occur only towards the end of the constant fresh weight period, indicating that these systems are inherently resistant to a certain degree of water stress (Kessler, Engelberg, Chen and Greenspan, 1964). Chen, Sarid and Katchalski (1968b) have shown that wheat embryos lose much of their drought-resistant character about 72 hours after the start of imbibition when, it is claimed, new non-resistant RNA's begin to be synthesised. Gates (1964) has pointed out that embryonic tissues are highly resistant to water loss, but their synthetic capacity may be affected by desiccation. In summary, changes in the physiological activity of the maturing seed are, as Klein and Pollock (1968) have argued, adaptations which render the cells resistant to desiccation. It would be interesting to know whether in other respects water stress in seeds resembles water stress in such organs as leaves. For example, it is known that hormone levels in leaves change during wilting (Wright and Hiron, 1969). Are hormone levels in seeds at least partly deter-mined by hormonal changes occurring in response to water stress during development?

The proteins synthesised by maturing seeds are very different from those made by germinating seeds (see p. 365 and p. 376); presumably, therefore, the mRNA's of the maturing seed are different from those of the germinating seed. It follows that at some stage during seed development there must be destruction of old messengers and a synthesis (and possible storage) of messengers essential for protein synthesis in germination. The breakdown of polysomes observed in desiccating seeds is an indication of messenger destruction (Öpik, 1968). There is evidence that, during late maturation, germination-specific mRNA's are synthesised and stored in 'masked' froms (see p. 381). Although there is no evidence that this qualitative change in RNA metabolism is directly related to seed desiccation it is perhaps significant that in water-stressed leaves the RNA synthesised has a base composition different from that of RNA synthesised in fully hydrated leaves (Kessler and Frank-Tishel, 1962; Kessler *et al.*, 1964). These workers point out that structural features of the RNA molecules of stressed leaves, such as a high (G + C) content, and possibly double-strandedness, make these molecules particularly stress resistant. That there is a relationship between the resistance of RNA to water stress and its stability in germinating seeds has been pointed out by Chen *et al.* (1968b).

There must obviously be changes at the gene level leading to these changes in mRNA metabolism during maturation. Furthermore, in some cases there is evidence that in dormancy, which develops as the seed matures, there is repression of chromatin activity (see p. 371). It is possible that desiccation, by directly affecting cell nuclear volume and the osmotic and ionic environment of chromatin, causes the ordered opening up or closing down of genetic areas, a control mechanism that appears to operate in heterokaryons and that has been suggested by Harris (1968) to operate for eukaryotes in general.

THE CONTROL OF THE SYNTHESIS OF RESERVE PROTEIN

Many of the mechanisms that control the synthesis of reserve protein may operate in the regulation of enzyme synthesis in developing seeds. For this reason, discussion of the accumulation of seed reserves is confined to reserve protein. The control of the synthesis of other reserves has been adequately covered elsewhere: Turner (1969a and b) and Jenner (1970) have reviewed the subject of starch synthesis, and Mudd (1967), Drennan and Canvin (1969), and Harris and James (1969) have considered in detail various aspects of lipid synthesis in seeds.

The subject of protein reserves in seeds has recently been reviewed by Altschul, Yatsu, Ory and Engleman (1966) and Mosse (1968). Studies with light and electron microscopes have shown the presence of protein bodies in seeds. Correlation of these studies with work on the chemistry of seed proteins has shown that much of the storage protein of seeds is located in these bodies. All such protein bodies will be referred to as 'aleurone grains' (Altschul *et al.*, 1966).

The development of aleurone grains has been studied by electron microscopy in several species. These studies have revealed that the cells of immature embryos are meristematic and contain no vacuoles. Vacuolation, which marks the end of meristematic activity and onset of cell maturation, is accompanied by the appearance of protein deposits associated with the developing tonoplasts. Vacuoles grow by increase in volume, or coalescence of smaller vacuoles, or both, and finally occupy a large proportion of total cell volume. Protein eventually fills the vacuole, which retains its membrane (Briarty *et al.*, 1969; Bain and Mercer, 1966a; Öpik, 1968; Engleman, 1966; Buttrose, 1963a and b; Jennings *et al.*, 1963).

Aleurone grains have been shown to contain a number of enzymes and inclusions other than storage protein, such as phytate (Lui and Altschul, 1967; Ory and Henningsen, 1969) and enzymes characteristic of lysosomes (Yatsu and Jacks, 1968; Matile, 1968).

A number of hypotheses have been advanced to account for the origins of aleurone grain protein (Buttrose, 1963b). Of these, only two have received support from detailed biochemical studies. Firstly, it is possible that the storage proteins are synthesised within developing vacuoles by a protein synthesising system independent of the cytoplasmic system. Graham, Morton and Raison (1963) showed that isolated aleurone bodies from wheat incorporated ^{35}S sulphate and ^{14}C-glycine. The unit synthesising storage protein was called the 'proteoplast' by Morton and Raison (1963, 1964), who demonstrated a possible function of aleurone grain phytate by isolating a phospho-transferase preparation from wheat grain catalysing the formation of ATP from ADP and phytate. They also showed that this ATP-generating system could supply energy for amino acid incorporation by cell free proteoplast preparations from wheat. It seems clear now, however, that aleurone bodies able to incorporate radioactive amino acids are contaminated by microorganisms. Wilson (1966) showed that sterile aleurone grains have almost no capacity for ^{14}C leucine incorporation.

The second suggested origin of aleurone protein is that the protein is synthesised on membrane-bound polyribosomes and transferred to developing vacuoles through the ER. Rough ER has been shown to increase during the period of accumulation of protein in vacuoles (Briarty, Coult and Boulter, 1969; Bain and Mercer, 1966a; Öpik, 1968). Payne and Boulter (1969a) separated free (f) and membrane-bound (mb) polyribosomes from developing *Vicia* seeds by sucrose density gradient centrifugation. They showed a high f/mb ratio at the phase of rapid cell division which fell during the accumulation of protein. This was due to a preferential synthesis of mb ribosomes rather than the conversion of f ribosomes to mb ribosomes. Both classes of ribosome are active in protein synthesis. Bailey, Cobb and Boulter (1970) described a cotyledon slice system for demonstrating the transfer of protein from rough ER into developing protein vacuoles by electron microscope autoradiography. Although there were difficulties in pulse-chasing the tissue they showed that *Vicia* cotyledon slices, when fed labelled amino acid, transferred label from ER to protein vacuoles. The transfer time was estimated to be about 25 minutes.

A coarse control of reserve protein synthesis may be effected by the supply of amino acids to the developing seed. Folkes (1970) has proposed that the supply of amino acids from the breakdown of leaf proteins, and the proportions in which the individual amino acids are supplied to the seed controls protein synthesis in the early stages of seed development. Another coarse control mechanism is suggested by the work of Wheeler and Boulter (1967) and Walbot (1970). During rapid embryonic growth there is a marked increase in the RNA and ribosome content of the seed, but it appears that cell division, RNA increase and the differentiation of major organs must be complete before storage products can be synthesised. A number of workers have suggested that this is due simply to a competition for precursors by systems concerned with differentiation and those concerned with the accumulation of reserves (Commoner, 1964; Grzesiuk, Mierzwinska and Sojka, 1962). A fall in total nucleic acid levels has been observed in wheat endosperm (Jennings and Morton, 1963) and in maize seeds (Ingle *et al.*, 1965) as storage protein synthesis commences. The latter workers showed that an RNase increases during endosperm development, and it might be that this increase regulates RNA levels in the seed.

The association of ribosomes with membranes during seed development may exert a controlling influence on the synthesis of reserve protein. Since there is a change from the synthesis of non-storage to the synthesis of storage protein during development there must be a switching-off of the synthesis of some kinds of RNA and a switching-on of the synthesis of others. Furthermore, since the phase of storage protein synthesis coincides with the phase of mb ribosome production then it is reasonable to assume that new RNA is associated with mb ribosomes. Rough ER has been observed connected with the nuclear membrane (Buttrose, 1963a; Briarty *et al.*, 1969) and there is electron microscopic evidence of considerable nuclear activity and changes in the nuclear membrane during the phase of storage protein synthesis (Öpik, 1968; Bain and Mercer, 1966a; Briarty *et al.*, 1969). The advantages of mb polyribosomes to the cell have been reviewed by Campbell (1970). Membrane-bound polyribosomes are particularly well adapted for the synthesis of protein that is to be moved elsewhere in the cytoplasm. In a number of animal tissues such as the mammary gland (Gaye and Denamur, 1970) and rat liver (Takagi, Tanaka and Ogata, 1970) it has been shown that the protein synthesised on mb polyribosomes are different from the proteins synthesised on f polyribosomes. An interesting observation has been reported by Fridlender and Wettstein (1970). They have demonstrated that there are at least two differences in the ensemble of ribosomal proteins from the f and mb polysomes of chick embryo cells. This suggests that there may well be a difference not only in the messenger RNA's translated by mb and f ribosomes but also in the structure of the ribosomes themselves. The role of membranes in the regulation of development is poorly understood, but the marked changes in membranes that occur in the cells of seeds during development indicate a fundamental role for these cell components (Bain and Mercer, 1966a).

METABOLISM AND METABOLIC SYSTEMS IN DRY SEEDS

For an enzymatic process to occur at a significant rate, substrate and product molecules must be free to diffuse to and from the active sites of the enzyme. Although there appears to be no severe limit imposed on metabolic activity in dry seeds by a lack of substrates or of functional enzymes, the very dryness of the seed would seem to rule out the possibility of metabolism in any accepted sense of the word. Changes do occur, however, in dry stored seeds. The process of after-ripening, for example (see p. 375) would be easier to understand if the condition of the seed was one that allowed metabolic changes to occur, but the process often seems to be hastened by desiccation. Before examining the question of biochemical organisation in the dry seed, therefore, it is necessary to establish just what is meant by the term 'dry'.

Aksenov, Askochenskaya and Petinov (1969) and Askochenskaya and Aksenov (1970) studied the state of water in the seeds of a number of species by nuclear magnetic resonance (NMR) techniques. They found three fractions of intracellular water in dry seeds: a tenaciously-bound fraction hydrating reserve starch, a more mobile fraction hydrating protein and a very strongly bound fraction again associated with seed protein but not detectable by NMR. These studies suggest a role for reserve substances in addition to their role as stores of nutrients for use in germination. They also indicate that the term 'dry' when applied to seeds is a relative one, but the question of whether any of the fractions of water in dry seeds are metabolically important remains unanswered.

The metabolite contents of dry seeds have been studied by a number of workers, especially as part of general studies of seed development. For example, Ingle *et al.* (1965) showed that the levels of total N, protein, fat, nucleic acid, soluble N, amino acids, sugars and soluble nucleotides in wheat embryos were higher at maturity than at any other time during maturation. If the embryo also contains enzymes able to function under conditions of extreme desiccation then presumably metabolic changes can take place in the dry seed. Enzymes such as those concerned with mobilisation or glycolysis (see p. 376) are synthesised or activated during imbibition; the dry seed appears to contain very low levels of extractable, functional enzymes of this type. On the other hand seeds also contain enzymes that are apparently not synthesised or activated during germination. For example, Quarles and Dawson (1969) showed that the level of phospholipase-D in peas declined during germination. In their view the enzyme, although actively breaking down cotyledonary phospholipids during germination and growth, has been 'left over' from seed maturation. Levels of functional enzymes of this type in dry seeds appear to be much higher than those of enzymes that become activated during imbibition; it might be argued, therefore, that if biochemical changes do occur in dry seeds it will be the 'left over' enzymes that catalyse them. The term 'after-ripening' applied to the dormancy-breaking process that occurs in dry storage would thus be biochemically as well as physiologically apt. The observation that enzymes characteristic of lysosomes are located in reserve deposits (see p. 365) which also contain a large proportion of the bound water of the dry seed, favours the view that enzymes and residual water come into contact in the dry seed.

Ultrastructural studies of dry and maturing seeds have established that the major cell organelles (nuclei, mitochondria) are present in the desiccated seed but are often in a diffuse, ill-defined state (Bain and Mercer, 1966a; Öpik, 1968; Yatsu, 1965; Klein and Pollock, 1968; Ovcharov, Doman and Popov, 1970). The absence of organised ER and mb ribosomes is a striking feature of the cells of dry seeds Öpik, 1968; Payne and Boulter, 1969a). The general impression gained from the results of studies on dry seeds with the electron microscope is one of biochemical inactivity but conventional methods of examining the activity of isolated cell organelles cannot be employed to test the validity of this impression since the conditions under which organelles are extracted and their activities assayed are usually aqueous and therefore liable in effect to convert them to the organelles of imbibing seeds.

Facts about the metabolism of dry seeds are obviously difficult to come by, and it is easier to think of changes occurring in dry storage as being of a non-enzymatic nature. Cytological and genetic changes are known to occur in dry stored seeds (see Chapter 9) and the falling activity of seed enzymes (presumably due to denaturation) observed in dry-stored seeds has been considered to be a direct indication of declining viability (see Barton, 1961 and Chapters 7, 8 and 9). It is possible to speculate almost endlessly on the relation of non-enzymatic changes in seed macromolecules to the processes of after-ripening or decreasing viability that occur during dry storage, but to devise a satisfactory explanation of these phenomena in biochemical terms is much more difficult.

Seed dormancy

There has been much discussion in the literature regarding the definition of the term 'dormancy' (Vegis, 1964; Amen, 1968; Wareing, 1969); see also in this volume pp. 11, 158, 181 and 321. The word is used here to describe 'instances where the seed of a given species fails to germinate under conditions of moisture, temperature and oxygen supply which are normally favourable for the later stages of germination and growth of that species' (Wareing, 1965). Dormancy in the seed of a particular species can often be overcome by treating the seed in any one of a number of different ways (Barton, 1965a; Stokes, 1965), but under what may be termed 'natural' conditions many forms of dormancy may be overcome by chilling or light treatment or by storage in dry conditions. The present discussion considers dormancy from these three aspects. The small measure of understanding we have of the metabolic aspects of dormancy indicate that this division is biochemically as well as ecologically justifiable.

DORMANCY IN SEEDS WITH A CHILLING REQUIREMENT

Seeds with a chilling requirement to overcome dormancy often contain both growth inhibitors (Wareing, 1965) and growth promoters (Frankland and Wareing, 1966), and a great deal of evidence now eixsts to support the view that dormancy of this type is controlled by an inhibitor/promoter balance that alters when the seed is chilled (Amen, 1968; Wareing and Saunders, 1971). The broader aspects of inhibitor/promoter interactions have been reviewed recently (Wareing, 1969; Wareing and Saunders, 1971); the present discussion is limited to a consideration of the metabolic consequences of inhibitor/promoter interactions in chilled seeds.

Frankland and Wareing (1966) found that chilling caused increases in gibberellin in hazel and beech seeds. Inhibitor levels did not change appreciably. A similar response to chilling has been found in other seeds with a chilling requirement, such as ash (Villiers and Wareing, 1965; Kentzer, 1966). A more detailed study of gibberellin synthesis in chilled hazel seeds has been made by Bradbeer (1968) and Ross and Bradbeer (1968; 1971a and b). Chilling appears to remove a block to gibberellin biosynthesis but gibberellins are apparently not made in quantities sufficient to promote germination until the seed is transferred to a higher temperature. Gas-liquid chromatography showed the presence of a number of gibberellins in hazel extracts and Ross and Bradbeer (1971a) point out that their results are consistent with a scheme of gibberellin biosynthesis proposed by Katsumi and Phinney (see Lang, 1970) in which GA_1 is an early product and is converted into other gibberellins such as GA_4, GA_5 and GA_8. Furthermore, the amount of gibberellin per unit weight of embryonic axis is very much greater than that of the cotyledons of chilled seeds and it seems likely that there is translocation from the axis to the cotyledons of a gibberellin (possibly GA_1) which is subsequently converted to other GA's. Ross and Bradbeer (1971b) used known inhibitors of gibberellin biosynthesis to test the hypothesis (Bradbeer, 1968) that chilling removes a block to gibberellin synthesis and that levels of gibberellins in chilled seeds do not increase to any great extent until the seed is transferred from the chilling temperature (5°C) to a higher temperature (20°C). A marked inhibition of germination and gibberellin accumulation was shown to occur in chilled seeds treated with inhibitors of gibberellin biosynthesis at 20°C. On the other hand Sínska and Lewak (1970) have shown that marked changes in GA_4 and GA_7 levels occur in apple seeds *during* the period of stratification.

A possible answer to the question of how GA brings about germination comes from the work of Jarvis, Frankland and Cherry (1968a and b) on gene repression in dormant hazel seeds. Tuan and Bonner (1964) found that chromatin extracted from buds of dormant potato tubers is almost incapable of supporting DNA-dependent RNA synthesis while the chromatin of non-dormant buds was a highly active template for RNA synthesis. Jarvis and co-workers found very much the same situation in hazel seeds: the chromatin of the embryonic axes of water-treated seeds was a much less effective template for RNA synthesis than chromatin from GA-treated seeds.

The problem of the nature of the block to GA synthesis in dormant seeds remains. The inhibitor present in dormant hazel seeds resembles abscisic acid (ABA) (Bradbeer, personal communication). ABA inhibits the germination of chilled hazel seeds in a similar way to known inhibitors of gibberellin biosynthesis such as Phosphon-D (Ross and Bradbeer, 1971b). Furthermore, the large increase in endogenous gibberellin that normally follows the transfer of spinach plants from short to long days is inhibited by ABA (Wareing *et al.*, 1968a) and ABA also reduces gibberellin levels in maize seedlings (Wareing *et al.*, 1968b). Two possible ways in which ABA might inhibit gibberellin synthesis in dormant seeds come to mind. Firstly, since both ABA and GA probably have a common biosynthetic origin in the mevalonate pathway (Lang, 1970) GA levels might be reduced by a diversion of precursors into ABA which is being turned over at a relatively rapid rate (Addicott and Lyon, 1969) but since only a small fraction of the total pool of mevalonate is converted into ABA this type of control seems unlikely; alternatively, there may be feedback inhibition of early enzymes in the biosynthetic sequence by ABA. Secondly, ABA might prevent the synthesis of enzymes of gibberellin biosynthesis. ABA is known to reduce chromatin activity in radish hypocotyls (Pearson and Wareing, 1969) and Villiers (1968) has produced autoradiographic evidence that ABA inhibits RNA synthesis in dormant ash seeds, though protein synthesis is not inhibited and is presumably supported by stable mRNA's similar to the postulated 'masked' messengers of germinating seeds (see p. 381). Changes in nucleotide and lipid metabolism have been observed during stratification (Bradbeer and Colman, 1967; Stobart and Pinfield, 1970). It is interesting to note that d'Apollonia and Bradbeer (1971) have observed that a fall in the extractability of hazel ribosomal RNA early in the chilling period can be accounted for by the binding of ribosomes to cell membranes, a process that appears to be particularly significant in the regulation of seed protein synthesis (see p. 366) and which the increasing potential of the chilled seed for gibberellin biosynthesis may reflect.

DORMANCY IN SEEDS WITH A LIGHT REQUIREMENT

The subject of light-controlled seed germination has been comprehensively reviewed recently (Evenari, 1965; Wareing, 1969). The ecological significance has been discussed in Chapter 11 (p. 333). Here, however, we shall be concerned with the possible mechanisms and consequently will need to go into greater physiological detail. Early work by the group at Beltsville (Borthwick, Hendricks, Parker, Toole and Toole, 1952) suggested that many of the effects of light on plant morphogenesis could be interpreted in terms of changes in a photoreceptive pigment which was called phytochrome. The phytochrome hypothesis explained previous observations of Flint and McAlister (1935, 1937) that light of wavelength 650 nm promoted and light at 750 nm inhibited the germination of lettuce seed, and has been the basis of most subsequent explanations of light-promoted seed germination. The present discussion is, therefore, much concerned with the mode of action of phytochrome; although the existence of other photosystems in seeds is suspected, next to nothing is known of the metabolic effects of their operation.

Phytochrome has been spectrophotometrically detected in seeds (Mancinelli and Tolkowsky, 1968). The active form of phytochrome appears to be P_{fr} (Hendricks and Borthwick, 1965). The P_{fr} molecule has an absorption maximum at 730 nm, that is, in the far-red (FR) region of the spectrum, and is reversibly formed by the action of red light (R) on the inactive P_r form which absorbs maximally at 660 nm. White light has the same action as R because the conversion of P_r to P_{fr} requires only a quarter of the amount of energy required by the reverse action.

Smith (1970) has summarised the present state of knowledge of the physical and optical properties of phytochrome. Three possible modes of action of P_{fr} are suggested by the results of work on light-stimulated seed germination, namely that P_{fr} affects gibberellin metabolism, that it activates previously inactive genes, and that it interacts in some way with cell membranes.

The observation that GA can substitute for light in breaking the dormancy of many light-sensitive seeds led Brian (1958) to suggest that the effect of P_{fr} was to increase gibberellin synthesis, but experiments by Bewley, Negbi and Black (1968) suggest that P_{fr} does not act in this way. It was found that R and GA act synergistically to stimulate lettuce seed germination and not additively as might be expected if the primary effect of P_{fr} is to increase gibberellin levels. Nevertheless, inhibitors of GA biosynthesis inhibit the light-induced germination of a number of species; it appears, therefore, that gibberellins play some part in the germination of light-sensitive seeds (see Black, 1969). An alternative (or additional) possibility is that P_{fr} stimulates the release of gibberellins from a bound form. This has been suggested to occur in the phytochrome-mediated unrolling of etiolated wheat leaves (Loveys and Wareing, 1971a and b). Bound forms of gibberellin seem to occur in peas (Barendse, Kende and Lang, 1968); presumably they occur in light-sensitive seeds also.

The idea that P_{fr} acts in some way to active genes (Mohr, 1966) is consistent with the view that germination is a genetically-programmed process. Mohr's group has examined a number of enzymes in relation to phytochrome action, for example enzymes of anthocyanin synthesis (Lange and Mohr, 1965), phenylalanine ammonia lyase (PAL) (Rissland and Mohr, 1967) and enzymes of lipid metabolism (Hock, Kühnert and Mohr, 1965). They found that there is a lag between the time when P_{fr} is formed by illumination with R and the time when enzyme activity increases. During this period inhibitors of RNA synthesis and translation, which are known to inhibit light-induced germination (Khan, 1967), inhibit the enzyme increase. Conclusions based on the use of inhibitors alone are open to objection (see p. 384). However, Schopfer and Hock (1971) have unequivocally shown the light-induced *de novo* synthesis of PAL in mustard seedlings by the density-labelling technique. This is strong evidence in favour of an effect, direct or indirect, of P_{fr} on the synthesis of specific proteins and taken with the results obtained with metabolic inhibitors indicates that the activation of certain genes may be under phytochrome control.

It has been suggested, without much evidence, that P_{fr} in some way affects cell membranes. Since there is a lack of detailed knowledge of the nature and possible consequences of changes in cell membranes it is impossible to say whether or not phytochrome could exert all its observed effects by causing such changes. If phytochrome is a hormone permease (Smith, 1970) or if it stimulates the release of (membrane?) bound gibberellins then the membrane hypothesis might explain observations on the role of gibberellins in light-induced seed germination. The observation of Unser and Mohr (1970) of a phytochrome-mediated increase in galactolipid synthesis is particularly interesting since it suggests a link between the membrane and gene activation hypotheses of P_{fr} action.

Embryo dormancy is rare in seeds with a light requirement (Wareing, 1969). The need for P_{fr} can be reduced or overcome altogether in many light-sensitive seeds if the seed coats are removed (Evenari and Neumann, 1952) and reinstated by subjecting the isolated embryo to osmotic stress (Scheibe and Lang, 1965). The presence of seed coats may impose a light requirement for germination by mechanical restriction (Ikuma and Thimann, 1963) or by restricting gaseous exchange (Koller, Poljakoff-Mayber, Berg and Diskin, 1963) or by being a source of inhibitors (Black and Wareing, 1959).

As it stands, the reversible photoreaction by which P_{fr} is formed by the action of R on P_r is an incomplete representation of phytochrome changes within plants. Work on light-inhibited seeds (Black and Wareing, 1960; Chen, 1968), on the so-called 'high energy reaction' (Borthwick, Hendricks, Schneider, Taylorson and Toole, 1969) and on the decay and non-photostimulated reversion of P_{fr} (Kendrick and Frankland, 1969; Taylorson and Hendricks, 1969) suggests that although P_{fr} is undoubtedly the active form of phytochrome there exists a number of forms in addition to P_r and P_{fr}. If the nature of some of these other forms of phytochrome were better known it might be easier to understand a little more of how phytochrome controls morphogenesis.

AFTER-RIPENING IN DRY STORAGE

The seeds of many plant species, having completed their development on the parent plant, will not germinate under apparently favourable conditions unless they have been stored for a period in dry conditions. The relationship between temperature and the rate of after-ripening of dry seeds has been discussed briefly in Chapter 11 (p. 326). This form of dormancy is found in the seeds of many cereals and wild grasses and a large number of other species (Barton, 1965a). However in many of these species, including the cereals, the dry after-ripening period may be omitted if it is substituted by a short stratification (i.e. cold, wet) treatment. In dormancy removed by stratification or light treatment the seed must be partially or fully imbibed if treatment is to be effective (Stokes, 1965). Under these conditions the seed is biochemically active and the block to germination is removed by a process that is apparently metabolic. On the other hand Belderok (1961) has shown that the duration of the dormant period in dry-stored wheat seeds is much reduced by exposing grains to desiccation. Since the loss of this form of dormancy is favoured by dry conditions, there is some doubt that the block to germination is removed by a metabolic process.

Biochemical explanations of after-ripening have, however, been proposed. Simpson (1965) has suggested that embryo dormancy in *Avena fatua* can be attributed to interference with the gibberellin-induced mobilisation and utilisation of seed reserves. The most serious objection to this explanation is that the synthesis of mobilisation enzymes is apparently a post-germination phenomenon (Drennan and Berrie, 1962). Roberts (1969) has proposed that coat-imposed dormancy is partly attributable to a barrier to gaseous diffusion imposed by the testa consequently leading to competition for O_2 between the respiratory systems of the seed. However, the nature of the after-ripening process that increases the ability of the seed to overcome blocks to oxidative metabolism is not clear.

It is difficult to explain the process of after-ripening in dry storage in biochemical terms and the tendency has been to regard after-ripening as the expression of purely physical changes in the seed. Dormancy relieved by dry storage appears, however, to have many aspects and it may well be that the primary event of after-ripening is not a change in growth regulators or the characteristics of the seed coat but is some yet unknown change in the ability of the embryo to overcome restrictions to germination that develop in seed maturation. This change may be biochemical in nature, but a lot more needs to be known about the content and state of enzymes and metabolites in dry seeds and the characteristics and activities of metabolic systems in very dry tissues before it becomes clear what exactly it is.

Seed germination

Metabolic changes in the earliest stages of seed germination are controlled by mechanisms that are either present in resting seeds or that are rapidly organised in predetermined ways during imbibition. The pattern of biochemical events in early germination can thus be seen to be a direct expression of the operation of regulatory systems present in the resting seed. This section examines how much of the metabolic apparatus of the germinating seed was already present before imbibition occurred. The question of the extent to which the enzymes of respiration and mobilisation are activated or synthesised during imbibition will be examined first; secondly the interesting possibility will be considered that the mRNA that codes for germination-specific enzymes is present in the resting seed in a stable or 'masked' form.

THE CONTROL OF ENZYME ACTIVATION DURING SEED GERMINATION

Soon after the seed begins to imbibe water there is a rapid increase in respiration and a mobilisation of reserves (Mayer and Poljakoff-Mayber, 1963). It has been established that the start of mobilisation and the increase in respiration are direct results of synthesis or activation or both of hydrolytic and respiratory enzymes. Increases occur in the enzymes of glycolysis, of the pentose phosphate pathway, of the TCA and glyoxylate cycles, of phosphate and nitrogen metabolism and of protein, carbohydrate and lipid breakdown (Filner, Wray and Varner, 1969). The present discussion will examine the activation of respiratory pathways in germination and will go on to consider, as an example of the control of reserve mobilisation, the fat-sugar conversion system in lipid-storing seeds (Beevers, 1961).

In the early hours of germination when the seed is imbibing water the rate of gaseous exchange rises rapidly (Yemm, 1965). During this phase there is a synthesis or activation or both of many of the enzymes of respiration including those of the glycolytic and pentose phosphate pathways (Marré, 1967; Brown and Wray, 1968; Chakravorty and Burma, 1959). The investigations of Brown and Wray (1968), Marré (1967) and co-workers, Bianchetti and Sartirana (1968) and Sartirana (1968) have shown that sugars, glycolytic intermediates, co-factors such as NAD and adenosine phosphates regulate the activities of many enzymes in the glycolytic and reverse glycolytic pathways. For example, the activity of fructose diphosphatase is decreased by glucose, hexose phosphates and triose phosphates (Bianchetti and Sartirana, 1968). This repression of enzyme activity by glycolytic intermediates occurs partly by a depression of enzyme synthesis and partly by a depression of enzyme function (Sartirana, 1968). Another fairly specific control of respiratory enzyme induction has been observed in the cotyledons of germinating castor beans. Glucose, fructose, galactose and sucrose markedly increase the rate of development of soluble hexokinases but not of other glycolytic enzymes (Marré, Cornaggia and Bianchetti, 1968).

The product of glycolysis is pyruvate (for details see Beevers, 1961). During the very early stages of seed germination most of the pyruvate is further oxidised to CO_2. In many seeds the initial rapid increase in the rate of respiration is followed by a period when the respiration rate stays the same for several hours and when the RQ rises to 2–3 (Yemm, 1965). In this phase, pyruvate from glycolysis is converted to ethanol (Yemm, 1965), and other fermentation products. Fermentative enzymes such as alcohol dehydrogenase (ADH) have been identified in a number of germinating seeds (Cossins and Turner, 1962). App and Meiss (1958) showed that ethanol could induce ADH activity in germinating rice seeds, and Hageman and Flesher (1960) have suggested that acetaldehyde induces ADH in germinating maize. A marked Pasteur effect is observed when the O_2 supplies to fermenting seeds are improved. One way in which the effect can be brought about is by removing the testa, which suggests that seed covering structures restrict the supply of oxygen to the seed (Yemm, 1965). An impressive body of evidence supports the idea that seed oxidation processes are controlled by the restrictive effect of seed coats (Roberts, 1969). When the oxygen supply is not restricted, ADH activity decreases (App and Meiss, 1958) and seeds are capable of very rapidly metabolising ethanol (Cameron and Cossins, 1967).

The control of glycolysis and fermentation in seeds bears a striking resemblance to the control of a number of metabolic sequences in microorganisms: the induction of respiratory enzymes by intermediates in glycolysis and fermentation during early germination is similar to the sequential induction of enzymes by intermediates observed in, for example, leucine biosynthesis in *Neurospora* (Gross, 1965). The activation of seed metabolism during imbibition and germination may be a 'cascade' process in which the unimbibed seed contains only a limited number of key enzymes: these enzymes are activated when the seed imbibes water, the products of the reactions they catalyse induce other enzymes, and the process continues until the metabolism of the seed is fully operational.

Studies of the ultrastructure of seeds have shown that mitochondria become organised and apparently active during germination

(Bain and Mercer, 1966b; Ovcharov *et al.*, 1970). In the early phase of germination, when the rate of respiration increases sharply, there is an increase in the activity of the electron transfer chains and TCA cycle oxidases while the activity of dehydrogenases increases less rapidly or not at all (Kollöffel, 1967; Kollöffel and Sluys, 1970). There appears to be mitochondrial protein synthesis at this stage, and the ratio of mitochondrial RNA to mitochondrial protein increases (Lado and Schwendimann, 1967). These increases may be accompanied by an increase in the number of mitochondria (Breidenbach *et al.*, 1966). During the fermentative phase of germination the activity of both oxidases and dehydrogenases increases (Kollöffel and Sluys, 1970). Ranson, Walker and Clark (1957) have observed an inhibition of succinic oxidase by high levels of CO_2. This control mechanism may operate in the fermentation period, when there appears to be a restriction of gaseous exchange.

The development of active cotyledonary mitochondria during the early stages of germination seems, in some seeds, to be under the control of the embryonic axis: in peas, although the presence of the axis is not necessary for the conversion of carbohydrate and protein to soluble products in the cotyledons, it is necessary for the normal organisation of mitochondria, ER and nuclei (Bain and Mercer, 1966c).

There is much evidence to suggest that although the cytochrome system and TCA cycle enzymes are present in germinating seeds, in many species only a small proportion of the carbon of acetyl CoA gets fully oxidised by these mitochondrial pathways (Canvin and Beevers, 1961). In castor bean endosperm and maize scutellum for example, the glyoxylate cycle competes with the TCA cycle for acetyl CoA and a large proportion of the carbon of the lipid reserves becomes converted to carbohydrate carbon because of the high activity of the enzyme isocitratase (Tanner and Beevers, 1965). Tanner and Beevers (1965) have identified an ADP-regulated 'catabolic glyoxylate cycle' in castor bean that produces ATP by the oxidation of acetyl CoA to CO_2. The enzymes of the glyoxylate cycle appear, in many seeds, to be located in glyoxysomes and are thus spatially separated from the mitochondrial TCA cycle. It therefore appears likely that there is compartmentation of acetyl CoA metabolism rather than direct competition between the TCA and glyoxylate cycles for acetyl CoA (Oaks and Bidwell, 1970).

Fifteen minutes after ^{14}C-ethanol is fed to pea slices, aspartate becomes heavily labelled and after 30 minutes other amino acids, such as methionine, are labelled (Cossins, 1962). Oaks and Beevers (1964) noted rapid labelling of glutamate in maize scutellum fed with ^{14}C acetate. It appears that in many germinating seeds TCA cycle intermediates are used more as precursors for amino acid synthesis and other metabolic pathways than as intermediates in the complete oxidation of acetyl CoA. Cameron and Cossins (1967) have shown a substantial drain of α-oxoglutarate to support amino acid biosynthesis in germinating peas.

A number of studies have indicated that respiratory electron transport chains may not be confined to the mitochondria of germinating seeds. Spragg and Yemm (1959) and Mapson and Moustaffe (1956) have examined the possibility that ascorbic acid and glutathione occur in soluble systems and may be responsible for the continuing increase in O_2 uptake late in germination when mitochondrial activity is falling.

Many seeds contain phytate, a store of phosphorus and cations for germination (Williams, 1970). Respiration may be controlled by a regulation of the availability of phosphate for respiratory phosphorylations. The hydrolytic enzyme phytase is synthesised during seed germination (Mandal and Biswas, 1970). Phytase synthesis is repressed by inorganic phosphate. Phosphate inhibition of enzyme synthesis is similar to the inhibition by actinomycin-D, and this has led to the suggestion that phosphate acts by repressing the synthesis of phytase-specific mRNA (Bianchetti and Sartirana, 1967). A similar end-product inhibition of enzyme activity appears to control lipid mobilisation in castor bean seeds.

The pathway of transformation of fat to carbohydrate in germinating seeds is visualised as comprising β-oxidation, the glyoxylate cycle and the EMP pathway (Fig. 12.1). Desveaux and Kogane-Charles

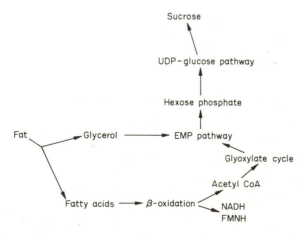

FIGURE 12.1

(1952) demonstrated the net conversion of fat to carbohydrate and its passage from endosperm to embryonic axis during germination. The conversion is marked by an RQ of less than 0.6 (Marré, 1967). Lipases and lipoxidases are known to occur in seeds and seedlings (Beevers, 1961; St Angelo and Altschul, 1964; Jacks, Yatsu and Altschul, 1967).

The enzymes of β-oxidation have been identified in seeds and are known to occur in cellular inclusions called glyoxysomes or peroxisomes (Hutton and Stumpf, 1969). The enzymes of the glyoxylate cycle are also located in the glyoxysomes of castor bean endosperm (Breidenbach *et al.*, 1968). Glyoxysomes have been demonstrated in a number of seeds, for example peanut (Longo and Longo, 1970a), maize (Longo and Longo, 1970a and b) and pine (Ching, 1970).

Glyoxysomes are rapidly synthesised during germination (Gerhardt and Beevers, 1970). The development of glyoxysomes and their subsequent decrease is closely paralleled by changes in the activity of the marker enzymes isocitrate lyase and malate synthetase (Hutton and Stumpf, 1969). The increase in these two enzymes is a result of *de novo* synthesis (Gientka-Rychter and Cherry, 1968; Longo, 1968). The electron microscope reveals that glyoxysomes are large membrane-bound inclusions which appear to be formed from smaller, denser bodies during the first three days of germination in maize (Longo and Longo, 1970a). Ching (1970) has shown that pine seed glyoxysomes develop in a similar fashion. The synthesis of glyoxysomes has been suggested (Ching, 1970) to be programmed by stable mRNA's and ribosomes in pre-existing glyoxysomes. It is interesting to note in this connection that RNA and, in pine seed, DNA have been found in association with glyoxysomes (Gerhardt and Beevers, 1969; Ching, 1970). Presley and Fowden (1965) could find no effect of an amino acid analogue on the development of acid phosphatase and isocitratase activity in mung beans, pea seeds, cucumber seeds, sunflower seeds and pumpkin seeds; they suggest that these enzymes develop from zymogen-like inactive forms during germination.

There is a decline in glyoxysomes and associated enzymes, paralleling the depletion of stored lipid, after 4–13 days of germination (Longo and Longo, 1970a; Ching, 1970; Vigil, 1970). The decrease of glyoxylate cycle enzymes is due to a combination of decreased synthesis and normal turnover, rather than to a specific breakdown. It is associated with breakdown of glyoxysomes and release of enzymes into the cytosol (Longo and Longo, 1970a).

The development and decline of glyoxysomes and their associated enzymes so closely follows the rise and fall of the level of fatty acids derived from the hydrolysis of lipid reserves that a causal relationship between the two sets of changes may exist. In other words, the formation of glyoxysome enzymes may be induced either by fatty acids or by some metabolite derived from fatty acids. Furthermore, the continued processing of fatty acids by this system is favoured both by the equilibrium constant of the sucrose phosphate synthetase reaction, which lies very strongly towards sucrose synthesis, and the sink effect of developing tissues to which the sucrose is exported. Clearly, then, in the control of lipid mobilisation in germination compartmentation, enzyme induction and feedback regulation are all important.

It is interesting that the synthesis of isocitratase in castor bean is inhibited by glucose (Marré, 1967). The inhibition of enzymes early in metabolic sequences by the end-products of these sequences is a classical control mechanism (Epstein and Beckwith, 1968). It appears, therefore, that the 'cascade' process, described in connexion with the induction of respiratory enzymes can be considered to be controlled by a very sensitive mechanism in which both substrate induction and end-product repression of enzyme activity participate.

The hormone or substrate-controlled induced synthesis of a number of other enzymes, has been demonstrated in germinating seeds. Many of these have been listed by Filner *et al.* (1969) in their review of enzyme induction in plants.

MASKED MESSENGER RNA AND SEED GERMINATION

The idea that 'masked' or long-lived forms of messenger RNA might exist in germinating seeds was developed to explain similarities between the early metabolic changes in seed germination and changes during early embryogenesis in a number of animals. Fertilisation of the egg of, for example, the sea urchin results in a very rapid increase in protein synthesis (Gross, 1967). This increase occurs in enucleate cells and in the presence of such inhibitors of RNA synthesis as actinomycin-D. The implication is that there is pre-formed mRNA present in the egg and that fertilisation stimulates the translation rather than the transcription of this mRNA. A number of particles and precursors believed to contain stored mRNA have been identified in animal cells and this has led Spirin (1966) to propose that protein synthesis in early embryogenesis is controlled by the regulated conversion of precursors to active polysomes.

Chen, Sarid and Katchalski (1968a) undertook a comparison of the RNA of dry wheat embryos with the RNA of embryos 48 hours and 48–72 hours after the start of imbibition. The technique they used was DNA-RNA hybridisation. It was found that about 1.15 per cent of the wheat embryo genome codes for RNA that is neither ribosomal RNA (rRNA) nor soluble RNA (sRNA). It was further shown that the RNA of embryos 0–48 hours after the start of imbibition is undistinguishable in hybridisation competition experiments from the RNA of dry embryos, but is different from the RNA of embryos imbibed for 48–72 hours. It was concluded from these results that new mRNA species appear during the 48–72 hours period and since no apparent increase in the size of the genome was observed then the synthesis of these new mRNA species must be switched on as the synthesis of old mRNA's is switched off. If the non-ribosomal, non-soluble RNA of wheat embryos is, as Chen and co-workers suggest, mRNA then the fraction present in the first stage of germination appears to be the same as the mRNA stored in the dry embryo and it is not until the second stage of germination, beginning 48–72 hours after imbibition, that new mRNA's begin to be synthesised.

A serious criticism of these studies is that the hybridisation techniques used are not specific or sensitive enough to allow Chen and his colleagues to draw the conclusions they have drawn. It is clear from experiments on animal RNA species, that one cannot designate as messenger the RNA coded for by the portion of the genome that hybridises with the non-ribosomal, non-soluble fraction of total cell RNA. Much of this RNA is probably nuclear RNA which never passes into the cytoplasm (Harris, 1968) and only a very small proportion of heterogeneous nuclear RNA, probably too small to be detected by hybridisation techniques at present, ever reaches the ribosomes and gets translated (Soeiro and Darnell, 1970). The technical difficulties that prevent the direct identification of a mRNA fraction in bulk RNA preparations from seeds are so serious that the best evidence for the existence of masked forms of seed mRNA has come from studies of mRNA by more indirect means.

Messenger RNA associated with ribosomes in polysomal aggregates can be identified by established methods such as sucrose density gradient centrifugation (Loening, 1968) and protein-synthesising cell free systems (Boulter, 1970). Marcus and Feeley (1964) studied protein synthesis in peanut cotyledons and wheat embryos during imbibition. They showed an activation of protein synthesis in the imbibition phase of seed germination. The capacity for incorporation of radioactive amino acid into protein by wheat embryos increases almost 130-fold in sixteen hours (Marcus and Feeley, 1965). This increase is inhibited by puromycin, an inhibitor of ribosome function. The ratio of monosomes to polysomes fell during this phase. Actinomycin-

D, an inhibitor of RNA synthesis, does not inhibit these changes in protein synthesis and polysome levels. Marcus and Feeley suggest that activation of mRNA during imbibition may be due to a change in the spatial separation of ribosomes and mRNA leading to polysome formation, or to synthesis of mRNA, or to the removal of an inhibitor of mRNA-ribosome association, or to a change in the ribosomes themselves allowing them to become competent to receive mRNA.

Similar observations of an increase in polysomes during germination have been made on peanut (Jachymczyk and Cherry, 1965), red pine seed (Sasaki and Brown, 1970), *Pisum arvense* (Barker and Rieber, 1967) and sycamore (*Acer pseudoplatanus*) (Thomas, unpublished) as well as on germinating pollen grains (Linskens, Schrauwen and Konings, 1970), fungal spores (Henney and Storck, 1964) and spores of bacteria (Woese, Langridge and Morowitz, 1960). There is evidence that the new polysomes that appear during germination are mb (Payne and Boulter, 1969b). The increase in polysomes may or may not be associated with an increase in ribosome synthesis (Marré, 1967). An increasing ability of ribosome preparations to incorporate amino acids into protein in cell-free systems parallels the increase in polysomes during the early stages of germination (Jachymczyk and Cherry, 1968; Sturani, 1968).

A second line of approach in the study of seed mRNA has been to use inhibitors of RNA and protein synthesis. The rationale behind this approach is that in a system where protein synthesis is inhibited by inhibitors of ribosome function (such as cycloheximide) but not by inhibitors of RNA synthesis (such as actinomycin-D) the mRNA must be long-lived. Dure and Waters (1965) and Waters and Dure (1966) extracted ^{32}P-labelled polysomes and polysomal RNA from cotton seed that had been imbibed for 4 hours in actinomycin-D and labelled for a further 12 hours. There was a 63 per cent inhibition of ^{32}P-incorporation into RNA by actinomycin-D at a concentration of 20 μg/ml but there was no effect on the incorporation of ^{14}C-amino acid into soluble protein. Actinomycin-D also decreased ^{32}P-incorporation into polysomes but there was no change in the monosome to polysome ratio. Similarly actinomycin-D reduced ^{32}P-incorporation into polysomal RNA by more than 60 per cent but there was no change in the amount of RNA judged by the absorbancy profile of RNA fractionated by sucrose density gradient centrifugation. Sucrose density gradient fractionation of the nuclear RNA of embryos imbibed for 12 hours and labelled for 30 minutes with ^{32}P-RNA showed the presence of a fraction with a high specific activity that had a base composition intermediate between that of DNA and rRNA and that was sensitive to inhibition by actinomycin-D. During the germination of cotton seed, then, as in the germination of wheat, there is synthesis of RNA but inhibition of this RNA synthesis by actinomycin-D inhibits neither protein synthesis nor germination. Polysome profiles (and therefore, presumably, mRNA) are stable under conditions of actinomycin-D inhibition.

The effects of inhibitors on seed metabolism tend to be variable. Cycloheximide has been shown to inhibit protein synthesis by peanut cotyledon ribosomes (Jachymczyk and Cherry, 1968) but not by wheat or castor bean ribosomes (Marcus and Feeley, 1966; Parisi and Ciferri, 1966). Actinomycin-D is reported to inhibit RNA synthesis in wheat (Marcus and Feeley, 1965) and cotton seed (Dure and Waters, 1965) but Marré (1967) has pointed out that in the experiments of these workers no clear evidence for any effect of actinomycin-D on mRNA synthesis was presented and, in general, when high enough concentrations of antibiotic are used, enzyme synthesis, growth and respiration are all inhibited. Even where the effects of actinomycin-D have been investigated on the development of the same enzyme in different seeds such as isocitratase in watermelon and peanut seeds (see Gientka-Rychter and Cherry, 1968) conflicting results are obtained. This suggests that there may be some species-specificity in the response to these metabolic inhibitors. A further complication limiting the usefulness of cycloheximide and actinomycin-D relates to the lack of specificity of these antimetabolites in plants. The interference of cycloheximide with energy transfer and ion uptake by non-green tissues has been demonstrated (Ellis and MacDonald, 1970) and actinomycin-D has been shown to inhibit respiration and ATP production in human leukemic leucocytes (Laszlo, Miller, McCarthy and Hochstein, 1966). Clearly, before reliable conclusions can be drawn from the results of experiments on the effects of inhibitors on seeds, it should be established that these effects are specifically on RNA and protein metabolism.

A third line of approach has been to study the synthesis of specific proteins during germination since it seems likely that the metabolism of a specific mRNA can be investigated by studying the synthesis of the protein for which it codes. In such investigations in order to relate protein synthesis to mRNA metabolism it is assumed that the protein being studied is synthesised *de novo* during germination. This has been unequivocably demonstrated for only a small number of enzymes, namely peanut isocitratase and malate synthetase (Gientka-Rychter and Cherry, 1968; Longo, 1968), barley α-amylase (Filner and Varner, 1967), barley protease (Jacobsen and Varner, 1967) and barley peroxidases (Austine, Jacobsen, Scandalios and Varner, 1970), by a density-labelling technique.

Ihle and Dure (1969, 1970) examined the effects of actinomycin-D and cycloheximide on the synthesis of a cotton cotyledon protease. During normal germination protease activity develops after 24 hours, rising to a maximum after about 3 days and thereafter gradually falls away. It was established that the 24 hour lag period is probably not an effect of an inhibitor of the protease, but is a period during which there seems to be no protease, active or inactive, present in the seed. If the seed is imbibed in cycloheximide then no enzyme activity appears at all, even if the seed is transferred to agar containing no cycloheximide. Cycloheximide applied at any time during the first 3 days of germination immediately inhibits further development of protease activity. The changes in response to cycloheximide appear to be a result of changes in enzyme synthesis since they are reflected in the incorporation of ^{14}C-amino acid into protein and the specific activity of protein synthesised during cycloheximide inhibition. On the other hand actinomycin-D, at concentrations inhibiting all RNA synthesis detectable by ^{32}P-incorporation, has little effect on the development of protease by whole seeds or by excised cotyledons on miracloth, except where long exposure to actinomycin-D causes necrosis in the tissue.

In an attempt to locate the time during embryo development when the mRNA for the protease was synthesised, Ihle and Dure (1969) treated embryos of various ages with actinomycin-D and studied the development of the protease during precocious germination on agar (Table 12.1).

Clearly the mRNA has been stored by the 100 mg stage, but not at the 95 mg stage. Therefore the mRNA for cotton seed protease must have been synthesised and stored when embryo development was about 60 per cent complete.

TABLE 12.1 *The effect of actinomycin-D on protease synthesis when applied to cotton embryos at various stages of development. (From Ihle and Dure, 1969.)*

Age of embryo as indicated by weight	Enzyme Units/Cotyledon Pair	
	−Actinomycin-D	+Actinomycin-D
115 mg Embryos	12.48	9.33
100 mg Embryos	6.23	8.65
95 mg Embryos	8.58	0

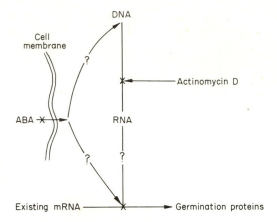

FIGURE 12.2

Ihle and Dure (1970) examined the nature of the mechanism that delays the translation of stored messengers until imbibition. They found that both ABA and an extract of the ovule wall inhibit the synthesis of protease in precociously germinated embryos at the stage of development when the protease-specific mRNA was known to be synthesised. This inhibition requires RNA synthesis since actinomycin-D overcomes the effect of both ABA and ovule extract. Ihle and Dure proposed the scheme shown in Fig. 12.2 for the hormonal control of the synthesis of proteins necessary for germination during the stage of cotton seed development when the protease mRNA is being synthesised.

The translation of the masked messenger of wheat seeds also appears to be regulated by plant hormones such as ABA and GA (Chen and Osborne, 1970) and Walton *et al.* (1970) have proposed that ABA inhibits protein synthesis in excised bean axes at the translation level.

Acknowledgement

I am very grateful to Professor P. F. Wareing for the benefit of his experience, advice and constructive criticism at every stage of preparation of this chapter and to the University of Wales for the grant of a Research Studentship.

References to Chapter 12

ADDICOTT, F. T., and LYON, J. L., 1969. Physiology of abscisic acid and related substances. *Ann. Rev. Pl. Physiol.*, **20**, 139–64.

AKSENOV, S. I., ASKOCHENSKAYA, N. A., and PETINOV, N. S., 1969. The fractions of water in wheat seeds. *Fiz. Rast.*, **16**, 58–64.

ALTSCHUL, A. M., YATSU, L. Y., ORY, R. L., and ENGLEMAN, E. M., 1966. Seed proteins. *Ann. Rev. Physiol.*, **17**, 113–36.

AMEN, R. D., 1968. A model of seed dormancy. *Bot. Rev.*, **34**, 1–31.

D'APOLLONIA, S. T., and BRADBEER, J. W., 1971. Effects of chilling on ribosomal RNA in hazel seeds. January S.E.B. Meeting, London.

APP, A. A., and MEISS, A. N., 1958. Effect of aeration on rice alcohol dehydrogenase. *Arch. Biochem. Biophys.*, **77**, 181–90.

ASKOCHENSKAYA, N. A., and AKSENOV, S. I., 1970. Characteristics of the hydration of seeds of different composition. *Fiz. Rast.*, **17**, 95–100.

AUSTINE, W., JACOBSEN, J. V., SCANDALIOS, J. G., and VARNER, J. E., 1970. Deuterium oxide as a density label of peroxidases in germinating barley embryos. *Pl. Physiol., Lancaster*, **45**, 148–52.

BAILEY, C. J., COBB, A., and BOULTER, D., 1970. A cotyledon slice system for the electron autoradiographic study of the synthesis and intracellular transport of seed storage protein of *Vicia faba*. *Planta, Berl.*, **95**, 103–18.

BAIN, J. M., and MERCER, F. V., 1966a. Subcellular organization of the developing cotyledons of *Pisum sativum* L. *Aust. J. biol. Sci.*, **19**, 49–68.

BAIN, J. M., and MERCER, F. V., 1966b. Subcellular organization of the cotyledons in germinating seeds and seedlings of *Pisum sativum* L. *Aust. J. biol. Sci.*, **19**, 69–84.

BAIN, J. M., and MERCER, F. V., 1966c. The relationship of the axis and the cotyledons in germinating seeds and seedlings of *Pisum sativum* L. *Aust. J. biol. Sci.*, **19**, 85–96.

BARENDSE, G. W. M., KENDE, H., and LANG, A., 1968. The fate of tritiated GA_1 in mature and germinating seeds of peas and Japanese morning glory. *Pl. Physiol., Lancaster*, **43**, 815–22.

BARKER, G. R., and HOLLINSHEAD, J. A., 1965. Ribosomes from the cotyledons of *Pisum arvense*. *Biochim. Biophys. Acta*, **108**, 323–5.

BARKER, G. R., and RIEBER, M., 1967. Formation of polysomes in the seed of *Pisum arvense*. *Biochem. J.*, 1195–201.

BARTON, L. V., 1961. *Seed preservation and longevity*. Leonard Hill, London.

BARTON, L. V., 1965a. Seed dormancy: general survey of dormancy types in seeds, and dormancy imposed by external agents. *Encycl. Pl. Physiol.*, **15** (2), 699–720.

BARTON, L. V., 1965b. Dormancy in seeds imposed by the seed coat. *Encycl. Pl. Physiol.*, **15** (2), 727–45.

BEEVERS, H., 1961. *Respiratory metabolism in plants*. Row, Peterson & Co., Evanston.

BELDEROK, B., 1961. Studies on dormancy in wheat. *Proc. int. Seed Test. Ass.*, **26**, 697–760.

BEWLEY, J. D., NEGBI, M., and BLACK, M., 1968. Immediate phytochrome action in lettuce seeds and its interaction with gibberellins and other germination promoters. *Planta, Berl.*, **78**, 351–7.

BIANCHETTI, R., and SARTIRANA, M. L., 1967. The mechanism of the repression by inorganic phosphate of phytase synthesis in the germinating wheat embryo. *Biochim. Biophys. Acta*, **145**, 484–90.

BIANCHETTI, R., and SARTIRANA, M. L., 1968. AMP-sensitive fructose diphosphatase in wheat embryos: changes in the levels of glycolytic intermediates during
enzyme depression. *Life Sci.*, **7**, 121–7.

BLACK, M., 1969. Light-controlled germination of seeds. *Symp. Soc. exp. Biol.*,
23, 193–217.

BLACK, M., and WAREING, P. F., 1959. The role of germination inhibitors and oxygen
in the dormancy of the light sensitive seed of *Betula* spp. *J. Exp. Bot.*, **10**, 134–45.

BLACK, M., and WAREING, P. F., 1960. Photoperiodism in the light-inhibited seed
of *Nemophila insignis*. *J. exp. Bot.*, **11**, 28–39.

BORTHWICK, H. A., HENDRICKS, S. B., PARKER, M. W., TOOLE, E. H., and TOOLE,
V. K., 1952. A reversible photoreaction controlling seed germination. *Bot. Gaz.*,
115, 205–25.

BORTHWICK, H. A., HENDRICKS, S. B., SCHNEIDER, M. J., TAYLORSON, R. B.,
and TOOLE, V. K., 1969. The high-energy light action controlling plant responses
and development. *Proc. Nat. Acad. Sci., US*, **64**, 479–86.

BOULTER, D., 1970. Protein synthesis in plants. *Ann. Rev. Pl. Physiol.*, **21**, 91–114.

BRADBEER, J. W., 1968. Studies in seed dormancy. IV. The role of endogenous
inhibitors and gibberellin in the dormancy and germination of *Corylus avellana* L.
seeds. *Planta, Berl.*, **78**, 266–76.

BRADBEER, J. W., and COLMAN, B., 1967. Studies in seed dormancy. I. The metabolism of ($2^{14}C$) acetate by chilled seeds of *Corylus avellana* L. *New Phytol.*, **66**,
5–15.

BREIDENBACH, R. W., and BEEVERS, H., 1967. Association of the glyoxylate cycle
enzymes in a novel subcellular particle from castor bean endosperm. *Biochem.
Biophys. Res. Commun.*, **27**, 462–9.

BREIDENBACH, R. W., CASTELFRANCO, P., and PETERSON, C., 1966. Biogenesis
of mitochondria in germinating peanut cotyledons. *Pl. Physiol., Lancaster*, **41**,
803–9.

BREIDENBACH, R. W., KAHN, A., and BEEVERS, H., 1968. Characterization of
glyoxysomes from castor bean endosperm. *Pl. Physiol, Lancaster*, **43**, 705–13.

BRIAN, P. W., 1958. The role of gibberellin-like hormones in regulation of plant
growth and flowering. *Nature, Lond.*, **181**, 1122–3.

BRIARTY, L. G., COULT, D. A., and BOULTER, D., 1969. Protein bodies of developing seeds of *Vicia faba*. *J. exp. Bot.*, **20**, 358–72.

BROWN, A. P., and WRAY, J. L., 1968. Correlated changes of some enzyme activities
and cofactor and substrate contents of pea cotyledon tissue during germination.
Biochem. J., **108**, 437–44.

BUTTROSE, M. S., 1963a. Ultrastructure of developing aleurone cells of wheat grain.
Aust. J. biol. Sci., **16**, 768–74.

BUTTROSE, M. S., 1963b. Ultrastructure of the developing wheat endosperm. *Aust.
J. biol. Sci.*, **16**, 305–17.

CAMERON, D. S., and COSSINS, E. A., 1967. Studies of intermediary metabolism in
germinating pea cotyledons. The pathway of ethanol metabolism and the role of
the tricarboxylic acid cycle. *Biochem. J.*, **105**, 323–31.

CAMPBELL, P. N., 1970. Functions of polyribosomes attached to membranes of
animal cells. *FEBS Letters*, **7**, 1–7.

CANVIN, D. T., and BEEVERS, H., 1961. Sucrose synthesis from acetate in the germinating castor bean; kinetics and pathway. *J. biol. Chem.*, **236**, 988–95.

CHAKRAVORTY, M., and BURMA, D. P., 1959. Enzymes of the pentose phosphate pathway in the mung bean seedling. *Biochem. J.*, **73**, 48–53.

CHEN, D., and OSBORNE, D. J., 1970. Hormones in the translational control of early germination in wheat embryos. *Nature, Lond.*, **226**, 1157–60.

CHEN, D., SARID, S., and KATCHALSKI, E., 1968a. Studies on the nature of messenger RNA in germinating wheat embryos. *Proc. Nat. Acad. Sci., US*, **60**, 902–9.

CHEN, D., SARID, S., and KATCHALSKI, E., 1968b. The role of water stress in the inactivation of messenger RNA of germinating wheat embryos. *Proc. Nat. Acad. Sci., US*, **61**, 1378–83.

CHEN, S. S. C., 1968. Germination of light-inhibited seed of *Nemophila insignis*. *Am. J. Bot.*, **55**, 1177–83.

CHING, T. M., 1970. Glyoxysomes in megagametophyte of germinating Ponderosa pine seeds. *Pl. Physiol., Lancaster*, **46**, 475–82.

COMMONER, B., 1964. Roles of DNA in inheritance. *Nature, Lond.*, **202**, 960–8.

COOPER, T. G., and BEEVERS, H., 1969. β-Oxidation in glyoxysomes from castor bean endosperm. *J. biol. Chem.*, **244**, 3515–20.

COSSINS, E. A., 1962. Utilization of ethanol-2-^{14}C by pea slices. *Nature, Lond.*, **194**, 1095–6.

COSSINS, E. A., and TURNER, E. R., 1962. Losses of alcohol and alcohol dehydrogenase activity in germinating seeds. *Ann. Bot.*, **26**, 591–7.

CRAFTS, A. S., 1968. Water deficits and physiological processes. In *Water Deficits and Plant Growth*, **2**, ed. T. T. Kozlowski, 85–133. Academic Press, New York and London.

DALBY, A., and DAVIES, I. ab I., 1967. Ribonuclease activity in the developing seeds of normal and opaque-2 maize. *Science, NY*, **155**, 1573–5.

DESVEAUX, R., and KOGANE-CHARLES, M., 1952. Germination of oleaginous seeds. *Ann. Inst. Recherches Agron. Ser. A., Ann. Agron.*, **3**, 385–416.

DOVE, L. D., 1967. Ribonuclease activity of stressed tomato leaflets. *Pl. Physiol., Lancaster*, **42**, 1176–8.

DRENNAN, C. H., and CANVIN, D. T., 1969. Oleic acid synthesis by a particulate preparation from developing castor oil seeds. *Biochim. Biophys. Acta*, **187**, 193–200.

DRENNAN, D. S. H., and BERRIE, A. M. M., 1962. Physiological studies of germination in the genus *Avena*. I. The development of amylase activity. *New Phytol.*, **61**, 1–9.

DURE, L., and WATERS, L., 1965. Long-lived mRNA: evidence from cotton seed germination. *Science, NY*, **147**, 410–12.

ELLIS, R. J., and MACDONALD, I. R., 1970. Specificity of cycloheximide in higher plant systems. *Pl. Physiol., Lancaster*, **46**, 227–32.

ENGLEMAN, E. M., 1966. Ontogeny of aleurone grains in cotton embryo. *Am. J. Bot.*, **53**, 231–7.

EPSTEIN, W., and BECKWITH, J. R., 1968. Regulation of gene expression. *Ann. Rev. Biochem.*, **37**, 411–36.

EVENARI, M., 1965. Light and seed dormancy. *Encycl. Pl. Physiol.*, **15** (2), 804–47.

EVENARI, M., and NEUMANN, G., 1952. The germination of lettuce seed. II. The influence of fruit coat, seed coat and endosperm upon germination. *Bull. Res. Coun. Israel*, **2**, 15–17.

FILNER, P., and VARNER, J., 1967. A simple and unequivocal test for *de novo* synthesis of enzymes: density labelling with H_2O^{18} of barley α-amylase induced by gibberellic acid. *Proc. Nat. Acad. Sci., US*, **58**, 1520–6.

FILNER, P., WRAY, J. L., and VARNER, J. E., 1969. Enzyme induction in higher plants. *Science, NY*, **165**, 358–67.

FLINT, L. H., and MCALISTER, E. D., 1935. Wavelength of radiation in the visible spectrum inhibiting the germination of light sensitive lettuce seed. *Smithsonian Inst. Misc. Coll.*, **94**, 1–11.

FLINT, L. H., and MCALISTER, E. D., 1937. Wavelength of radiation in the visible spectrum promoting germination of light sensitive lettuce seed. *Smithsonian Inst. Misc. Coll.*, **96**, 1–8.

FOLKES, B. F., 1970. The physiology of the synthesis of amino acids and their movement into the seed proteins of plants. *Proc. Nutr. Soc.*, **29**, 12–20.

FRANKLAND, B., and WAREING, P. F., 1966. Hormonal regulation of seed dormancy in hazel (*Corylus avellana* L.) and beech (*Fagus sylvatica*). *J. exp. Bot.*, **17**, 596–611.

FRIDLENDER, B. R., and WETTSTEIN, F. O., 1970. Differences in the ribosomal protein of free and membrane-bound polysomes of chick embryo cells. *Biochem. Biophys. Res. Commun.*, **39**, 247–53.

GATES, C. T., 1964. The effect of water stress on plant growth. *J. Australian Inst. agric. Sci.*, **30**, 3–22.

GAYE, P., and DENAMUR, R., 1970. Preferential synthesis of β-lactoglobulin by the bound polyribosomes of the mammary gland. *Biochem. Biophys. Res. Commun.*, **41**, 266–72.

GERHARDT, B. P., and BEEVERS, H., 1969. Occurrence of RNA in glyoxysomes from castor bean endosperm. *Pl. Physiol., Lancaster*, **44**, 1475–7.

GERHARDT, B. P., and BEEVERS, H., 1970. Developmental studies on glyoxysomes in *Ricinus endosperm. J. Cell Biol.*, **44**, 94–102.

GIENTKA-RYCHTER, A., and CHERRY, J. H., 1968. *De novo* synthesis of iso-citratase in peanut (*Arachis hypogaea* L.) cotyledons. *Pl. Physiol., Lancaster*, **43**, 653–9.

GRAHAM, J. S. D., MORTON, R. K., and RAISON, J. K., 1963. Isolation and characterization of protein bodies from developing wheat endosperm. *Aust. J. biol. Sci.*, **16**, 375–83.

GROSS, P. R., 1967. The control of protein synthesis in embryonic development and differentiation. In *Current Topics in Developmental Biology*, **2**, ed. A. Monroy and A. A. Moscana, 1–46. Academic Press, New York and London.

GROSS, S. R., 1965. The regulation of synthesis of leucine biosynthetic enzymes in *Neurospora. Proc. Nat. Acad. Sci., US*, **54**, 1538–46.

GRZESIUK, S., MIERZWINSKA, T., and SOJKA, E., 1962. On the physiology and biochemistry of seed development in broad bean. *Fiz. Rast.*, **9**, 544–52.

HAGEMAN, R. H., and FLESHER, D., 1960. The effect of an anaerobic environment on the activity of alcohol dehydrogenase and other enzymes of corn seedlings. *Arch. Biochem. Biophys.*, **87**, 203–9.

HARRIS, H., 1968. *Nucleus and cytoplasm*. Clarendon Press, Oxford.

HARRIS, P., and JAMES, A. T., 1969. Effect of low temperature on fatty acid biosynthesis in seeds. *Biochim. Biophys. Acta*, **187**, 13–18.

HENDRICKS, S. B., and BORTHWICK, H. A., 1965. The physiological functions of phytochrome. In *Chemistry and Biochemistry of Plant Pigments*, ed. T. W. Goodwin, 405–36. Academic Press, New York.

HENNEY, H. R., and STORCK, R., 1964. Polyribosomes and morphology in *Neuro-*

spora crassa. Proc. Nat. Acad. Sci., US, **51**, 1050–5.

HOCK, B., KÜHNERT, E., and MOHR, H., 1965. Die regulation von Fettbau und Atmung bei Senfkeimlingen durch Licht (*Sinapis alba* L.). *Planta, Berl.*, **65**, 129–38.

HSIAO, T. C., 1970. Rapid changes in levels of polyribosomes in *Zea mays* in response to water stress. *Pl. Physiol., Lancaster*, **46**, 281–5.

HUTTON, D., and STUMPF, P. K., 1969. Fat metabolism in higher plants. XXXVII. Characterization of the *β*-oxidation systems from maturing and germinating castor bean seeds. *Pl. Physiol., Lancaster*, **44**, 508–16.

IHLE, J. N., and DURE, L., 1969. Synthesis of a protease in germinating cotton cotyledons catalyzed by masked mRNA synthesized during embryogenesis. *Biochem. Biophys. Res. Commun.*, **36**, 705–10.

IHLE, J. N., and DURE, L., 1970. Hormonal regulation of translation inhibition requiring RNA synthesis. *Biochem. Biophys. Res. Commun.*, **38**, 995–1001.

IKUMA, H., and THIMANN, K. V., 1963. The role of seed coats in germination of photosensitive lettuce seeds. *Plant and Cell Physiol.*, **4**, 169–85.

INGLE, J., BEITZ, D., and HAGEMAN, R. H., 1965. Changes in composition during development and maturation of maize seeds. *Pl. Physiol., Lancaster*, **40**, 835–9.

JACHYMCZYK, W. J., and CHERRY, J. H., 1968. Studies on mRNA from peanut plants: *in vitro* polyribosome formation and protein synthesis. *Biochim. Biophys. Acta*, **157**, 368–77.

JACKS, T. J., YATSU, L. Y., and ALTSCHUL, A. M., 1967. Isolation and characterization of peanut spherosomes. *Pl. Physiol., Lancaster*, **42**, 585–97.

JACOBSEN, J. V., and VARNER, J. E., 1967. Gibberellic acid-induced synthesis of protease by isolated aleurone layers of barley. *Pl. Physiol., Lancaster*, **42**, 1596–600.

JARVIS, B. C., FRANKLAND, B., and CHERRY, J. H., 1968a. Increased nucleic acid synthesis in relation to the breaking of dormancy of hazel seed by gibberellic acid. *Planta, Berl.*, **83**, 257–66.

JARVIS, B. C., FRANKLAND, B., and CHERRY, J. H., 1968b. Increased DNA template and RNA polymerase associated with the breaking of seed dormancy. *Pl. Physiol., Lancaster*, **43**, 1734–6.

JENNER, C. F., 1970. Relationship between levels of soluble carbohydrate and starch synthesis in detached ears of wheat. *Aust. J. biol. Sci.*, **23**, 991–1003.

JENNINGS, A. C., and MORTON, R. K., 1963. Changes in nucleic acids and other phosphorus containing compounds of developing wheat grain. *Aust. J. biol. Sci.*, **16**, 332–41.

JENNINGS, A. C., MORTON, R. K., and PALK, B. A., 1963. Cytological studies of protein bodies of developing wheat endosperm. *Aust. J. biol. Sci.*, **16**, 366–74.

KENDRICK, R. E., and FRANKLAND, B., 1969. The *in vivo* properties of *Amaranthus* phytochrome. *Planta, Berl.*, **86**, 21–32.

KENTZER, T., 1966. Gibberellin-like substances and growth inhibitors in relation to the dormancy and after-ripening of ash seeds (*Fraxinus excelsior* L.). *Acta Soc. Bot. Pol.*, **35**, 575–85.

KESSLER, B., ENGELBERG, N., CHEN, D., and GREENSPAN, H., 1964. Studies on physiological and biochemical problems of stress in higher plants. *Volcani Inst. Agr. Res. Unit Plant Physiol. Biochem., Rehovot, Israel. Special Bull.*, **64**, Project ALO–CR–7.

KESSLER, B., and FRANK-TISHEL, J., 1962. Dehydration-induced synthesis of nucleic acids and changing composition of RNA: a possible protective reaction in drought-resistant plants. *Nature, Lond.*, **196**, 542–5.

KHAN, A. A., 1967. Dependence of lettuce seed germination on actinomycin-D resistant RNA synthesis. *Physiol. Plant.*, **20**, 1039–44.

KLEIN, S., and POLLOCK, B. M., 1968. Cell fine structure of developing lima bean seeds related to seed desiccation. *Am. J. Bot.*, **55**, 658–72.

KOLLER, D., POLJAKOFF-MAYBER, A., BERG, A., and DISKIN, T., 1963. Germination-regulating mechanisms in *Citrullus colocynthis*. *Am. J. Bot.*, **50**, 597–603.

KOLLÖFFEL, C., 1967. Respiration rate and mitochondrial activity in the cotyledons of *Pisum sativum* L. during germination. *Acta Bot. Neerl.*, **16**, 111–22.

KOLLÖFFEL, C., 1970. Oxidative and phosphorylative activity of mitochondria from pea cotyledons during maturation of the seed. *Planta, Berl.*, **91**, 321–8.

KOLLÖFFEL, C., and SLUYS, J. V., 1970. Mitochondrial activity in pea cotyledons during germination. *Acta Bot. Neerl.*, **19**, 503–8.

LADO, P., and SCHWENDIMANN, M., 1967. Changes of mitochondrial RNA level during the transition from rest to growth in the endosperm of germinating castor beans. *Life Sci.*, **6**, 1681–90.

LANG, A., 1970. Gibberellins: structure and metabolism. *Ann. Rev. Pl. Physiol.*, **21**, 537–70.

LANGE, H., and MOHR, H., 1965. Die Hemmung der Phytochrom-induzierten Anthocyansynthese durch Actinomycin-D und Puromycin. *Planta, Berl.*, **67**, 107–21.

LASZLO, J., MILLER, D. S., MCCARTHY, K. S., and HOCHSTEIN, P., 1966. Actinomycin-D: inhibition of respiration and glycolysis. *Science, NY*, **151**, 1007–10.

LINSKENS, H. F., SCHRAUWEN, J. A. M., and KONINGS, R. N. H., 1970. Cell-free protein synthesis with polysomes from germinating *Petunia* pollen grains. *Planta, Berl.*, **90**, 153–62.

LOENING, U. E., 1968. The occurrence and properties of polysomes in plant tissues. In *Plant Cell Organelles*, ed. J. B. Pridham, 216–27. Academic Press, London and New York.

LONGO, C. P., 1968. Evidence for *de novo* synthesis of iso-citratase and malate synthetase in germinating peanut cotyledons. *Pl. Physiol., Lancaster*, **43**, 660–4.

LONGO, C. P., and LONGO, G. P., 1970a. The development of glyoxysomes in peanut cotyledons and maize scutella. *Pl. Physiol., Lancaster*, **45**, 249–54.

LONGO, G. P., and LONGO, C. P., 1970b. The development of glyoxysomes in maize scutellum. Changes in morphology and enzyme compartmentation. *Pl. Physiol., Lancaster*, **46**, 599–604.

LOVEYS, B. R., and WAREING, P. F., 1971a. The red light controlled production of gibberellin in etiolated wheat leaves. *Planta, Berl.*, in press.

LOVEYS, B. R., and WAREING, P. F., 1971a. The red light controlled production rolling. *Planta, Berl.*, in press.

LUI, N. S. T., and ALTSCHUL, A. M., 1967. Isolation of globoids from cotton seed aleurone grain. *Arch. Biochem. Biophys.*, **121**, 678–84.

MANCINELLI, A. L., and TOLKOWSKY, A., 1968. Phytochrome and seed germination. V. Changes of phytochrome content during germination of cucumber seeds. *Pl. Physiol., Lancaster*, **43**, 489–94.

MANDAL, N. C., and BISWAS, B. B., 1970. Metabolism of inositol phosphates. I. Phytase synthesis during germination in cotyledons of mung beans, *Phaseolus aureus. Pl. Physiol., Lancaster*, **45**, 4–7.

MAPSON. L. W., and MOUSTAFA, E. M., 1956. Ascorbic acid and glutathione as respiratory carriers in the respiration of pea seedling. *Biochem. J.*, **62**, 248–59.

MARCUS, A., and FEELEY, J., 1964. Activation of protein synthesis in the imbibition phase of seed germination. *Proc. Nat. Acad. Sci., US*, **51**, 1075–9.

MARCUS, A., and FEELEY, J., 1965. Protein synthesis in imbibed seeds. II. Polysome formation during imbibition. *J. biol. Chem.*, **240**, 1675–80.

MARCUS, A., and FEELEY, J., 1966. Ribosome activation and polysome formation *in vitro*: requirement for ATP. *Proc. Nat. Acad. Sci., US*, **56**, 1770–7.

MARRÉ, E., 1967. Ribosome and enzyme changes during maturation and germination of the castor bean seed. In *Current Topics in Developmental Biology*, **2**, eds. A. Monroy and A. A. Moscana, 75–105. Academic Press, London and New York.

MARRÉ, E., CORNAGGIA, M. P., and BIANCHETTI, R., 1968. The effects of sugars on the development of hexose phosphorylating enzymes in the castor bean cotyledons. *Phytochem.*, **7**, 1115–23.

MATILE, P., 1968. Aleurone vacuoles as lysosomes. *Z. Pflanzenphys.*, **58**, 365–8.

MAYER, A. M., and POLJAKOFF-MAYBER, A., 1963. *The germination of seeds*. Pergamon Press, London.

MAYER, A. M., and SHAIN, Y., 1968. Zymogen granules in enzyme liberation and activation in pea seeds. *Science, NY*, **162**, 1283–4.

MOHR, H., 1966. Differential gene activation as a mode of action of phytochrome 730. *Photochem. Photobiol.*, **5**, 469–83.

MORTON, R. K., and RAISON, J. K., 1963. A complete intracellular unit for incorporation of amino acid into storage protein utilizing ATP generated from phytate. *Nature, Lond.*, **200**, 429–33.

MORTON, R. K., and RAISON, J. K., 1964. The separate incorporation of amino acids into storage and soluble proteins catalyzed by two independent systems isolated from developing wheat endosperm. *Biochem. J.*, **91**, 528–38.

MOSSE, J., 1968. Cereal (grain) proteins. *Actual. Sci. Ind. No. 1305*, 47.

MUDD, J. B., 1967. Fat metabolism in plants. *Ann. Rev. Pl. Physiol.*, **18**, 229–52.

NIR, I., POLJAKOFF-MAYBER, A., and KLEIN, S., 1970a. The effect of water stress on mitochondria of root cells. *Pl. Physiol., Lancaster*, **45**, 173–7.

NIR, I., POLJAKOFF-MAYBER, A., and KLEIN, S., 1970b. Effect of water stress on polysome population and ability to incorporate amino acids in maize root tips. *Israel J. Bot.*, **19**, 451–62.

OAKS, A., and BEEVERS, H., 1964. The glyoxylate cycle in maize scutellum. *Pl. Physiol., Lancaster*, **39**, 431–4.

OAKS, A., and BIDWELL, R. G. S., 1970. Compartmentation of intermediary metabolites. *Ann. Rev. Pl. Physiol.*, **21**, 43–66.

ÖPIK, H., 1965. Respiration rate, mitochondrial activity and mitochondrial structure in the cotyledons of *Phaseolus vulgaris* L. during germination. *J. exp. Bot.*, **16**, 667–82.

ÖPIK, H., 1968. Development of cotyledon cell structure in ripening *Phaseolus vulgaris* seeds. *J. exp. Bot.*, **19**, 64–76.

ORY, R. L., and HENNINGSEN, K. W., 1969. Enzymes associated with protein bodies isolated from ungerminated barley seeds. *Pl. Physiol.*, **44**, 1488–98.

OVCHAROV, K. E., DOMAN, N. N., and POPOV, B. A., 1970. Correlation between mitochondrial ultrastructure and germinating capacity of maize seeds. *Fiz. Rast.*,

17, 329–36.

PARISI, B., and CIFERRI, O., 1966. Protein synthesis by cell-free extracts from castor bean seedlings. I. Preparation and characteristics of the amino acid incorporation system. *Biochem.*, **5**, 1638–45.

PAYNE, P. I., and BOULTER, D., 1969a. Free and membrane bound ribosomes of the cotyledons of *Vicia faba* L. I. Seed development. *Planta, Berl.*, **84**, 263–71.

PAYNE, P. I., and BOULTER, D., 1969b. Free and membrane-bound ribosomes of the cotyledons of *Vicia faba* L. II. Seed germination. *Planta, Berl.*, **87**, 63–68.

PEARSON, J. A., and WAREING, P. F., 1969. Effect of abscisic acid on activity of chromatin. *Nature, Lond.*, **221**, 672–73.

PRESLEY, H. J., and FOWDEN, L., 1965. Acid phosphatase and iso-citratase production during seed germination. *Phytochem.*, **4**, 169–76.

QUARLES, R. H., and DAWSON, R. M. C., 1969. The distribution of phospholipase-D in developing and mature plants. *Biochem. J.*, **112**, 787–94.

RANSON, S. L., WALKER, D. A., and CLARK, L. D., 1957. The inhibition of succinic oxidase by high CO_2 concentrations. *Biochem. J.*, **66**, 57P.

RISSLAND, I., and MOHR, H., 1967. Phytochrom-induzierte Enzymbildung (Phenylalanindesaminase), ein schnell ablafender Prozess. *Plant Berl.*, **77**, 239–49.

ROBERTS, E. H., 1969. Seed dormancy and oxidation processes. In *Symp. Soc. exp. Biol.*, **23**, 161–92.

ROSS, J. D., and BRADBEER, J. W., 1968. Concentrations of gibberellin in chilled hazel seeds. *Nature, Lond.*, **220**, 85–86.

ROSS, J. D., and BRADBEER, J. W., 1971a. Studies in seed dormancy. V. The concentrations of endogenous gibberellins in seeds of *Corylus avellana* L. *Planta, Berl.*, in press.

ROSS, J. D., and BRADBEER, J. W., 1971b. Studies in seed dormancy. VI. The effects of inhibitors of gibberellin synthesis on the gibberellin content and germination of chilled seeds of *Corylus avellana* L. *Planta, Berl.*, in press.

ST ANGELO, A. J., and ALTSCHUL, A. M., 1964. Lipolysis and the free fatty acid pool in seedlings. *Pl. Physiol., Lancaster*, **39**, 880–3.

SARTIRANA, M. L., 1968. *In vivo* glucose induced acceleration of the inactivation of fructose-diphosphatase in wheat embryos. *Giorn. Bot. Ital.*, **102**, 261–6.

SASAKI, S., and BROWN, G. N., 1970. Changes in polysomes in red pine embryos during early stages of seed germination. *Pl. Physiol., Lancaster*, **45**, Suppl. xvii.

SCHEIBE, J., and LANG, A., 1965. Lettuce seed germination: Evidence for a reversible light-induced increase in growth potential and for phytochrome mediation of the low temperature effect. *Pl. Physiol., Lancaster*, **40**, 485–92.

SCHNEPF, E., 1961. Über Veränderungen der plasmatischen Feinstrukturen während des Welkens. *Planta, Berl.*, **57**, 156–75.

SCHOPFER, P., and HOCK, B., 1971. Nachweis der Phytochrom-induzierten *de-novo*-synthese von Phenylalaninammonium lyase (PAL EC 4.3.1.5) in Keimlingen von *Sinapis alba* L. durch Dichtmarkiertung mit Deuterium. *Planta, Berl.*, **96**, 248–53.

SIMPSON, G. M., 1965. Dormancy studies in seed of *Avena fatua*. IV. The role of gibberellin in embryo dormancy. *Can. J. Bot.*, **43**, 793–816.

SIMPSON, G. M., and NAYLOR, J. M., 1962. Dormancy studies in the seed of *Avena fatua*. III. A relationship between maltase, amylase and gibberellin. *Can. J. Bot.*, **40**, 1659–73.

SİNSKA, I., and LEWAK, S., 1970. Apple seed gibberellins. *Physiol. Vég.*, **8**, 661–7.

SMITH, H., 1970. Phytochrome and photomorphogenesis in plants. *Nature, Lond.*, **227**, 665–8.

SOEIRO, R., and DARNELL, J. E., 1970. A comparison between heterogeneous nuclear RNA and polysomal RNA in HeLa cells by RNA-DNA hybridization. *J. Cell Biol.*, **44**, 467–75.

SPIRIN, A. S., 1966. On 'masked' forms of messenger RNA in early embryogenesis and in other differentiating systems. In *Current Topics in Developmental Biology*, **1**, eds. A. Monroy and A. A. Moscana, 1–38. Academic Press, London and New York.

SPRAGG, S. P., and YEMM, E. W., 1959. Respiratory mechanisms and the changes of glutathione and ascorbic acid in germinating peas. *J. exp. Bot.*, **10**, 409–25.

STOBART, A. K., and PINFIELD, N. J., 1970. Glycerol utilization in seeds of *Corylus avellana* L. *New Phytol.*, **69**, 939–49.

STOKES, P., 1965. Temperature and seed dormancy. *Encycl. Pl. Physiol.*, **15** (2), 746–803.

STONER, C. D., and HANSON, J. B., 1966. Swelling and contraction of corn mitochondria. *Pl. Physiol., Lancaster*, **41**, 255–66.

STURANI, E., 1968. Protein synthesis activity of ribosomes from developing castor bean endosperm. *Life Sci.*, **7**, 527–37.

TAKAGI, M., TANAKA, T., and OGATA, K., 1970. Functional differences in protein synthesis between free and bound polysomes of rat liver. *Biochim. Biophys. Acta*, **217**, 148–58.

TANNER, W., and BEEVERS, H., 1965. The competition between the glyoxylate cycle and the oxidative breakdown of acetate in *Ricinus* endosperm. *Z. Pflanzenphys.*, **53**, 126–39.

TAYLORSON, R. B., and HENDRICKS, S. B., 1969. Action of phytochrome during prechilling of *Amaranthus retroflexus* L. seeds. *Pl. Physiol., Lancaster*, **44**, 821–5.

TUAN, D. Y. H., and BONNER, J., 1964. Dormancy associated with repression of genetic activity. *Pl. Physiol., Lancaster*, **39**, 768–72.

TURNER, J. F., 1969a. Physiology of pea fruits. VI. Changes in uridine diphosphate glucose pyrophosphorylase and adenosine diphosphate glucose pyrophosphorylase in the developing seed. *Aust. J. biol. Sci.*, **22**, 1145–51.

TURNER, J. F., 1969b. Starch synthesis and changes in uridine diphosphate glucose pyrophosphorylase and adenosine diphosphate glucose pyrophosphorylase in the germinating wheat grain. *Aust. J. biol. Sci.*, **22**, 1321–7.

UNSER, G., and MOHR, H., 1970. Phytochrome mediated increase of galactolipids in mustard seedlings. *Naturwiss.*, **57**, 358.

VAADIA, Y., RANEY, F. C., and HAGAN, R. M., 1961. Plant water deficits and physiological processes. *Ann. Rev. Pl. Physiol.*, **12**, 265–72.

VEGIS, A., 1964. Dormancy in higher plants. *Ann. Rev. Pl. Physiol.*, **15**, 185–224.

VIGIL, E. L., 1970. Cytochemical and developmental changes in microbodies (glyoxysomes) and related organelles of castor bean endosperm. *J. Cell Biol.*, **46**, 435–54.

VILLIERS, T. A., 1968. An autoradiographic study of the effect of the plant hormone abscisic acid on nucleic acid and protein metabolism. *Planta, Berl.*, **82**, 342–54.

VILLIERS, T. A., and WAREING, P. F., 1965. The growth substance content of dormant fruits of *Fraxinus excelsior*. *J. exp. Bot.*, **16**, 534–44.

WALBOT, V., 1970. RNA metabolism during *Phaseolus* embryo development (Abstr.). *J. Cell Biol.*, **47**, 218a.

WALTON, D. C., SOOFI, G. S., and SONDHEIMER, E., 1970. The effects of abscisic acid on growth and nucleic acid synthesis in excised embryonic bean axes. *Pl. Physiol., Lancaster*, **45**, 37–40.

WAREING, P. F., 1965. Endogenous inhibitors in seed germination and dormancy. *Encycl. Pl. Physiol.*, **15** (2), 909–924.

WAREING, P. F., 1966. Ecological aspects of seed dormancy and germination. In *Reproductive biology and taxonomy of vascular plants. BSBI Conf. Reports*, **9**, 103–21.

WAREING, P. F., 1967. Natural inhibitors as growth hormones. In *Trends in Plant Morphogenesis*. Longmans Green, London.

WAREING, P. F., 1969. Germination and Dormancy. In *Physiology of Plant Growth and Development*, ed. M. B. Wilkins, 605–44. McGraw-Hill, London.

WAREING, P. F., GOOD, J., and MANUEL, J., 1968a. Some possible physiological roles of abscisic acid. In *Biochemistry and Physiology of Plant Growth substances*, eds. F. Wrightman and G. Setterfield, 1561–79. Runge Press, Ottawa.

WAREING, P. F., GOOD, J., POTTER, H., and PEARSON, J. A., 1968b. Preliminary studies on the mode of action of abscisic acid. In *Plant Growth Regulators* Monograph, **31**, 191–207. Soc. Chem. Ind., London.

WAREING, P. F., and SAUNDERS, P. F., 1971. Hormones and dormancy. *Ann. Rev. Pl. Physiol.*, **22**, in press.

WATERS, L., and DURE, L., 1966. Ribonucleic acid synthesis in germinating cotton seeds. *J. mol. Biol.*, **19**, 1–27.

WHEELER, C. T., and BOULTER, D., 1967. Nucleic acids of developing seeds of *Vicia faba* L. *J. exp. Bot.*, **18**, 229–40.

WILLIAMS, S. G., 1970. The role of phytic acid in the wheat grain. *Pl. Physiol., Lancaster*, **45**, 376–81.

WILSON, C. M., 1966. Bacteria, antibiotics and amino acid incorporation into maize endosperm protein bodies. *Pl. Physiol., Lancaster*, **41**, 325–7.

WOESE, C. R., LANGRIDGE, R., and MOROWITZ, H. J., 1960. Microsome distribution during germination of bacterial spores. *J. Bact.*, **79**, 777–82.

WRIGHT, S. T. C., and HIRON, R. W. P., 1969. (+)-Abscisic acid, the inhibitor induced in detached wheat leaves by a period of wilting. *Nature, Lond.*, **224**, 719–20.

YATSU, L. Y., 1965. The ultrastructure of cotyledonary tissue from *Gossypium hirsutum* L. seeds. *J. Cell Biol.*, **25**, 193–8.

YATSU, L. Y., and JACKS, T. J., 1968. Association of lysomal activity with aleurone grains in plant seeds. *Arch. Biochem. Biophys.*, **124**, 466–71.

YEMM, E. W., 1965. The respiration of plants and their organs. In *Plant Physiology – A Treatise*, **IVA**, ed. F. C. Steward, 231–310. Academic Press, New York and London.

Organisation of the United States National Seed Storage Laboratory

Edwin James

Since early colonial times plant introductions have contributed materially to the progress of agriculture in the United States, but until 1898 this activity was poorly organised and rarely, if at all, funded. In 1901 the Section of Seed and Plant Introduction was established in the Bureau of Plant Industry to put plant introduction in its proper perspective and on a systematic basis. Beginning in 1898, however, an inventory of plant introductions was begun wherein each introduction received a number. Up to 1970 over 350,000 introductions have been numbered.

Unfortunately, there was no effective organisation established for the preservation of various collections. The seeds were sent either to specialists in the Bureau of Plant Industry or to research workers in state experiment stations, none of whom had adequate storage facilities. If plant introductions exhibited no outstanding attributes the remnants of seeds were either discarded or tossed in a drawer or box and left there until all viability was lost. It has been estimated that before the late 1940s as much as 66 per cent for some and as much as 98 per cent of other plant introductions were lost. Such losses led to repeated requests for the introduction of the same material. In many cases it has been impossible to enter many fruitful areas to reintroduce germ plasm that would possibly contribute to our plant breeding activity.

Scientists finally awakened to the fact that the preservation of germ plasm was sadly neglected. In 1944 the National Research Council recommended that the United States Department of Agriculture establish a facility for the preservation of valuable germ plasm. This proposal received the support of practically all organisations involved in plant breeding activities but funds for the construction of a National Seed Storage Laboratory did not become available until 1957. Construction was begun at once and the Laboratory was ready for operation in the autumn of 1958. It was the first such facility of its kind and is still the only one with almost unlimited storage facilities (Fig. A1.1).

FIGURE AI.I   Exterior of the United States National Seed Storage Laboratory.

The National Seed Storage Laboratory for the United States is located at Fort Collins, Colorado, which abuts to the foothills of the Rocky Mountains. The climate in this area is characterised by moderate winters and temperate summers. The relative humidity varies from as low as 5 per cent during the summer to occasional days in rainy spells during the winter and spring when it is as high as 95 per cent. The atmospheric humidity has little effect on the humidity in the closed system of the storage rooms, but moderate temperatures do obtain economy of refrigeration costs.

The Laboratory is on three levels. Figure AI.2 shows the arrangement of the three levels. All refrigeration and air conditioning equipment is on the ground floor, administrative offices on the second, seed storage rooms and the germination laboratory (Fig. AI.3) on the third floor. The equipment and seed storage rooms are constructed with heavily reinforced concrete and will reputedly withstand an atomic blast at three miles. The only windows in the building are in the ground floor room across the corridor from the garage, the entire front of the office area, and at the ends of the germination laboratory.

The ten storage rooms are accessible from a common corridor and have capacities of 42,000 450-gm accessions for the larger ones, approximately 12,000 each for those at the top of Fig. AI.2, and about

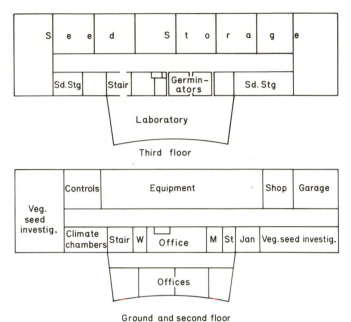

Third floor

Ground and second floor

FIGURE AI.2   Floor plan of the United States National Seed Storage Laboratory (the main part of the structure measures 41 × 10.65 m).

FIGURE AI.3   Germination laboratory in the United States National Seed Storage Laboratory. (Walk-in germinators not visible on left.)

4,000 for the one small room. In all, the storage area has a capacity of 180,000 450-gm samples. If it ever becomes necessary, the storage capacity could be expanded to many times the 180,000 by placing small quantities of seeds in foil envelopes. For example, 4–5 gm of tobacco seeds now take as much storage space as 500 gm of wheat. By placing tobacco seeds in small foil envelopes, 20 accessions would require no more space than one tin can.

The accessions are arranged in numbered steel trays, and the trays placed in numbered steel racks, Fig. A1.4. Any risk from fire is practically eliminated except for the 4 inches of cork insulation on the periphery of the rooms. The cork, however, is protected by a half-inch of plaster.

Storage conditions are maintained at about 4°C with an average relative humidity of 32 per cent. The humidity is maintained at this level by the reheat method, wherein the air is delivered from the

FIGURE AI.4   Typical seed storage room in the United States Seed Storage Laboratory.

evaporator at a temperature at least 12–15°C lower than 4°C and reheated to the specified 4°C. The temperature of 4°C was selected as being suitable for most seeds. Three rooms, however, are equipped so that a temperature of −12°C can be maintained.

For research purposes, three small chambers have been installed on the ground floor where different climatic conditions can be simulated. These have temperatures of 10°, 21° and 32°C, and cabinets inside the chambers have relative humidities of 50, 70 and 90 per cent, thus providing nine climatic analogues ranging from the prevailing conditions in our northern states to those found in our deep south.

Seeds are accepted from all public agencies, commercial seed firms, and individuals involved in plant breeding or seed research. When received all seeds are tested for viability and given accession numbers. If the germination is satisfactory the seeds are then placed in tin cans and transferred to the storage rooms. The cans are not hermetically sealed because of the necessity of frequent sampling. With a relative humidity of 32 per cent, sealing is not required. The seeds equilibriate with the 32 per cent relative humidity which along with the temperature of 4°C eliminates insect infestation and fungal invasion (see Fig. 2.6, p. 51).

Each sample is given two accession numbers. One of these is in serial order and the other is coded for genus and species. Using oat seed as an example, the first lot of seed received is designated as A–1 and the hundredth as A–100. *Avena* is punched numerically on cards as 054 and *sativa* as 407. Throughout the coding system *Avena*, for instance, is always punched as 054 regardless of species and *sativa* as 407 for any genera as a *sativa* species.

The germination laboratory (Fig. A1.3) is equipped with two walk-in germinators which are electronically controlled to provide either constant or alternating temperatures between 15° and 30°C. Six additional small germinators with the same temperature range provide a wide range of conditions for research purposes.

When seeds are accepted for storage, the Laboratory has the responsibility of future maintenance. A 5-year retest programme is followed and if during the storage period seed viability drops to 50 per cent, the Laboratory is authorised, through contracts, to produce a new generation with the same genetic composition as the original accessions. This, of course, eliminates the practicability of storing hybrids. We do, however, attempt to obtain inbred parents of the hybrids. Seeds which lose viability relatively rapidly are tested every 2 years.

Any seeds of present or potential value are accepted for storage but

adequate documentation is required. Documentation consists of descriptions of the varieties or breeding lines, or references to publications in which descriptions can be found. Each crop has its own code
for horticultural or agronomic characteristics and these are recorded
on punched cards for retrieval purposes. All documentation will
eventually be converted to tapes for computer retrieval.

The specified quantities of seeds for storage are dependent on the
kind and the difficulty of production. For most kinds, such as named
varieties, we prefer to store between 5,000 and 10,000 seeds. This
amount provides for many retests and for seed requests. In the case
of genetic materials, of which it is often difficult to obtain large
amounts of seeds, 500 seeds are acceptable and the donor then
assumes the responsibility of any possible future increases. The
number of seeds used for germination tests of such genetic stocks is
scaled down to give us only a rough estimate of viability.

When the Laboratory was first established, an inter-disciplinary
group consisting of commercial plant breeders, seed producers and
marketers, experiment station representatives, and personnel of the
US Department of Agriculture promulgated certain policies under
which the Laboratory operates. When seeds are accepted for storage
they become public property and are subject to disbursement. The
only exception to this rule is a provision that the donor can request
a 5-year moratorium on distribution.

Only seeds are stored and they are expected to be clean and have
viabilities above 60 per cent, preferably above 75 per cent. When seeds
of low viability are received they are stored on a temporary basis until
seeds of higher viability are received as a replacement. When large
collections are received the donor is provided with a report on the
germinabilities of the entire collection. With such a report he is able
to plan for seed replacements according to his increase or breeding
programme.

Although the Laboratory was not intended to be a distributing
agency, any *bona fide* research worker in the United States or its
possessions can obtain nominal amounts of seeds without charge provided supplies are unobtainable elsewhere. We are not a procurement
agency for research workers but do cooperate by advising individuals
as to sources of seeds stored in the Laboratory.

The Laboratory has no responsibility in relation to commitments
with foreign countries. All deposits and requests for seeds from
foreign countries are channelled through the New Crops Research
Branch, Agricultural Research Service, US Department of Agriculture, Beltsville, Maryland. This policy does not imply that foreign

countries cannot avail themselves of the services of the Laboratory. We do accept from and disburse seeds to foreign scientists but only with the approval of the New Crops Research Branch.

The Laboratory is charged with the publishing of inventories of seed stocks in storage. These are published every 2 years and distributed throughout the United States so that all research workers are advised regarding the stocks available. The inventories are broken down into various crop groups and contain only a listing of seed stocks and donors. The last inventories, published in March 1970, listed over 73,000 accessions.

It is recognised that the definition of valuable germ plasm, which the Laboratory was designed to store, will vary greatly depending on the significance attached to the present commercial value of the crop involved and the individual research worker's evaluation, whether he be a geneticist, horticulturist, agronomist, physiologist or plant pathologist. In accordance with the policy statement, valuable germ plasm covers many kinds of seeds.

The largest quantities of seeds in storage are world collections. These are composed of small grains, cotton, soya beans and other oil seeds, sorghum and tobacco. World collections are stored for the most part as a reserve in case collections held by other agencies are, for unforeseen circumstances, destroyed. The Laboratory has no responsibility for disbursements of world collections.

In the United States we have four Regional Plant Introduction Stations where plant introductions are increased and evaluated. Seeds produced at these stations are stored in their own long-term storages and made available to research workers. When the working stocks of these stations exceed their needs a portion is transferred to the National Seed Storage Laboratory as a reserve. These introductions are the second largest of our collections.

Most of the balance of the 73,000 accessions now held consist of varietal material. These include varieties as they are released, those approaching obsolescence, and those already obsolete. The latter are as important as newly released varieties because no one can say that a variety which is susceptible to certain prevailing pathogens would not be resistant to a new race of the future. Our crop history in the United States shows this to be true. In fact, we frequently receive requests for seeds of varieties that have disappeared from the agricultural community for 30 to 40 years.

Open-pollinated varieties are being rapidly replaced by hybrids in both the agronomic and horticultural crops. It is important that the open-pollinated material be preserved to give us a broad base for the development of inbred lines. The inbred lines, whether in present use or not, are also considered valuable germ plasm and all available ones are stored.

Other valuable germ plasm which is stored in limited quantities include certain genetic stocks which have no commercial value but have pronounced interest academically. One example of this type of material is the classic *Datura* collection by A. H. Blakeslee. Also eligible for storage are seeds of those varieties or strains being used as differential hosts for pathogens, those used for indexing viruses, and those used in physiological studies or assays.

The present value of the stored seed stocks may not be immediately apparent but as time goes on they will become increasingly important. Modern improved varieties are gradually encroaching on centres of crop origins so that pockets where disease resistances are found are becoming more and more confined. If material from these pockets is not collected and preserved it will be lost forever to plant breeders. By placing all plant introductions in appropriate storage they will be available to plant breeders indefinitely. Furthermore, areas for fruitful exploration in Red China are no longer accessible to all plant explorers. Consequently, it is extremely important that plant materials from this area be preserved. One valuable plant introduction may save an agricultural industry and the benefits realised from one introduction would conceivably pay for the construction of a store and the cost of maintaining one for many years.

Organisation of the National Seed Storage Laboratory for Genetic Resources in Japan

Hiroshi Ito

The National Seed Storage Laboratory for genetic resources was opened in the Division of Genetics, Department of Physiology and Genetics, National Institute of Agricultural Sciences, Hiratsuka, Kanagawa, Japan in 1966. The city of Hiratsuka is located about 60 km to the south-west of Tokyo. This Laboratory is the principal establishment of the Genetic Resources Project for Crops, Ministry of Agriculture and Forestry and was started in 1965. Thus the Laboratory is part of the Institute but it is operated under the supervision of the Agriculture, Forestry and Fishery Research Council, Ministry of Agriculture and Forestry.

Before the establishment of this laboratory the permanent maintenance and distribution of genetic resources had been undertaken by individual breeders or geneticists. But nowadays, the demands for genetic resources are not only very wide but also stringent because of the rapid development of breeding and genetics, and so it became necessary to establish a specialised organisation for this purpose. The Laboratory is becoming the centre for seed storage and the distribution of genetic resources for the whole of Japan. The main roles of the Laboratory are first to preserve genetic resources for future use in breeding and genetical studies; secondly to collect and maintain a large stock of different types of adequately described genetic material for use by breeders and geneticists; and thirdly to prevent the loss of valuable genetic resources.

The number of different types of seeds accepted for storage is steadily increasing, making a total of 12,991 samples from 61 species at the end of March, 1970, comprising 10,321 samples from 12 species of food crops, 1,192 samples from 31 species of horticultural crops, 1,299 samples from 6 species of industrial crops and 179 samples from 12 species of fodder crops. During the last three years, 3,782 seed samples were distributed for breeding and genetic studies.

PRINCIPLES OF LONG-TERM SEED STORAGE

Before the development of long-term seed storage methods, it was very difficult to satisfy the above roles at a genetic resources centre because it is impossible to raise seed free from the effect of natural selection acting in successive generations raised in the field. In 1956 I started to study ways of developing systems of long-term seed storage, and these principles, together with a new breeding system based on long-term seed storage, have been presented in a series of publications (Ito and Hayashi, 1960a and b, 1961; Ito, 1965; Ito and Kumagi, 1969; Ito, 1970).

Owing to the development of the air-conditioning and freezing industry, long-term seed storage has recently become easy and popular, and it now is the basic method of the maintenance of plant materials for genetic resources.

At the FAO/IBP Joint Conference for Plant Genetic Resources held in Rome, Harrington (1967) suggested 'it appears possible to keep seed of most species for hundreds of years if the storage conditions are ideal'. He continued 'ideal storage conditions for seed of most species include a relative humidity of about 15 per cent, a temperature as cold as possible (preferably at a deep freeze temperature of −20°C or lower), an atmosphere low in oxygen and high in carbon dioxide, an absence of light, and a storage room or container that minimises radiation damage'.

Under the ideal condition cited above, moisture contents of seed are about 4–6 per cent. Roberts (1960, 1961) studied the relationships between the longevity of seeds and storage conditions, concluding that the storage temperature and moisture content of seeds are the most important factors affecting their longevity. We have used the empirical equations (2.9, 2.10 and 2.11) he used to describe these relationships in conjunction with the viability constants he published for rice (see Appendix 3) for calculating the periods during which germinability would not fall below 90 per cent. Tables based on these values (which can be constructed from the rice nomograph in Appendix 3) have agreed well with our results obtained in practice under various conditions since 1956 (Ito, 1965, and unpublished data).

SPECIFIC FEATURES OF THE STORAGE METHOD

The seed storage methods employed in the Laboratory have the following three features. First the adoption of a 'two-step seed-storage system', secondly the use of a low humidity room for drying and packing seed for storage, and thirdly the use of special kinds of

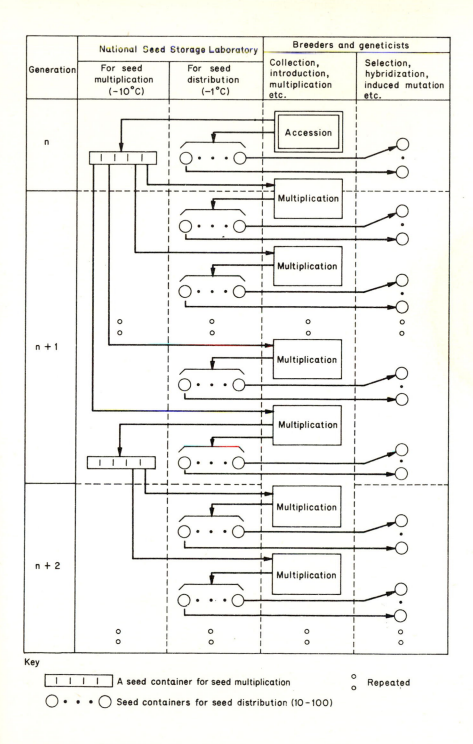

FIGURE A2.1 Explanatory diagram of the two-step seed-storage system.

407

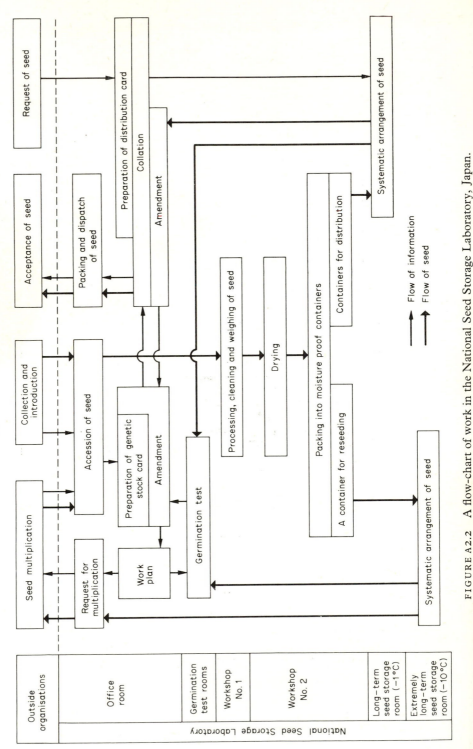

FIGURE A2.2   A flow-chart of work in the National Seed Storage Laboratory, Japan.

moisture-proof containers for distribution as well as for storage.

The main priority for seed storage in the Laboratory is to maintain the genetic constitution of the material at the time of accession by the Laboratory. The best way to keep the genetic constitution of a material is to store a sufficient quantity of seeds at the beginning, to mix them well, and to maintain the viability of these seeds for a long time. This method cannot be employed for practical purposes without some modification because of the limitation of the capacity of storage room, but almost the same effects can be expected if the number of generations of stored seeds are kept to a minimum.

An effective way to avoid the use of an unnecessary number of generations is to adopt the 'two-step seed-storage system' illustrated in Fig. A2.1. In this Laboratory, two separate seed samples are maintained – one for multiplication of stored seeds and the other for distribution. The storage room for multiplication provides better storage conditions than that for the distribution of seeds. If the Laboratory meets a demand for seed distribution, a sample from a container for distribution is given to a breeder or a geneticist. This procedure is continued until the seed samples for distribution are about to be exhausted or the high viability has begun to decline. Then some seed is taken out from the multiplication store, and the Laboratory will ask a breeder or a geneticist to multiply it in the ecological region from which the seed originated.

The multiplied seeds are stored for further distribution. These cycles are repeated till the stored seeds for multiplication are about to be exhausted or the viability of seed has declined. Thus the number of generations through which the seeds is passed before distribution is minimised and consequently the undesirable changes associated with the maintenance of seed stocks are prevented.

When the seed for multiplication is about to be exhausted or is beginning to decline in viability, the fresh seeds produced by multiplication are used not only to recover a fresh stock for distribution but also for a further stock for multiplication.

WORK IN THE LABORATORY

The work of the Laboratory is shown schematically in Fig. A2.2. The essential work of the Laboratory is the maintenance and exchange of seed samples and relevant information. Figure A2.2 shows that the information services are carried out only in the office room and the other rooms are used for seed storage, reprocessing, cleaning, weighing and germination tests.

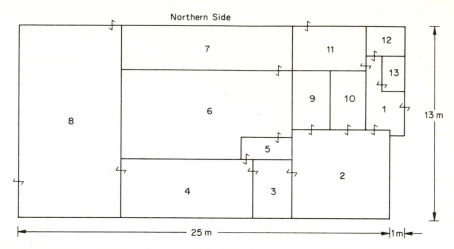

FIGURE A2.3   Ground plan of the National Seed Storage Laboratory, Japan.

In order to facilitate the functions shown in Fig. A2.2, the ground plan of the Laboratory was designed as shown in Fig. A2.3. Figure A2.4 shows a general view of the Laboratory. In this Laboratory, the rooms are used not only for storage but also for research work on seed storage. The Laboratory is located on the main road of the Institute and the position of the storage rooms and germination-test rooms are arranged so as to avoid the influence of sunshine. Further details of the design and organisation are as follows.

FIGURE A2.4   General view of the National Seed Storage Laboratory, Japan.

Room No. 1, Entrance
To be free from dust, all visitors are requested to take special shoes for exclusive use in the Laboratory.

Room No. 2, Office room
The main occupations in the office room are information and seed-exchange services. Hand-sorted punch cards are used for registration and records of seed distribution. The control board is installed in this room in order to check and control the status of all the air-conditioning machines.

Room No. 3, Workshop No. 1
Seeds originate from all over Japan and abroad and the level of grading and cleaning of the seeds which are received is not always the same. Consequently, in some cases it is necessary to grade and clean the seeds again in the Laboratory and this room is used for these purposes. The room is constructed with sound-proof walls and ceiling and a strong ventilation fan is equipped to exhaust the dust from the room.

Room No. 4, Workshop No. 2
This is one of the most important rooms in the Laboratory. Japan is one of the most humid countries in the world during summer and so

FIGURE A2.5   Moisture-proof seed containers for genetic resources.
At the front: Three different sizes of containers used for seed distribution.
At the rear: Food can used as the container for seed-multiplication stocks.

FIGURE A2.6   Moisture-proof seed containers for seed distribution.
At the rear: left, view of container from the bottom; right, view inside
At the front: left, view of container from the top; right, view inside.

it is very difficult to dry seed down to 4–6 per cent moisture content for long-term storage because, in order to be sure of not damaging the seeds, the drying temperature of seed should not exceed 40°C. However, since this room is intensely dehumidified to less than 30 per cent relative humidity, it is possible to dry seed well and also the seed will not absorb moisture during the subsequent packing.

To cope with the difficulty of maintaining such a dried condition for a long period at low temperature, dried seeds are packed in moisture-proof containers. Suitable containers which are used for distribution and extremely long-term storage for seed multiplication are shown in Figs. A2.5 and A2.6.

Room No. 5, Preparation room
This is a general-purpose room which is attached to the cold storage room.
Room No. 6, Long-term seed-storage room
This room contains many cardboard book boxes arranged systematically on shelves (Fig. A2.7). 10–80 small moisture-proof containers for seed distribution are packed in each book box (Fig. A2.8). About 300 seeds are packed in each small container. Three different sizes

FIGURE A2.7   An inside view of the seed storage room for distribution.

FIGURE A2.8   Arrangement in the shelf of book boxes and moisture-proof seed containers for distribution.

of containers and book boxes are used for different kinds of crops according to the size of their grains. When the Laboratory is requested to send a sample, an officer in the Laboratory only needs to take out a single container from a book box. The containers for seed distribution can easily be opened for inspection by plant quarantine officers.

Room No. 7, Extremely long-term seed-storage room
This room is similar to Room No. 6 except that it is kept at a very low temperature. Because the aim is to maintain the viability of the seed stored for multiplication for longer than the seed for distribution, the standard container used in this room is a vacuum-sealed tin food can of 120 cc capacity (Fig. A2.5).

Room No. 8, Machinery room servicing rooms Nos. 4–7
This room has wide windows on three sides in order to facilitate the cooling of the machinery. An oil tank of 2,000 litres and a cooling tower are attached to this room.

Rooms Nos. 9 and 10, Germination-test rooms
These rooms are air-conditioned ranging from 15–35°C. Within this range, the temperature of Room No. 9 is kept constant, and in Room No. 10 one constant temperature is maintained during the daytime and another during the night. Both rooms are illuminated at a minimum light intensity of 2,000 lux. Preparation of seed for germination tests and observation of germination are carried out in these rooms but, in some cases, the office room is also used for the same purposes.

Room No. 11, Machinery room servicing rooms Nos. 9 and 10
In this room a package-type air-conditioner is installed for Room No. 9 and another for Room No. 10.

Rooms Nos. 12 and 13, Lumber room and lavatory
The lumber room is used for cleaning petri-dishes after germination tests with an automatic washer.

FUTURE PROBLEMS

Four years' experience of operating the Laboratory has disclosed several problems that are important for planning future seed-storage laboratories for genetic resources.
(1) Ideal storage conditions for seed are not suitable for human beings. Workers in the Laboratory have to work under the uncomfortably dry and cold conditions for a long time. It would be better if the

workers could store or take out seed without entering the storage rooms. A future seed-storage room should incorporate a rotary, ring-type room system.

(2) The workshop space in the Laboratory is not enough. The space of Room No. 3 (Workshop No. 1) for seed cleaning should be determined by the kinds of seeds stored, because on this depends the number of seed-cleaning machines required. Accordingly it is suggested that if many kinds of seeds are to be stored, the room area for seed cleaning should be at least 30 m², but at present this room only has a floor area of 10 m². Similarly, the space of Room No. 4 (Workshop No. 2) for seed drying and packing is not sufficient for smooth working and the present space of 39 m² should be enlarged to 70 m².

(3) Systematic networks for continuous seed multiplication for the Laboratory should have been organised among breeders and geneticists when the new seed-storage laboratory was established. In general the quantities of seed obtained at the time of accession are not sufficient for the seed storage laboratory.

References to Appendix 2

HARRINGTON, J. F., 1967. Seed and pollen storage for conservation of plant gene resources. *FAO Technical Conference of Exploration, Utilization, and Conservation of Plant Gene Resources*, 1–22 WM/63264 FAO.

ITO, H., 1965. Studies on maintenance of genetic stocks and a breeding system for rice plants based on long-term seed storage. *Bull. Natn. Inst. agric. Sci.*, D, **13**, 163–230. [Japanese with English summary.]

ITO, H., 1970. A new system of cereal breeding based on long term seed storage. *SABRAO (Society for the Advancement of Breeding Researches in Asia and Oceanea) Newsletter*, **2** (1), 65–70.

ITO, H., and HAYASHI, K., 1960a. Studies on the storage of rice seeds. I. Influences of temperature and moisture content on the longevity of rice seeds and methods of drying seeds. *Proc. Crop. Sci. Soc., Japan*, **28**, 363–64. [Japanese with English summary.]

ITO, H., and HAYASHI, K., 1960b. Studies on the storage of rice seeds. II. The moisture content, viability, and longevity of rice seeds in different methods of processing. *Proc. Crop Sci. Soc., Japan*, **29**, 97–99. [Japanese with English summary.]

ITO, H., and HAYASHI, K., 1961. Tentative plans for improvement of the permanent maintenance of genetic stocks and breeder's seed of rice, and the procedure of rice breeding based on long-term storage of seeds. *Japan J. Breed.*, **11**, 59–64. [Japanese with English summary.]

ITO, H., and KUMAGAI, K., 1969. The National Seed Storage Laboratory for genetic resources in Japan. *Japan agric. Res. Quart.*, **4**, 32–38.

ROBERTS, E. H., 1960. The viability of cereal seed in relation to temperature and moisture. *Ann. Bot.*, **24**, 12–30.

ROBERTS, E. H., 1961. The viability of rice seed in relation to temperature, moisture content, and gaseous environment. *Ann. Bot.*, **25**, 381–90.

Viability Nomographs

E. H. Roberts and Dorothy L. Roberts

The following nomographs have been constructed as a guide to the prediction of viability under hermetic storage for all species for which there is sufficient information at present. To some extent these nomographs can also be used as a guide to loss of viability under open storage; but see Chapter 2 for a discussion of their limitations. The nomographs have been constructed from equations (2.9), (2.10) and (2.11) using the following values for the four viability constants (for mean viability period in days):

Fig.	Species	Viability constants				Source (references to Chapter 2)
		K_v	C_1	C_2	K_a	
A3.1	Rice (*Oryza sativa*)	6.531	0.159	0.069	0.210	Roberts (1961b)
A3.2	Wheat (*Triticum aestivum*)	5.067	0.108	0.050	0.350	Roberts (1960, 1961a)
A3.3	Barley (*Hordeum distichon*)	6.745	0.172	0.075	0.301	Roberts and Abdalla (1968)
A3.4	Broad beans (*Vicia faba*)	5.766	0.139	0.056	0.379	Roberts and Abdalla (1968)
A3.5	Peas (*Pisum sativum*)	6.432	0.158	0.065	0.384	Roberts and Abdalla (1968)

METHOD OF USING THE NOMOGRAPHS

The nomographs may be used in various ways; two of the most useful are as follows:

(1) To estimate the time taken for viability to fall to any given level at any given temperature and moisture content

Put a ruler on the required temperature (scale *a*) and moisture content (scale *b*). Note the value indicated on scale *c* (this gives the mean viability period). Using this point on scale *c* as a pivot, move the ruler to indicate any required percentage viability on scale *e*. The value now indicated on scale *d* is the time taken for viability to drop to the percentage viability chosen.

(2) To find the various combinations of temperature and moisture content necessary to maintain viability above a given value for a given period

Select the minimum level of viability required on scale *e*. Select required storage period on scale *d*. Put a ruler through both points and note the value it indicates on scale *c*. Using this point on scale *c* as a pivot, move the ruler through scales *a* and *b*. Any position of the ruler indicates a combination of values for temperature (scale *a*) and moisture content (scale *b*) which, during the required storage period, viability would be expected to fall to the chosen value.

Erratum

On pp. 419–423 (Figs. A.3.1–A.3.5), in column (c), for 'Moisture' read 'Mean'.

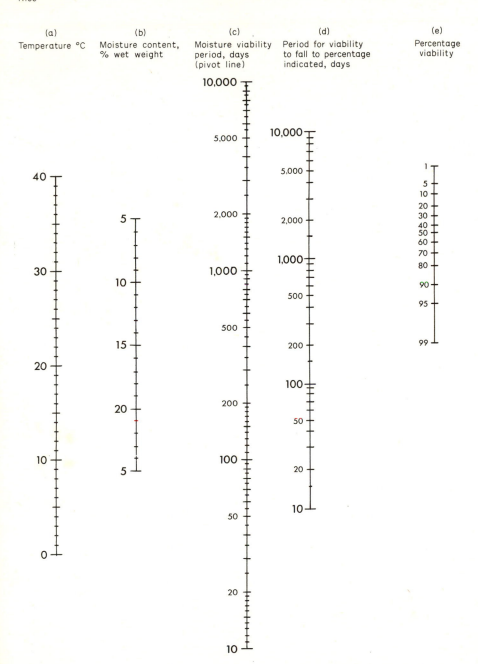

FIGURE A3.1 Viability nomograph for Rice (*Oryza sativa*)

Wheat

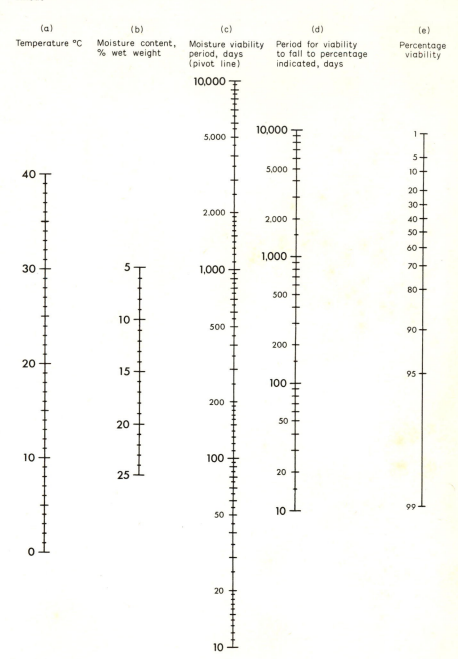

FIGURE A3.2   Viability nomograph for Wheat (*Triticum aestivum*)

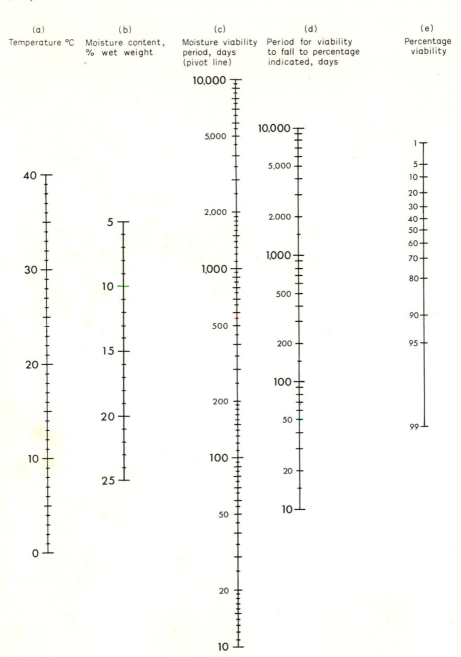

FIGURE A3.3   Viability nomograph for Barley (*Hordeum distichon*)

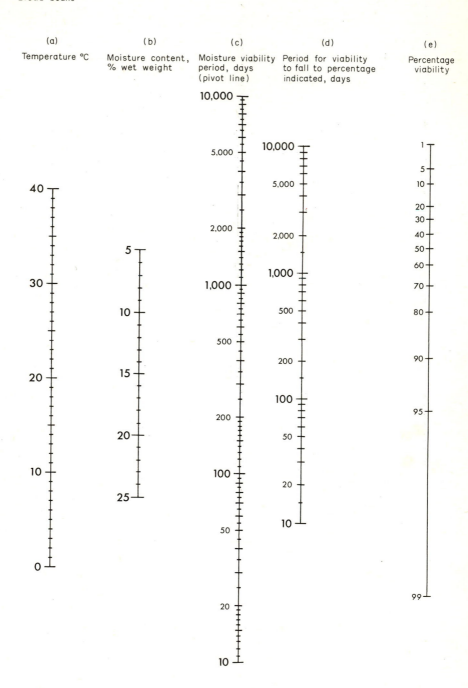

FIGURE A3.4   Viability nomograph for Broad beans (*Vicia faba*)

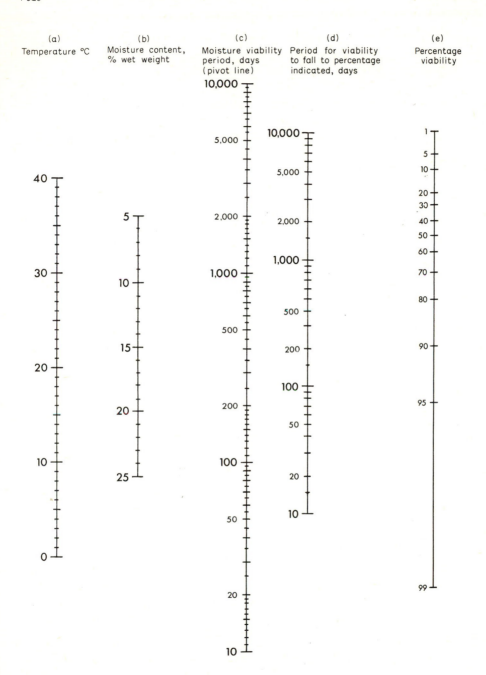

FIGURE A3.5. Viability nomograph for Peas (*Pisum sativum*)

Moisture Content of Seeds

E. H. Roberts and Dorothy L. Roberts

Extracts from the International Seed Testing Association rules for determining moisture contents [a]

The air-oven method at 130°C can be applied to most kinds of seeds. For tree seeds and other seeds containing volatile oils, a temperature of 130°C is excessive; therefore, the air-oven method at 105°C, or in some cases the toluene distillation method, must be applied.

AIR-OVEN 130°C METHOD

The air-oven method at 130°C is applicable to seeds of the following species:

Agrostis spp.
Ageratum spp.
Alopecurus pratensis
Alyssum spp.
Anethum graveolens
Anthoxanthum odoratum
Anthriscus spp.
Antirrhinum spp.
Apium graveolens
Arachis hypogaea [b]
Arrhenatherum spp.
Asparagus officinalis
Aster spp.
Avena spp. [b]
Bellis spp.
Beta vulgaris
Brassica spp.
Bromus spp.
Camelina sativa
Campanula spp.
Cannabis sativa
Carum carvi
Celosia spp.

Cheiranthus spp.
Chloris gayana
Chrysanthemum spp.
Cicer arietinum
Cichorium spp.
Citrullus vulgaris [b]
Cucumis spp.
Cucurbita spp.
Cuminum cyminum
Cynodon dactylon
Cynosurus cristatus
Dactylis glomerata
Dahlia spp.
Daucus carota
Delphinium spp.
Deschampsia spp.
Dianthus spp.
Digitalis spp.
Fagopyrum esculentum [b]
Festuca spp.
Godetia spp.
Gossypium spp. [b]
Holcus lanatus

Hordeum vulgare [b]
Lactuca sativa
Lathyrus spp.
Lepidium sativum
Linum usitatissimum
Lolium spp.
Lotus spp.
Lupinus spp. [b]
Lycopersicon esculentum
Medicago spp.
Melilotus spp.
Myosotis spp.
Nasturtium spp.
Nemesia spp.
Nicotiana tabacum
Onobrychis viciifolia
Ornithopus sativus
Panicum spp.
Papaver somniferum
Paspalum dilatatum
Pastinaca sativa
Penstemon spp.
Petroselinum crispum

[a] *Proc. int. Seed Test. Ass.*, **31**, 128–34 (1966).
[b] Seeds must be ground before drying.

Petunia spp.
Phacelia tanacetifolia
Phalaris spp.
Phaseolus spp.[b]
Phleum pratense
Phlox spp.
Pimpinella spp.
Pisum sativum[b]
Poa spp.
Portulaca spp.
Pyrethrum spp.

Ricinus communis[b]
Salvia spp.
Scorzonera hispanica
Secale cereale[b]
Sesamum indicum
Sinapis spp.
Solanum tuberosum
Sorghum spp.
Spergula sativa
Spinacia oleracea
Tagetes spp.

Trifolium spp.
Trisetum flavescens
Triticum spp.[b]
Valerianella locusta
  var. *olitoria*
Vicia spp.[b]
Viola spp.
Zea mays[b]
Zinnia spp.

AIR-OVEN 105°C METHOD

The air-oven method at 105°C is applicable to seeds of the following species:

Allium ascalonicum
Allium cepa
Allium porrum
Allium sativum

Capsicum spp.
Ceratonia siliqua
Glycine max
Raphanus sativus

Solanum melongena
  var. *esculentum*
Trigonella foenum-graecum

and
all tree seeds except the following which are listed under the toluene distillation method.

TOLUENE DISTILLATION METHOD

The toluene distillation method is applicable to seeds of the following species:

Abies spp.
Cedrus spp.

Fagus spp.
Picea spp.

Pinus spp.
Tsuga spp.

GRINDING

The grinding mill should grind evenly and should not be operated at such a high speed that the ground material is heated. Air currents that might cause loss of moisture must be reduced to a minimum.

Some kinds of seed, particularly the larger-seeded kinds, must be finely or coarsely ground before drying, but generally this is not necessary for the smaller-seeded kinds. The following general rules apply:

(1) Thorough grinding is necessary in the case of cereals (including maize, rice, sorghum) and cotton seed. The mill-setting for these

[b] Seeds must be ground before drying.

kinds of seed should be such that for hard wheat at least 50 per cent of the ground material passes through a wire sieve with meshes of 0.5 mm, and not more than 10 per cent remains on a wire sieve with meshes of 1.0 mm.

(2) Coarse grinding is necessary in the case of large leguminous seeds such as *Vicia*, *Phaseolus*, *Pisum* and *Lupinus*, and tree seeds such as *Quercus* and *Fagus*. The mill-setting should be such that at least 50 per cent of the ground material passes through a sieve with meshes of 3.4 mm.

(3) For most other kinds of seeds, grinding should be omitted. It is usually best not to grind seeds of high oil content because they are difficult to grind properly and also because oxidation of the oil during drying may result in a gain in weight and thus cause errors in the moisture determination. Oxidation of the oil is a particularly serious problem in seeds containing oil of high iodine number, such as flax seed.

DRYING GROUND SEEDS OF HIGH MOISTURE CONTENT

For seeds of all species for which grinding is necessary, two-stage drying must be applied when the moisture content is higher than 18 per cent (leguminous seeds higher than 20 per cent).

AIR-OVEN 130°C METHOD

Equipment

(1) Dishes of noncorrosive metal (thickness approximately 0.5 mm) with side rounded at base and a flat bottom, fitted with covers which seat so snugly that loss of moisture is reduced to a minimum. The dimensions of the dishes must be such that not more than 0.3 gm of material is applied per cm². To ensure close fitting of the cover, the rim of the dish should be levelled by rubbing with an abrasive.

(2) An electrically-heated oven with adequate ventilation and thermostatic control which permits the temperature to be maintained at 130 ± 3°C. The heating capacity of the oven must be such that after pre-heating to a temperature of 130°C, followed by opening and loading with dishes, the oven will again reach 130°C within 45 minutes (preferably within 30 minutes).

(3) A desiccator (with a suitable desiccant), preferably fitted with a thick metal plate to promote rapid cooling of the dishes.

(4) A balance on which accurate weighings can be made in grammes to 3 decimal places.

Procedure

Weigh the dish with its cover. Weigh out 4–5 gm of the sample, previously mixed. Place the working sample in the dish, distributing it evenly over the bottom surface. Put the cover on the dish and weigh again. Place the dish on top of its cover in an oven heated beforehand to 130°C. In order to limit the loss of heat, the dishes must be placed in the oven rapidly. From the time that the oven again reaches 130°C, the drying period should be 60 minutes at this temperature (for hard wheats, 120 minutes). After termination of the drying period, cover the dishes immediately and place them in a desiccator to cool for 30–45 minutes. Weigh the dishes with their contents and covers. All weighings should be made to an accuracy of 1 mg.

If: M_1 is the weight in grammes of the dish and its cover,

M_2 is the weight in grammes of the dish, its cover and its contents, and

M_3 is the weight in grammes of the dish, cover and contents after drying, then the moisture content calculated on wet basis and expressed in percentage is:

$$(M_2 - M_3)\frac{100}{M_2 - M_1}$$

The determination must be made in duplicate. The results of duplicate determinations must not differ by more than 0.2 per cent. Should the difference be greater than this, the determination must be repeated in duplicate.

Two-stage drying

Weigh out approximately 50 gm of the sample. Transfer this working sample to a suitable weighed container and place in an air-oven at a temperature of 130°C for 5–10 minutes. The length of this preliminary drying period will depend on the amount of moisture it may be necessary to remove, the objective being to reduce the moisture content to 12–15 per cent. Spread the partly-dried seeds in an open tray and leave exposed in the laboratory for 2 hours. Transfer the material to the container in which it was oven-dried, and weigh. Calculate the loss of moisture as stated above. The determination must be made in duplicate, with the weighings being made to an accuracy of 10 mg.

Grind separately the two partly-dried working samples, and, on the ground material from each, make a single moisture determination as stated above, and calculate the loss of moisture.

From the results obtained in the first and second stages of the procedure, calculate the moisture content of the sample. If S_1 is the moisture lost in stage 1, and S_2 is the moisture lost in stage 2, each expressed as a percentage, then the original moisture content of the sample, calculated on the wet basis and expressed as a percentage, is:

$$S_1 + S_2 - \frac{S_1 \times S_2}{100}$$

AIR-OVEN 105°C METHOD

This method is similar to the air-oven 130°C method. The differences are:

(a) The oven temperature must be maintained at 105 ± 2°C.
(b) The drying period must be 16 hours at 105°C.
(c) The work should be conducted in a laboratory having a relative humidity below 70 per cent because at 105°C a high relative humidity influences considerably the result of the moisture determination.

TOLUENE DISTILLATION METHOD

Use a 250 ml distilling flask connected to a 20-inch Liebig condenser by a distilling trap of the Bidwell and Sterling type graduated in tenths of a millilitre. All connections should be of ground glass. Clean the trap and condenser with a solution of potassium dichromate in sulphuric acid, rinse thoroughly in water, then in alcohol, and dry in an oven to prevent water adhering to inner surfaces during the determination.

If the sample is likely to bump on boiling, add dry sand to cover the bottom of the flask. Add enough toluene to cover the sample completely (about 65 m.). Weigh and introduce into the toluene enough sample to give 2–5 ml of water and connect the apparatus. Fill the trap with toluene, pouring it through the top of condenser. Bring to boil and distil slowly at about 2 drops per second, until most of the water passes over, then increase rate of distilling to about 4 drops per second.

When all of the water is apparently over, wash down the condenser by pouring toluene in at the top, continuing distilling for a short time to see whether any more water will distil over; if it does, repeat the washing-down process. If any water remains in the condenser, remove it by brushing down with a tube brush attached to copper wire and saturated with toluene, washing down the condenser at the same time. The entire process is usually completed within one hour. Allow the trap to come to room temperature. If any drops adhere to sides of the trap, force them down, using copper wire with its end wrapped with a rubber band. Read volume of water and calculate the percentage. The determination must be made in duplicate.

Conversion of moisture-content values

FIGURE A4.1   Scale for converting moisture-content values between dry-weight basis and wet-weight basis.

$$\text{per cent moisture content, dry-weight basis} = \frac{\text{weight of water}}{\text{weight of dry matter}} \times 100$$

$$\text{per cent moisture content, wet-weight basis} = \frac{\text{weight of water}}{\text{weight of water} + \text{weight of dry matter}} \times 100$$

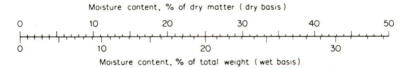

Moisture content, % of dry matter (dry basis)

Moisture content, % of total weight (wet basis)

TABLE A4.1 *Hygroscopic Equilibria of Seeds. Moisture content (% wet basis) in equilibrium with various relative humidities.*

Species	Source of data*	Conditions of absorption or desorption where known	Temp. °C	Relative humidity, %																		
				10	15	20	25	30	35	40	45	50	55	60	65	70	75	80	85	90	95	100
Allium cepa, Onion	3		25 approx.	4.6		6.8		8.0			9.5			11.2			13.4					
Allium fistulosum, Spring Onion	3		25 approx.	3.4		5.1		6.9			9.4			11.8			14.0					
Apium graveolens, Celery	3		25 approx.	5.8		7.0		7.8			9.0			10.4			12.4					
Arachis hypogaea, Groundnuts	3		25 approx.		2.6			4.2			5.6						9.8			13.0		
Groundnuts	1		30											7.2		7.0	8.0	9.3	11.3	14.3	20.0	>30.5
Avena sativa, Oats	5		fluctuating 12–25	5.5		7.2		8.8		10.2		11.4		12.5		14.0	15.2	17.0			22.6	
Beta vulgaris, Garden beetroot	7	Absorption	25 approx.	4.7		7.0		8.6		9.8		10.1		12.7		14.6						
Garden beetroot	3		25 approx.	2.1		4.0		5.8			7.6			9.4			11.2					
Sugar beet	5		fluctuating 12–25	4.4		6.3		8.0		9.8		10.7		12.0		13.3	14.5	16.6		20.5	22.5	
Brassica juncea, Mustard	3		25 approx.	1.8		3.2		4.6			6.3			7.8			9.4					
Brassica napus, Rape	5		fluctuating 12–25	3.1		3.9		4.5		5.2		6.0		6.9		8.0	8.6	9.3		12.1	15.3	
Brassica oleracea, Cabbage	3		25 approx.	3.2		4.6		5.4			6.4			7.6			9.6					
Cabbage	6	Absorption		3.4		4.7		5.5		6.3		7.1		8.1		9.7						
Brassica rapa, Turnip	3		25 approx.	2.6		4.0		5.1			6.3			7.4			9.0					

	No. of samples	Temperature										
Citrullus vulgaris, Watermelon	3	25 approx.	3.0	4.0	5.1	6.3	7.4	9.0				
Cucumis sativus, Cucumber	3	25 approx.	2.6	4.3	5.6	7.1	8.4	10.1				
Cucurbita maxima, Winter squash	3	25 approx.	3.0	4.3	5.6	7.4	9.0	10.8				
Daucus carota, Carrot	3	25 approx.	4.5	5.9	6.8	7.9	9.2	11.6				
Carrot	7	Absorption	4.2	5.8	7.0	7.9	8.9	10.0	11.9	16.0		
Fagopyrum esculentum, Buckwheat	3	25 approx.		6.7	9.1	10.8	12.7	15.0	19.1		24.5	
Festuca rubra, Red Fescue	7	Absorption	3.8	7.2	8.9	10.0	11.7	13.4	16.5	18.8		
Fraxinus elatior, Ash	7	Absorption	4.1	6.0	7.4	8.8	10.3	12.0	13.9			
Glycine max, Soya beans	3	25 approx.		4.3	6.5	7.4	9.3	13.1	18.8			
Hibiscus esculentus, Okra	3	25 approx.	3.8	7.2	8.3	10.0	11.2	13.1				
Hordeum vulgare, Barley	3	25 approx.		6.0	8.4	10.0	12.1	14.4	19.5		26.8	
Lactuca sativa, Lettuce	3	25 approx.	2.8	4.2	5.1	5.9	7.1	9.6				
Lettuce	6	Absorption	3.1	4.2	5.0	5.9	6.7	7.6	9.1			
Linum usitatissimum, Flax	3	25 approx.		4.4	5.6	6.3	7.9	10.0	15.2		21.4	
Lolium perenne,	7	Absorption	4.3	6.9	8.7	10.3	11.9	13.8	16.1	18.6		
Ryegrass	5	fluctuating 12–25	4.6	7.0	9.1	10.9	12.3	13.2	14.3	15.0	16.0	22.5
Lupinus hirsutus(?), Lupin	5	fluctuating 12–25	4.2	6.2	7.8	9.1	10.5	11.7	13.4	14.5	16.7	
Lycopersicon esculentum, Tomato	3	25 approx.	3.2	5.0	6.3	7.8	9.2	11.1				
Medicago sativa, Lucerne	6	Absorption 30	4.8	6.4	7.8	9.0	10.0	11.7	14.0	15.0		

TABLE A4.1 Hygroscopic Equilibria of Seeds. Moisture content (% wet basis) in equilibrium with various relative humidities (continued).

Species	Source of data*	Conditions of absorption or desorption where known	Temp. °C	Relative humidity, %																		
				10	15	20	25	30	35	40	45	50	55	60	65	70	75	80	85	90	95	100
Oryza sativa, Rice																						
Rice	2	Desorption	25	4.6		6.5		7.9		9.4		10.8		12.2		13.4		14.8		16.7		
	2	Absorption	25	3.9		5.3		6.8		7.9		9.2		10.4		11.8		13.6		16.6		
Pastinaca sativa, Parsnip	3		25 approx.	5.0		6.1		7.0			8.2			9.5			11.2					
Phaseolus limensis, Lima bean	3		25 approx.	4.6		5.6		7.7			9.2			11.0			13.8					
Phaseolus vulgaris, Kidney bean	6	Absorption		4.2		7.1		8.7		10.3		12.2		14.5		17.9						
Picea excelsa, Spruce	7	Absorption		2.5		4.2		5.5		6.7		7.8		9.0		10.4						
Pisum sativum, Peas	6	Absorption below 15% Desorption above 15%	30	5.5		7.3		8.6		9.9		11.3		13.1		15.4		19.0				
Peas	7	Absorption		4.0		7.0		8.8		10.2		12.0		13.9		16.2		20.5		28.4		
Peas	5		fluctuating 12–25	5.3		7.0		8.6		10.3		11.9		13.5		15.0	15.9	17.1		22.0	26.0	
Raphanus sativus, Radish	3		25 approx.	2.6		3.8		5.1			6.8			8.3			10.2					
Secale cereale, Rye	3		25 approx.		7.0			8.7			10.5			12.2			14.2			20.6		
Rye	5		fluctuating 12–25	6.9		8.2		9.6		10.9		12.2		13.5		15.1	16.2	17.5		21.6	24.5	26.7
Solanum melongena, Egg plant	3		25 approx.	3.1		4.9		6.3			8.0			9.8			11.9					

432

Species	Ref.	Equilibrium	Temp. (°C)											
Sorghum vulgare, Sorghum	3		25 approx.	6.4	8.6	10.5	12.0	15.2	18.8	21.9				
Spinacia oleracea, Spinach	3		25 approx.	4.6	6.5	7.8	9.5	11.1	13.2					
Triticum vulgare, Wheat	4	Desorption	35	5.5	7.2	8.5	9.8	11.0	12.2	13.4	15.1	19.5		
Wheat	4	Desorption	30	5.8	7.8	9.1	10.4	11.8	13.0	14.2	16.0	21.5		
Wheat	4	Desorption	25	6.0	8.0	9.3	10.6	12.0	13.2	14.7	16.3	21.5		
Wheat	4	Absorption	35	4.0	5.6	7.0	8.3	9.8	11.1	12.8	14.5	19.5		
Vicia faba, Broad bean	6	Absorption below 15%, Desorption above 15%		8.5	10.0	11.5	13.2	15.0	19.7					
Broad bean	5	fluctuating	12–25	4.7	6.8	8.5	10.1	11.6	13.1	14.8	15.9	17.2	22.6	27.2
Zea mays, Maize	4	Desorption	30	5.6	7.5	8.9	10.2	11.3	12.6	14.0	15.8	20.0		
Maize	4	Absorption	30	4.5	6.3	7.6	8.9	10.1	11.5	13.0	14.9	19.5		
Maize	5	fluctuating	12–25	6.2	7.9	9.3	10.7	11.9	13.1	14.6	15.5	16.5	20.7	25.0

*(1) AUSTWICK, P. K. C., and AYERST, G., 1963. Groundnut microflora and toxicity. *Chem. and Ind.*, 12 Jan., 1963, 55–61.

(2) BREESE, M. H., 1955. Hysteresis in the hygroscopic equilibria of rough rice at 25°C. *Cereal Chem.*, **32**, 481–7.

(3) HARRINGTON, J. F., 1959. Drying, storing, and packaging of seeds to maintain germination and vigor. *Proc. 1959 Mississippi Course for Seedsmen*, 89–107. [Mimeographed publication.]

(4) HUBBARD, J. E., EARLE, F. R., and SENTI, F. R., 1957. Moisture relations in wheat and corn. *Cereal Chem.*, **34**, 422–33.

(5) KREYGER, J., cited by OWEN, E. B., 1956. *The Storage of Seeds for the Maintenance of Viability*. Commonwealth Agricultural Bureaux, Farnham Royal, England.

(6) LUBATTI, O. F., and BUNDAY, G., 1960. The water content of seeds. I. The moisture relations of seed peas, etc. *J. Sci. Food Agric.*, **11**, 685–90.

(7) TOUZARD, M. J., 1961. La conservation des semences. *Congr. Natl. Semences, 1961*, 133–52.

TABLE A4.2 *Recommended maximum 'safe' drying temperatures for seed in commercial practice. Temperatures are quoted to the nearest 0.5°C.*

Species	Source of data★	Moisture content (% wet basis)	Max. drying temp. °C	
			Seed	Air
Allium cepa, Onion	10	12–20		32
		20+		21
Avena sativa, Oats	1	30		43
		28		45.5
		26		49
		24		53
		22		62
		18		67
	3	over 21	35	60
		below 21	40	70
Beta vulgaris, Mangels and beets	10	14–22		50
		22+		32
Brassicas, Cabbage, Brussels sprouts, marrow-stemmed kale	10	10–18		38
		18+		27
Dactylis glomerata, Cocksfoot grass	9	20		54
		25		40.5
		30		21
	10	13–21		49
		21+		32
	15	about 13		43.5 alternating with a cold blow
Fagopyrum esculentum, Buckwheat	3	below 21	40	70
		over 21	35	60
Festuca rubra, Chewings Fescue grass	6	5		90 for $\frac{3}{4}$ hr
		5		80 for 1 hr
		5		70 for $2\frac{1}{2}$ hr
		5		60 for 21 hr
		5		50 for 120 hr
		8		80 for $\frac{1}{2}$ hr
		8		70 for $1\frac{1}{2}$ hr
		8		60 for $3\frac{1}{2}$ hr
		8		50 for 80 hr
		11		70 for $\frac{1}{2}$ hr
		11		60 for 1 hr
		11		50 for 8 hr
		13		50 for 5 hr
Helianthus annuus, Sunflower	3	below 21	40	70
		over 21	35	60
Hordeum vulgare, Barley	1	30		43
		28		45.5
		26		49
		24		53
		22		62
		18		67

Species	Source of data*	Moisture content (% wet basis)	Max. drying temp. °C	
			Seed	Air
	16	25		59
		20		62.5
		16.8		59.5
	14	23.5		60.5
		20.4		65
		18.3		67
	3	below 21	40	70
		over 21	35	60
	8	28		55
		24		65
		20		75
Linum usitatissimum, Linseed	12	9–11		65.5
		15–18		79.5
Lolium perenne, Ryegrass	9	25		71
		30		65.5
		35		54.5
		40		48.5
		45		37.5
	10	14–22		49
		22+		32
	15	about 13.5		43.5 alternating with a cold blow
Onobrychis sativa, Sainfoin	11	30		21–42 'germination remained pratically unchanged'
Oryza sativa, Rice	13	above 20		60
		15–20		52
		15 or lower		50
Phalaris arundinacea, Reed grass	6	30+		38
Phleum pratense, Timothy grass	11	14–22		38
		22+		27
	15	13.5		43.5 alternating with a cold blow
	9	25		57.5 ⎫
		30		54 ⎬ hulled
		35		43 ⎭
	9	25		54 ⎫
		30		51.5 ⎬ unhulled
		35		48.5 ⎪
		40		40.5 ⎭
Pisum sativum, Peas	10	16–24		38
		24+		26.5
Secale cereale, Rye	3	over 21	35	60
		below 21	40	70
Setaria italica(?), Millet	3	below 21	40	70
		over 21	35	60
Trifolium, Clover	10	12–20		38
		20+		26.5

Species	Source of data*	Moisture content (% wet basis)	Max. drying temp. °C		
			Seed	Air	
Triticum aestivum, Wheat	1	18		67	
		22		62	
		24		53	
		26		49	
		28		45.5	
		30		43	
	4	14		60	for 30 min
		30		48	
	3	below 21	40	70	
		over 21	35	60	
	9	28		65	
Zea mays, Maize	7	approaching 50		40.5	
		up to 35		42	
		35		42	

*(1) CASHMORE, W. H., 1942–44. Temperature control of farm grain driers. *J. Minist. Agric.*, **49** (3), 144.

(2) DIMMOCK, F., 1947. *The effects of immaturity and artificial drying upon the quality of seed corn.* Tech. Bull. 58, Canada Dept Agric.

(3) GERZHOI, A. P., and SAMOCHETOV, V. F., 1960. *Grain Drying and Grain Driers.* Natn. Sci. Found., Washington. [Transl. of original Russian publ., 1958.]

(4) GREER, E. N., 1953. Risks in the storage of grain. *Agric. Merch.*, **33** (12), 369–71 and 379.

(5) GRIFFITH, W. L., and HARRISON, C. M., 1954. Maturing and curing temperatures and their influence on germination of reed canary grass seed. *Agron. J.*, **46**, 163–7.

(6) HYDE, E. O. C., 1935. Chewings fescue seed. The influence of temperature and moisture content upon the rate of loss of its germinating capacity. *N.Z.J. Agric.*, **51**, 40–2.

(7) KIESSELBACH, T. A., 1939. Effect of artificial drying upon the germination of seed corn. *J. Amer. Soc. Agron.*, **31**, 489–96.

(8) KREYGER, J., 1960. Invloed van de luchttemperatuur op het verloop van de korreltemperatuur en op de kwaliteit bij verticale doorstroomdrogers voor graan. (Effect of air temperature upon grain temperature and quality in vertical through-flow driers.) *Jversl. Inst. Bewar. Verwerk. LandbProd., Wageningen*, 1960, p. 25: Transl. 117, natn. Inst. agric. Engng, Silsoe.

(9) Nat. Inst. agric. Engng, Silsoe. *Open Day Cyclostyled Publ., 1969.*

(10) NORTH, C., 1948. Artificial drying of vegetable and herbage seeds. *Agriculture, Lond.*, **54**, 462–6.

(11) ŠINKAREV, BORSC, and HOLUBINSKAJA, 1936. [Drying sainfoin seeds.] [Russian.] Selek Semenovod. No. 6, 90. [Cited in Herb Abstr., 7, 64c.]

(12) SORENSON, J. W., and DAVENPORT, M. G., 1951. Drying and storing flaxseed in Texas. *Agric. Engng, St Joseph, Mich.*, **32**, 379–82.

(13) Texas Agricultural Experimental Station, 1950. Agricultural research in Texas, 1947–49. [Cited by E. B. Owen. *The Storage of seeds for maintenance of viability.* Commonwealth Agric. Bureaux, Farnham Royal, England.]

(14) WARNER, M. G. R., and BROWNE, D. A., 1962. Apparatus for the assessment of drying damage and some initial results with barley and wheat. *J. agric. Engng Res.*, **7**, 359.

15) WILLIAMS, M., 1938. The moisture content of grass seed in relation to drying and storing. *Welsh J. Agric.*, **14**, 213–32.

16) WOODFORDE, J., and LAWTON, P. J., 1965. Drying cereal grain in beds six inches deep. *J. agric. Engng Res.*, **10**, 146.

Index

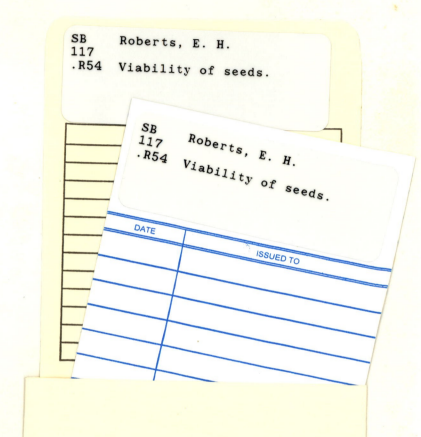